T0074995

CLIMATE VARIABILITY AND TROPICAL CYCLONE ACTIVITY

This book presents a comprehensive summary of research on tropical cyclone variability at various time scales, from intraseasonal and interannual to interdecadal and centennial. It covers the fundamental theory, statistics, and numerical modelling techniques used when considering climate variability in relation to tropical cyclone activity. Major climate oscillations including the Madden–Julian, El Niño, Atlantic Meridional Mode, Atlantic Multidecadal Osciilation, and Pacific Decadal Oscillations are covered, and their impacts on tropical cyclone activity in the Pacific and Atlantic oceans are discussed. Hurricane landfalls in the United States, Caribbean, and East Asia are also considered. Climate models and numerical simulations are used to show how prediction models of tropical cyclones are developed. Looking to the future, particular attention is paid to predicting how tropical cyclones will change in response to increased concentrations of greenhouse gases. This book is ideal for researchers and practitioners in atmospheric science, climatology, oceanography, and civil and environmental engineering sciences.

PAO-SHIN CHU is a Professor and Hawaii State Climatologist in the Department of Atmospheric Sciences at the University of Hawaii. He was the NASA/ASEE Summer Faculty Fellow at NASA Goddard Space Flight Center in 1994/5 and was appointed as Chair Professor in the Department of Atmospheric Sciences at the National Taiwan University in 2019 and at the National Central University during 2014/7. His research interests lie in the areas of tropical climate variability and change, tropical cyclones, statistical modelling, and extreme events, and he teaches several related graduate-level courses including Tropical Climate and Weather, Monsoon Meteorology, and Statistical Meteorology.

HIROYUKI MURAKAMI is a research scientist at the NOAA Geophysical Fluid Dynamics Laboratory/University Corporation for Atmospheric Research, Plainsboro, New Jersey. He was a contributing author for the Intergovernmental Panel on Climate Change 5th Scientific Assessment (AR5, 2013) and has been identified as one of the 500 most influential climate scientists by Reuters. His research interests include tropical cyclone climate projections, extreme events, and numerical weather prediction.

CLIMATE VARIABILITY AND TROPICAL CYCLONE ACTIVITY

PAO-SHIN CHU

University of Hawaii–Manoa

HIROYUKI MURAKAMI

NOAA GFDL/University Corporation for Atmospheric Research

CAMBRIDGE
UNIVERSITY PRESS

CAMBRIDGE
UNIVERSITY PRESS

University Printing House, Cambridge CB2 8BS, United Kingdom

One Liberty Plaza, 20th Floor, New York, NY 10006, USA

477 Williamstown Road, Port Melbourne, VIC 3207, Australia

314–321, 3rd Floor, Plot 3, Splendor Forum, Jasola District Centre, New Delhi – 110025, India

103 Penang Road, #05-06/07, Visioncrest Commercial, Singapore 238467

Cambridge University Press is part of the University of Cambridge.

It furthers the University's mission by disseminating knowledge in the pursuit of education, learning, and research at the highest international levels of excellence.

www.cambridge.org
Information on this title: www.cambridge.org/9781108480215
DOI: 10.1017/9781108586467

© Cambridge University Press 2022

First published 2022

Printed in the United Kingdom by TJ Books Limited, Padstow Cornwall

A catalogue record for this publication is available from the British Library.

ISBN 978-1-108-48021-5 Hardback

Contents

*Solutions to the exercises and Python codes for the Genesis Potential Index
are available online at www.cambridge.org/9781108480215
The plate section is to be found between pages 144 and 145*

Preface

Tropical climate has received increased attention over the last 40 years mainly because of the El Niño–Southern Oscillation (ENSO) phenomenon and the Madden–Julian Oscillation (MJO), and their associated impacts on a local, regional, and global scale. While the MJO is the most prominent disturbance that operates on the subseasonal time scale (less than 90 days but longer than 10 days), the ENSO is a powerful interplay between the tropical ocean and atmosphere on interannual time scales with a preferred recurrence interval of 2–7 years. Studies show that the MJO and ENSO can have a profound effect on global weather systems, such as shifting tropical cyclone (TC) formation location, altering frequency of occurrence, storm tracks, landfall locations, intensity, and lifespan in various ocean basins. There are at least two types of El Niño: the Eastern Pacific and Central Pacific types, which modulate regional TC activity in a different manner. In addition to the ENSO, other climate modes that also influence TC activity on the interannual time scale include the North Atlantic Oscillation, Pacific Meridional Mode, and Atlantic Meridional Mode. On a longer time scale, TC activity is modulated by the decadal to interdecadal oscillations in the Atlantic and Pacific. Aside from TCs, the aforementioned climate modes also alter precipitation and temperatures variations, resulting in drought, flooding, extreme cold or warm conditions, and public health issues in many parts of the world. TC attributes are projected to change over the next 50–100 years under an anthropogenic warming scenario, although uncertainty remains.

For the first time, this book brings together TC variability over the Pacific and Atlantic Ocean basins at all four time scales ranging from subseasonal, interannual, decadal/interdecadal, to centennial. This book also aims to present an overview of storm activity in each basin over the Pacific and Atlantic conditional on major climate modes, as well as hurricane landfalls in the United States, Caribbean, and East Asia, where a large portion of the world's population reside. Emphasis is placed on dynamical controls and physical understanding that determine TC variability across a variety of scales. The understanding of the aforementioned climate modes, together with methods applied, is critical to the development of modern methods for the prediction of TCs, not only for scientific curiosity, but also for emergency planning and disaster mitigation purposes. Because of the enormous financial and social impacts brought by TC-induced torrential rainfall and flooding, an understanding of the long-term changes in heavy precipitation associated with TCs is critically important. While directly observed long-term TC rainfall is scanty, a case study based on an island with reliable and high temporal resolution is presented. How TCs will change in response to the increased concentrations of greenhouse gases in the future is of great interest to our society. Physical mechanisms for projected changes in TCs are explored. This book provides essential reading for senior undergraduate and graduate students in atmospheric sciences or related fields (e.g., oceanography, geophysics, hydrology, geography, civil, and environmental engineering). It also targets researchers, policy planners, and faculty who are concerned about tropical cyclones and climate, and seek an updated and comprehensive reference book.

PSC wishes to acknowledge Stefan Hastenrath for mentoring him on tropical climate variability at the University of Wisconsin-Madison. Professor Hastenrath published a number of books on the subject, which has long inspired my interest to continue to work on a similar research arena. In 2004, PSC contributed a chapter entitled "ENSO and tropical cyclone activity" in a book published by Columbia University Press. This book – *Hurricanes and Typhoons: Past, Present, and Future* – was edited by R. J. Murnane and K.-B. Liu. The experience gained from this book chapter and the team-teaching of "Tropical Climate and Weather" at the University of Hawaii has gradually broadened my knowledge and scope of tropical climate and tropical cyclones, and enabled me to seriously contemplate authoring the current book. As anticipated, the long journey of book writing has been an extraordinary and fulfilling experience. In this regard, PSC is grateful to Hui-Ling Chang of the Central Weather Bureau (CWB) in Taiwan, who provided constant encouragement.

HM would like to express his gratitude to the numerous scientists all over the world. HM's efforts to understand the climate variability in TCs have been generously supported over the years by the research projects in Japan and the United States. HM is especially grateful to Dr. Akio Kitoh (Meteorological Research Institute [MRI]), Dr. Masato Sugi (MRI), Professor Bin Wang (University of Hawaii), Professor Tim Li (University of Hawaii), and Professor Gabrie A. Vecchi (Princeton University) for their kind support of HM's research career. HM's deepest appreciation also goes to Professor Pao-Shin Chu, who is one of the authors for this book, for his kind encouragement, discussion, and edition for writing the current book.

We are indebted to Phil Klotzbach, Pang-chi Hsu, Tianyi Wang, Leishan Jiang, Sen Zhao, Cheng-Ku Yu, Ming-Yue Tang, Hui-Ling Chang, Michelle Luo, Huang-Hsiung Hsu, and Malte Stuecker for their outstanding comments on specific chapters of the initial draft manuscript. Excellent editing was provided by May Izumi of the School of Ocean and Earth Science and Technology (SOEST), University of Hawaii. We are deeply grateful for her meticulous and timely work. We would also like to thank Boyi Lu and Haley Okun, graduate students from the Department of Atmospheric Sciences, University of Hawaii, and Lingwei Meng from the Geophysical Fluid Dynamics Laboratory of the Princeton University, for their technical assistance. Meng-Shih Chen and Ching-Teng Lee of the CWB Forecast Center kindly provided TC forecast verification statistics. Some financial support from SOEST is acknowledged.

Abbreviations

AC	anomaly correlation
ACE	accumulated cyclone energy
ADT-HURSAT	Advanced Dvorak Technique-Hurricane-Satellite
AEJ	African easterly jet
AGCM	atmospheric general circulation model
AMM	Atlantic meridional mode
AMO	Atlantic multidecadal oscillation
AMOC	Atlantic meridional overturning circulation
APCC	APEC Climate Center
BMA	Bayesian model averaging
BSISO	boreal summer intraseasonal oscillation
BSS	Brier skill score
CA	constructed analogues
CCA	canonical correlation analysis
CDD	annual or seasonal maximum length of consecutive dry days
CDF	cumulative distribution function
CFSv2	climate anomalies from the NCEP Coupled Forecast System Model Version 2
CGCM	coupled general circulation model
CHIPS	Coupled Hurricane Intensity Prediction System
CISK	conditional instability of the second kind
CLIPER	climatology plus persistence
CLIVAR	climate variability and predictability
CMIP	Climate Model Intercomparison Project
CMR	central mountain range
CNP	central North Pacific
CP	Central Pacific
CPC	Climate Prediction Center

CWB	Central Weather Bureau
DJF	December to February
ECHAM-SIT	an AGCM Coupled with Snow-Ice-Thermocline One-Column Ocean Model
ECMWF	European Center for Medium-range Weather Forecast
ECS	East China Sea
EEOF	extended empirical orthogonal function
EIO	East Indian Ocean
EMDR	Eastern Main Development Region
ENP	eastern North Pacific
ENSO	El Niño–Southern Oscillation
EOF	empirical orthogonal function
EP	Eastern Pacific
EPC	Eastern Pacific Cooling
ER	Equatorial Rossby
ERA-Interim	ECMWF Re-Analysis Data
ERSST	extended reconstructed sea surface temperature
ETCCDI	Expert Team on Climate Change Detection and Indices
FLOR	GFDL Forecast-oriented Low Ocean Resolution and Atmospheric Model
FLOR-FA	GFDL FLOR with Flux Adjustment
FMA	February to April
FSU	Florida State University
GCM	general circulation model
GEV	generalized extreme value
GFDL	Geophysical Fluid Dynamics Laboratory
GPI	genesis potential index
H	hurricanes
HiFLOR	GFDL Forecast-oriented Low Ocean Resolution and Higher Resolution Atmospheric Model
HD	hurricane days
HURDAT2	Atlantic Hurricane Database
IBTrACS	International Best Track Archive for Climate Stewardship
IID	independent and identically distributed
IPCC	Intergovernmental Panel on Climate Change
IPCC AR5	Intergovernmental Panel on Climate Change Fifth Assessment Report
IR	intensification rate
IRI	International Research Institute for Climate and Society
ISO	intraseasonal oscillation

ITCZ	Intertropical Convergence Zone
IWTC-III	WMO Third International Workshop on Tropical Cyclone
IWTC-VI	WMO Sixth International Workshop on Tropical Cyclone
IWTC-VII	WMO Seventh International Workshop on Tropical Cyclone
JASO	July to October
JJA	June to August
JJASO	June to October
JJASON	June to November
JTWC	Joint Typhoon Warning Center
K	Kelvin
LAD	least absolute deviation
LHF	latent heat flux
LMI	lifetime maximum intensity
LOOCV	leave one out cross validation
LSE	least square errors
MC	maritime continent
MCA	maximum covariance analysis
MCMC	Markov Chain Monte Carlo
MDR	main development region
ME	mixed El Niño
MGL	Mean Genesis Location
MH	major hurricanes
MHD	major hurricane days
MJJASON	May to November
MJO	Madden–Julian Oscillation
MKV	Markov model
MME	multimodel ensemble
MPI	maximum potential intensity
MRG	mixed Rossby gravity
MSE	moist static energy
MT	a combination of MRG Wave and TD-type disturbance
NA	North Atlantic
NAM	Northern Annular Mode
NAO	North Atlantic Oscillation
NCEP	National Centers for Environmental Prediction
NGEV	non-stationary generalized extreme value
NOAA	National Oceanic and Atmospheric Administration
NOAA/OGP	NOAA Office of Global Programs
NPI	North Pacific Index
NPO	North Pacific Oscillation

NS	named tropical storms and hurricanes
NTC	net tropical cyclone
NWS	National Weather Service
OLR	outgoing longwave radiation
ONI	ocean Niño index
PCA	principal component analysis
PDF	probability density function
PDI	power dissipation index
PDO	Pacific Decadal Oscillation
PMM	Pacific Meridional Mode
QBO	quasi-biennial oscillation
QBW	quasi-biweekly
QBWO	quasi-biweekly oscillation
R	largest single 24-h value in each of n years or seasons
R50	annual or seasonal count of days with daily rainfall $\geq$50 mm
R5d	annual or seasonal maximum consecutive 5-day rainfall
RCM	regional climate model
RCP	representative concentration pathway
RI	rapid intensification
RMSE	root-mean-squared errors
RPSS	ranked probability skill score
RSMC	Regional Specialized Meteorological Center
S2S	subseasonal to seasonal
SCS	South China Sea
SDII	Simple Daily Intensity Index
SLP	sea level pressure
SO	Southern Oscillation
SP	South Pacific
SPCZ	South Pacific Convergence Zone
SST	sea surface temperature
SSTA	sea surface temperature anomalies
SUB	subtropical
SVM	support vector machine
TC	tropical cyclone
TCTS	tropical cyclone translation speed
TD	tropical disturbance
TPW	tropospheric precipitable water
TRMM	Tropical Rainfall Measuring Mission
TRP	tropical
UK	United Kingdom

US	United States
VWS	vertical wind shear
WCRP	World Climate Research Program
WES	wind-evaporation–sea surface temperature
WMDR	Western Main Development Region
WMO	World Meteorological Organization
WNP	Western North Pacific

1

Introduction

Interest in understanding how and why tropical cyclone (TC) activity in various ocean basins is modulated by climate variability has grown substantially over the last four decades. This interest stems from the fact that the TC is one of the most destructive natural catastrophes and causes loss of lives and enormous property damage on a global scale. The large-scale circulation patterns conducive to TC activity during an extreme climate mode differ profoundly from those of an opposite climate mode. In this book, climate modes encompass a suite of time scales, ranging from the shortest ones within a season (90 days) such as the quasi-biweekly oscillation (QBW) and the Madden–Julian Oscillation (MJO), to the interannual variability such as the El Niño-Southern Oscillation (ENSO), North Atlantic Oscillation, and Atlantic and Pacific Meridional Mode, to the multi-decadal variability such as the Pacific Decadal Oscillation and the Atlantic Multidecadal Oscillation. TC activity includes formation location, frequency of storm occurrence, life span, tracks, landfall rate, and/or storm intensity.

The ocean basins considered here cover the western North Pacific (WNP) and the South China Sea, eastern and central North Pacific, South Pacific, and North Atlantic. Because reliable TC records for the Indian Ocean are relatively short, TC activity in the North and South Indian Oceans are excluded. This book is organized as follows: description of the intraseasonal oscillation in Chapter 2; interannual to interdecadal variability in Chapter 3; modulation of TC activity in each ocean basin in Chapter 4; discussion on the subseasonal to seasonal TC prediction in Chapter 5; typhoon rainfall variations under changing climate in Chapter 6; followed by Chapter 7 for future TC projections.

This book begins with the subject of intraseasonal oscillation (ISO), a time scale that is longer than 10 days but shorter than 90 days (Chapter 2). For the ISO, two distinct modes stand out clearly: the Madden–Julian Oscillation (MJO) and quasi-biweekly oscillation. The MJO phenomenon generally spans a time window of 30–60 days while the latter has a typical 14-day time scale. The MJO is fascinating and characterized by a planetary zonal wavenumber one structure in the global tropics. It exhibits a baroclinic vertical structure in winds and pronounced seasonality in its propagation direction. In the boreal winter, the MJO convection propagates eastward fast from the Indian Ocean to the equatorial central Pacific. Associated with the MJO deep convection are fluctuations in surface pressure and zonal winds at both lower and upper troposphere. In the boreal summer, the MJO system takes a different route, tracking northward in the Indian monsoon region and northwestward over the western North Pacific and South China Sea. Major aspects of the MJO to be discussed are the initiation process of the convection, periodicity, mechanisms for eastward propagation, role of the Maritime Continent in perturbing the eastward propagation, and mechanisms for the northward propagation in the boreal summer. We will also discuss three major theories to account for many aspects of the MJO. They include the coupled Kelvin–Rossby wave theory, moisture-mode theory, and unified dynamic moisture-mode theory. The recent moisture-mode theory that emphasizes the asymmetry of the column-integrated moist static energy tendency and related processes that are instrumental to the propagation and intensification of MJO convection will be discussed.

In Chapter 3, several climate feedback mechanisms involving the tropical ocean and atmosphere system are first introduced. Such feedbacks include the wind-evaporation process, Bjerknes hypothesis, footprinting mechanism, sea surface temperature–cloud feedback, and sea surface temperature–water vapor feedback. This is followed by the El Niño phenomenon, which is the leading mode of tropical climate variability with a recurring period of three to seven years. El Niño is manifested by the anomalous warming of the eastern and central tropical Pacific. La Niña is the opposite of El Niño and refers to anomalous cooling and steady and strong easterly trade winds in the tropical Pacific. Coupled to the oceanic El Niño and La Niña events is the atmospheric Southern Oscillation, a large-scale pressure seesaw between the tropical eastern and western Pacific. Taken together, the term El Niño–Southern Oscillation (ENSO) describes the coupled atmosphere–ocean interaction in the vast tropical Pacific basin. Two prevailing theories that provide foundation for understanding ENSO dynamics are presented. They are the delayed oscillation theory and recharge–discharge mechanism. Studies over the last 15 years show that there are at least two types of El Niño: Eastern Pacific and Central Pacific events, which modulate regional TC activity in a different manner. ENSO forecasts and a new approach called the Bayesian model averaging are followed. Other climate modes on the interannual time scale of interest include: North Atlantic Oscillation (NAO); Pacific Meridional Mode (PMM); and Atlantic Meridional Mode (AMM). On the decadal to interdecadal time scale well-known climate modes include the Pacific Decadal Oscillation (PDO) and the Atlantic Multidecadal Oscillation (AMO). The Aleutian low, atmospheric teleconnections from the tropics, and midlatitude ocean dynamics appear to play key roles in driving the PDO variability. The AMO variability is attributed to oceanic meridional overturning circulation in the Atlantic, stochastic forcing from the atmosphere, cloud feedback, and aerosol effects.

Chapter 4 discusses how TC activity in various ocean basins is influenced by climate variability of the ocean–atmosphere system. Indeed, marked ISO, interannual and multidecadal variations are intrinsic to tropical climate and they exert a pronounced impact on TC changes. For example, the 2020 Atlantic hurricane season was extremely active and set numerous records for overall activity. It featured a record 30 named storms, of which 13 developed into hurricanes and 6 further intensified into major hurricanes. Twelve out of these 30 storms made landfall in the contiguous United States, in stark contrast to the mean rate of less than two landfalls per year during 1900–2000. Besides these staggering numbers, accumulated cyclone energy (ACE), which measures the strength and duration of tropical storms and hurricanes, was 75% above the long-term mean of 1981–2010. In 2020, La Niña, vigorous African easterly waves, a positive AMM state, and abnormally warm sea surface temperatures over the North Atlantic main development region all fueled the extremely active North Atlantic hurricane season. Also noteworthy is the consecutive above average Atlantic hurricane seasons since 2016. Whether this above normal activity over the last 5 years will continue in the next 5–10 years is of a great scientific and socio-economic interest.

In Chapter 4, background climatology for each basin is first introduced. Large-scale flow patterns and equatorial waves, which are regarded as TC precursors, are then followed. For the WNP, eastern North Pacific, and North Atlantic, the tropical depression-type disturbances appear to be the most common equatorial waves, trailed by the equatorial Rossby waves. For the South Pacific, it is just the opposite. Regional TC changes in each ocean basin as influenced by various climate modes are the major themes of this chapter. Central to TC changes are the genesis frequency, location of formation, tracks, and landfalls. Recent studies show that regional TC activity is modulated by a combination of two or three climate modes, such as the MJO and ENSO. Modulation of the Eastern Pacific El Niño, Central Pacific El Niño, PMM, and AMM on TC activity over the Pacific and Atlantic Oceans from recent studies is described. Subsequently, the attention is focused on decadal and interdecadal TC variability. The chapter concludes with a section on the observed variations in TC attributes such as frequency of occurrence, translation speed, intensity, and meridional migration of the latitude of lifetime maximum intensity based on historical or modern data.

Because of the huge socioeconomic repercussions incurred by TCs, developing a sound and modern method for predicting TC activity from weeks, months, or seasons in advance is becoming increasingly important. Better forecasts, advance warnings, and proper emergency management efforts would all lead to a reduction of loss of life and property damage. Chapter 5 provides a review of the current status regarding subseasonal to seasonal TC prediction and the corresponding methods used therein by various researchers and government agencies. Broadly speaking, prediction methods can be classified into three approaches: purely statistical; dynamical; and statistical-dynamical. Statistical methods utilize logistic regression, least absolute deviation regression, Poisson regression, and/or multiple linear regression techniques. A Poisson generalized regression model cast in the Bayesian framework is also applied to probabilistic forecast TC activity. Dynamical forecast methods rely on dynamical climate models. Dynamical seasonal TC forecasts were regularly issued by many organizations throughout the world in the last 20 years using either atmospheric general circulation models forced by sea surface temperature anomalies or by coupled atmosphere–ocean models. Many models were constantly upgraded with increasing higher horizontal resolution, improved model physics, and ensemble forecasting techniques. Statistical-dynamical methods refer to the use of forecast information (e.g., predictors) from dynamical models in predicting TC metrics using statistical methods. This hybrid approach takes advantage of the future, yet-to-be-observed values from dynamical models and statistical relationships between TC metrics and environmental parameters to forecast future TC changes. For the subseasonal to seasonal TC, an international effort was initiated in the last few years to develop the subseasonal to seasonal (S2S) prediction data set. Specifically, the S2S dataset contains dynamical subseasonal forecasts and reforecasts with leads up to 60 days.

Heavy rainfall and flooding resulting from TCs result in devastating consequences on society, human and animal life, and economics. Besides natural variability, anthropogenic climate change elevates the water vapor capacity in the atmosphere. A warmer atmosphere can hold more moisture in the air – about 7% more per 1°C of warming. With increasing moisture under global warming, TCs are expected to produce heavier rainfall, a notion consistent with climate models that show a projected increase in TC precipitation rate. Chapter 6 describes how TC rainfall may have changed over the last several decades. Taiwan is used as an example because of the availability of reliable and long-term hourly and daily rainfall records, its unique geographic location with regard to the prevailing typhoon tracks, and abundance of published studies relevant to the subject. Here, changes in rainfall frequency and intensity, storm duration and translation speed associated with typhoons and typhoon tracks are considered. This chapter concludes with a section on long-term variations in return levels during the typhoon season modulated by time and an ENSO index using a non-stationary generalized extreme value distribution.

Chapter 7 is aimed toward providing the state-of-the-art knowledge of future TC projections based on the numerical simulation from a suite of climate models. Historically, low and medium-resolution (commonly 120–300 km) global climate models have been used to simulate TC numbers in a warmer climate. Because the horizontal resolutions used in these models are not high enough to estimate intensity changes in TCs, a "dynamical downscaling" or "statistical-dynamical downscaling" method is used to yield finer details of TCs with a regional or hurricane model. Independently, a high-resolution (~20 km) global atmospheric model is used to improve fidelity of TC simulations. Modeling results commonly indicate a global decrease in TC numbers in the future. A few hypotheses relevant to the projected decrease in TC frequency are discussed including: weakening of tropical overturning circulation; increasing of entropy deficit; and increase in ventilation index. More recently, climate models with higher resolution also demonstrated an increase in storm numbers in the future, a stark contrast to the earlier view. Hypotheses are also advanced to help explain this new result. At the end of this chapter is an updated review of the expert opinions on the potential future changes in TC in the context of the annual frequency of TCs, annual frequency of category 4 and 5 TCs, storm intensity, and mean precipitation rate induced by TCs.

2

Climate Variability. Part I: Intraseasonal Oscillation

2.1 Introduction

Pioneered by Madden and Julian (1971, 1972), the intraseasonal oscillation (ISO) did not receive much attention until mid to late 1980s. Since then there has been a surge of studies based on observations, numerical modeling, and theoretical treatment of its phenomenon. According to the two seminal papers by Madden and Julian, the ISO is marked by a planetary zonal wavenumber one structure and a fast ($\sim$5 m s^{-1}) eastward-propagating feature from the Indian Ocean to the central Pacific along the equator within a time window of 40–50 days. Associated with the ISO deep convections are fluctuations in surface pressure and zonal wind component at both lower and upper troposphere. A slower poleward propagation in the Pacific is also found. The discovery of tropical intraseasonal variability can be traced back in Chinese literature to Xie et al. (1963), who published eight years prior to the first paper by Madden and Julian (1971). Because of its publication in a non-English journal, it has eluded the international research community working in the field (Li et al., 2018).

Because of the remarkable discovery of this fascinating phenomenon in the global tropics, the Madden–Julian Oscillation (MJO) became synonymous with the ISO in the field, and the original 40–50-day oscillation band has broadened to periods of 20–90 days. The unique ISO time scale is different from synoptic-scale variability, which is usually less than 10 days, and the seasonal variations that run longer than 90 days. The MJO exhibits multiscale structure of the convective complex and motions (Nakazawa, 1988). It also exerts a profound influence on tropical cyclone activity, the onset of the Asian and Australian monsoon (e.g., Webster, 1986; Hendon and Liebmann, 1990; Hung and Yanai, 2004), extreme precipitation events in California (Jones, 2000), and modulate precipitation variations in South America (e.g., de Souza and Ambrizzi, 2006; Julia, 2012). For example, the onset of the Indian summer monsoon is primarily determined by two factors: the annual march of the sun and the phase of the MJO wave (Webster, 1986). Only when monsoon westerlies were enhanced by the right phase of the MJO wave did the monsoon onset occurs.

In the following sections, we will first introduce physical descriptions about the MJO phenomenon, followed by a discussion of the periodicity and mechanisms for eastward propagation during the boreal winter (austral summer), the initiation mechanism of major convection over the Indian Ocean, mechanisms for the northward propagation over the Asian monsoon regions during the boreal summer, and the role of the Indonesian Maritime Continent (MC) on the propagation of the MJO convection. Three major theories relevant to many MJO phenomena will be reviewed. A brief introduction of another kind of the ISO, namely, the 10–20-day oscillation, will also be provided. We will focus on the MJO phenomenon because of its dominance in the ISO variability. An excellent review of the MJO phenomenon can be found in Zhang (2005), Waliser (2006), Li (2014), Wang et al. (2016), Li and Hsu (2018), and Li et al. (2020) among others.

2.2 Physical Description of the Madden–Julian Oscillation (MJO) from Observations

Figure 2.1 displays the time–longitude cross section of outgoing longwave radiation (OLR) anomalies and MJO-filtered outgoing longwave radiation (OLR) anomalies between 10°N and 10°S from October 1, 2011 to April 2, 2012 (Kiladis et al., 2014). To isolate the MJO signal, the bandpass-filtered data which generally retain 30–60 or 20–90 days are commonly used. In the tropics, the OLR mainly reflects cloud-top temperature, with low OLR values corresponding to cold and high clouds, which generally denote enhanced convection. Thus, an inverse relationship holds between OLR and tropical convection. A succession of eastward-propagating OLR anomalies (or enhanced tropical convection) from Africa or west Indian Ocean to the western Pacific or even to the international dateline is evident (Fig. 2.1). While the convection of the MJO vanishes in the eastern equatorial Pacific, its signal in surface pressure and wind continues to propagate farther eastward as free waves, decoupled from convection, at much higher speeds (Zhang, 2005). The typical zonal extent of an MJO event is approximately 12,000–20,000 km, so a planetary-scale tropospheric circulation is evident in its footprint.

At lower troposphere, the MJO exhibits a slower eastward propagation ($\sim$5 m s^{-1}) over the warm Indian Ocean and the western Pacific, but continues to speed up ($\sim$15 m s^{-1}) over the cold eastern Pacific. The evolution of the bandpass-filtered upper-level velocity potential anomalies between 30°N and 30°S is shown in Fig. 2.2 (Higgins and Shi, 2001). The horizontal wind vector can be decomposed

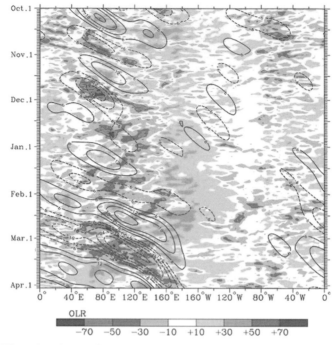

Fig. 2.1 Time–longitude diagram of OLR anomalies (shading) and MJO-filtered OLR (contour) averaged between 10°S and 10°N from October 1, 2011 to April 2, 2012. Negative anomalies are in blue at 20 W m^{-2} interval. The contour interval is 6 W m^{-2}. A black and white version of this figure will appear in some formats. For the color version, refer to the plate section.
Source: Kiladis et al., 2014; ©American Meteorological Society. Used with permission.

into the sum of divergent and rotational wind components and the velocity potential refers to the divergent part of the wind. In meteorology, the velocity potential describes the large-scale convergent or divergent component of the horizontal wind and the center of the velocity potential implies rising or sinking motion in the atmosphere. In the tropics, the upper-level divergence center derived from the velocity potential field is usually associated with deep convection. In Fig. 2.2, the time interval for each panel is five days apart and day 0 denotes MJO convection over the Central America. The upper-level velocity potential center clearly moves eastward and a global eastward circumferential propagation along the equator is evident. It should be noted that a continuous global propagation of the MJO only exists in the upper level as an atmospheric response to the convective heating perturbation.

The longitude–vertical cross section along the equator when the MJO center is located near 125°E is shown in Fig. 2.3 (Sperber, 2003). Upper-level divergence and low-level convergence are seen over the maximum convection center

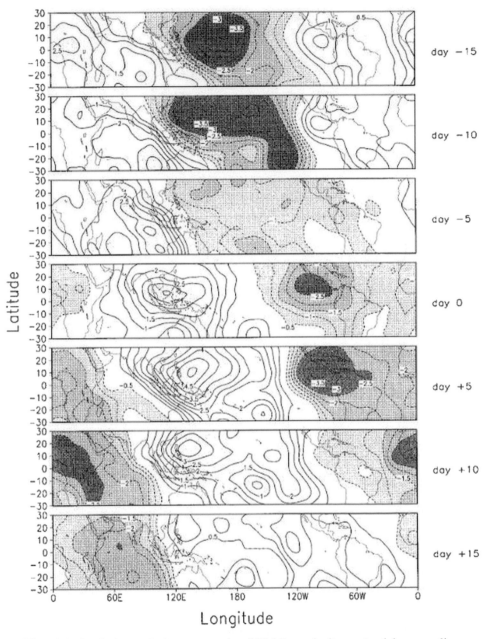

Fig. 2.2 Evolution of the composite 200-hPa velocity potential anomalies (10^6 m^2 s^{-1}) in MJO from day −15 to day 15. Negative contours are dashed.
Source: Higgins and Shi, 2001; ©American Meteorological Society. Used with permission.

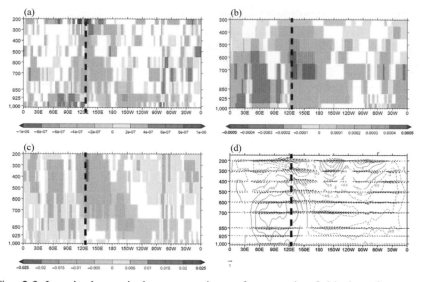

Fig. 2.3 Longitude–vertical cross sections of composite fields in reference to MJO center near 125°E (thick dashed line). (a) divergence (s^{-1}), (b) specific humidity (kg kg^{-1}), (c) vertical velocity (Pa s^{-1}), and (d) zonal wind and vertical velocity. In (a) convergence (divergence) is green or blue (yellow or red). In (b), positive (negative) humidity is green or blue (yellow or red), In (c), rising (sinking) motion in green or blue (yellow or red). Note that the vertical velocity (Pa s^{-1}) is multiplied by -100 to yield scaling compatible with the v (meridional) wind (m s^{-1}). A black and white version of this figure will appear in some formats. For the color version, refer to the plate section.
Source: Sperber, 2003; ©American Meteorological Society. Used with permission.

(Fig. 2.3a). Below 850 hPa, there is clear zonal asymmetry, with a juxtaposition of convergence to the east and divergence to the west of convection center. The rising motion tilts westward with increasing height and the maximum ascending motion occurs near 300–400 hPa (Fig. 2.3c). The specific humidity anomalies roughly follow the vertical velocity profile (Fig. 2.3b) and maximum humidity coincides in phase with the convection in middle-lower troposphere, with positive moisture anomalies leading the convection (Maloney and Hartmann, 1998). Low-level westerlies appear to the west of the convection center over the Indian Ocean while low-level easterlies, with longer extent, appear to the east of the convection center (Fig. 2.3d).

The empirical orthogonal function (EOF) analysis is then applied to the OLR data and the leading mode is chosen to represent the pattern and evolution of the MJO convection (Sperber, 2003). Both the EOF and principal component analysis (PCA) are commonly used in climate analysis and refer to the same set of procedures. Briefly, EOF (or PCA) reduces a large number of variables to a set of fewer new variables; however, the newer variables represent the maximum possible portion of the variability contained in the original data. By regressing the leading mode time series of OLR against other bandpass-filtered variables such as winds, rainfall, vertical velocity, shortwave radiation, and latent heat flux at lag 0 when the convection is centered over the Maritime Continent (MC), Sperber (2003) noted that the 200-hPa wind is dominated by easterlies while westerlies at 850 hPa prevail at and west of convection, indicative of a baroclinic vertical wind structure. A close correspondence between the reduced OLR and enhanced rainfall is seen over the MC, together with the enhanced upward motion and reduced shortwave radiation being collocated with rainfall estimates. Over the Indian Ocean and the MC, low-level westerly anomalies enhance the climatological mean wind, leading to enhanced latent heat flux to the atmosphere. Between 150°E and the dateline, the equatorial easterly anomalies weaken the climatological westerly flows, resulting in a weaker latent heat flux ahead of the MJO convection. The pattern of the net surface heat flux is similar to that in the latent heat flux, with the latter accounting for up to 70% of the net heat flux anomalies.

2.3 The Periodicity of the MJO

The periodicity of the MJO is commonly known as the time required for the upper tropospheric disturbance to navigate around the global tropics from the western Indian Ocean and returning to the same domain. It is hypothesized that the return of the eastward-propagating upper-level divergence would trigger deep convection for the next MJO cycle, although it is unclear how this process would work. Moreover, while the circumnavigating phenomenon may be evident in some composite studies, it is not clear in other case studies (e.g., Hsu et al., 1990; Li et al., 2015) or numerical experiments (Zhao et al., 2013). Because the MJO convection tends to remain quasi-stationary over the MC, its fast eastward propagation together with the slow development of convection over the MC could determine the periodicity of the MJO cycle (Hsu and Lee, 2005). Another mechanism, the so-called recharge–discharge paradigm of the moist static energy (MSE) or moist entropy, is also postulated to explain the periodicity of the MJO (Blade and Hartmann, 1993; Maloney, 2009).

The MSE is defined as $m = C_pT + gz + L_vq$, where T is temperature, z the height, q the specific humidity, C_p the specific heat at constant pressure, g the gravitational acceleration, and L_v the latent heat of vaporization. According to this paradigm, the low-level MSE slowly builds up (i.e., recharging) in the Indian Ocean and destabilize the atmosphere before the formation of the MJO deep convection. This is subsequently discharged by vertical motions during and after MJO convection (Kemball-Cook and Weare, 2001; Maloney, 2009). In other words, the MJO period is determined mainly by the slow recharge time for column MSE and the duration for the convective episode. In the lower troposphere, the MSE anomalies in the area where SST is warm are regulated by water vapor or SST variations (Maloney, 2009; Wang et al., 2016).

Maloney (2009) diagnosed the MSE budget of a composite MJO in a climate model. The leading terms in the vertically integrated MSE budget over the western Pacific warm pool (155°E) are the horizontal advection of MSE and surface latent heat flux (LHF). While the horizontal advection consistently recharges (discharges) column MSE within low-level easterly (westerly) anomalies before (during and after) peak precipitation, the behavior of the LHF acts reversely in sign to that of the horizontal advection term, yielding a smaller MSE tendency. As a result, the recharge–discharge time of MSE become more gradual than it would be when only the horizontal advection is considered. This has implication for the time scale of the intraseasonal oscillation.

2.4 Initiation of the MJO Convection

The initiation of the MJO convection occurs in the western Indian Ocean and is closely connected to its periodicity (e.g., Madden and Julian, 1971, 1972; Hendon and Salby, 1994; Maloney and Hartmann, 1998; Kemball-Cook and Weare, 2001). The convection in the Indian Ocean could have been triggered by an extratropical or midlatitude event. Hsu et al. (1990) first noted that Rossby wave train from the Northern Hemisphere propagates into the tropics to initiate convection. However, Zhao et al. (2013) argued that wave source from the Southern Hemisphere is more likely to trigger MJO convection. Based on the composite of OLR data when maximum MJO convection is located in the east Indian Ocean during the boreal winter, Zhao et al. (2013) found that nine days prior to that date, the convection formed in the southwestern Indian Ocean and then propagated eastward toward the equator while it strengthened. The low-level easterly anomaly, which is commonly observed over the equatorial Indian Ocean in the early stages of the MJO life cycle, is likely to provide lifting and frictional effects against the east African highland. By doing so, near-surface moisture convergence tends to occur on the windward slope of the topography, which preconditions the lower troposphere for the later development of deep heating anomalies (Hsu and Lee, 2005; Wu and Hsu, 2009). Recently, Hung and Sui (2018) showed that when the convection is suppressed in the central Indian Ocean, moistening by wave-induced boundary-layer convergence and advection by low-level anomalous easterlies contributes to the initiation of the MJO convection in the western Indian Ocean.

Defining a region between 20°S–0° and 50°E–70°E as the MJO initiation region, Zhao et al. (2013) examined the time evolution of the ISO (20–90 days) anomalies over this region, up to 30 days prior to the major convection in the eastern Indian Ocean. They found that the OLR anomaly transitions from positive (e.g., suppressed phase) to negative (active phase) 15 days prior to the major convection in the eastern Indian Ocean, at a time when the vertical motion also changes synchronously from an anomalous descending motion to an anomalous ascending motion (Fig. 2.4). Therefore, day –15 was regarded as the initiation date. One week

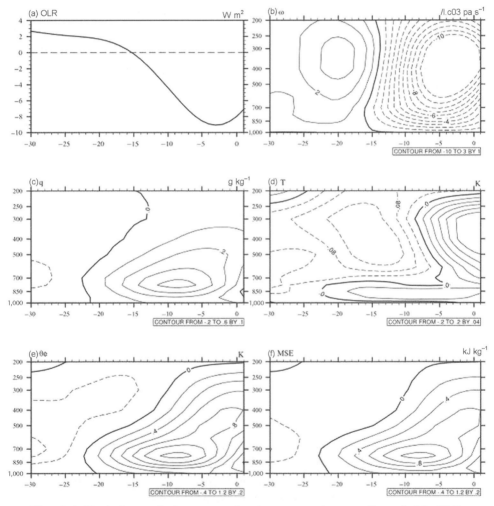

Fig. 2.4 Evolutions of the composite intraseasonal anomalies of (a) OLR, (b) vertical motion, (c) specific humidity, (d) temperature, (e) equivalent potential temperature, and (f) moist static energy, averaged over the MJO initiation region (20°S–0°, 50°E–70°E). Time (in days) is on the *x*-axis.

Source: Zhao et al., 2013; ©American Meteorological Society. Used with permission.

prior to the initiation date, the low-level specific humidity and temperature anomaly transition from negative to positive values. The marked increase in both specific humidity and temperature leads to an increased equivalent potential temperature and MSE. As a result, the atmosphere between 850 hPa and 500 hPa are already convectively unstable approximately one week prior to the initiation date. Thus, low-level moistening prior to the convection initiation (day −15) is crucial for the establishment of the convectively unstable atmosphere.

Applying an MJO-filtering operation to the moisture budget equation, Zhao et al. (2013) found that the positive moisture tendency during the initiation period, averaged from day −25 to day −15, is mainly contributed by the horizontal moisture advection. This process is different from that during its eastward propagation, in which the low-level moistening is caused by vertical moisture advection associated with the boundary-layer convergence. By decomposing the specific humidity and wind into three components, namely, the low-frequency background state with a period longer than 90 days, MJO with a period of 20–90 days, and high-frequency component with a period less than 20 days, the low-level MJO flow is dominated by anomalous easterlies and two anticyclonic Rossby gyres (Fig. 2.5). Such a

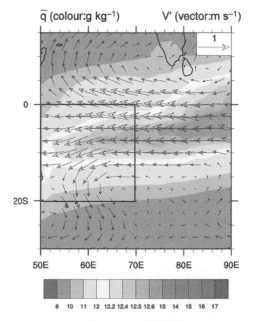

Fig. 2.5 Vertically integrated (1,000-hPa to 700-hPa) intraseasonal wind and low-frequency background state specific humidity averaged during the initiation period (day −25 to day −15). A black and white version of this figure will appear in some formats. For the color version, refer to the plate section.
Source: Zhao et al., 2013; ©American Meteorological Society. Used with permission.

pattern is typically observed when the suppressed MJO convection occurs in the eastern Indian Ocean. The anomalous easterlies advect the slowly varying background state high moisture content near 10°S and increase the low-level moisture over the western Indian Ocean for the initiation process. This moisture advection mechanism for the MJO initiation can also be regarded as a "recharge" process (Zhao et al., 2013). Based on the composited OLR and wind fields during the initiation period, the easterly anomaly over the equatorial Indian Ocean was mainly driven by the negative heating anomaly over the eastern Indian Ocean, implying the initiation of new convection over the west Indian Ocean is a direct Rossby wave response to the preceding, downstream suppressed MJO convection over the eastern Indian Ocean (Zhao et al., 2013). The next attention will discuss the mechanisms for the eastward MJO propagation.

2.5 Mechanisms for Eastward MJO Propagation

For any model that can simulate the MJO movement, an essential test is its ability to realistically demonstrate the eastward propagation of the major convection and accompanying zonal wind component over the equatorial Indian Ocean and western Pacific. Lau and Peng (1987) suggest that the tropical wave-conditional instability of the second kind (CISK) is responsible for the eastward propagation of the MJO convection. CISK refers to a concept in which cumulus convection and the large-scale environmental motion interact (Charney and Eliassen, 1964; Ooyama, 1964). The latent heat released by cumulus convection drives the large-scale circulation and maintains the low-level convergence. In turn, convergence moistens and destabilizes the environment through boundary-layer ascent and thus maintains favorable conditions for development of cumulus convection. Wave-CISK (Lindzen, 1974) considers the feedbacks between deep convection and low-level moisture convergence in the presence of the large-scale equatorial waves.

Low-level convergence in advance of the major convection is suggested as a key component for the eastward MJO propagation through the frictional wave-CISK mechanism, which includes the frictional effect on moisture convergence (Wang, 1988; Wang and Rui, 1990; Hendon and Salby, 1994; Maloney and Hartmann, 1998). Frictional convergence refers to wind convergence due to the Ekman effect when easterlies straddle the equator in the atmospheric boundary layer. In the presence of equatorial easterlies, the Ekman transport in the atmospheric boundary layer is to the left (right) of the wind in the Northern (Southern) Hemisphere; therefore, boundary-layer convergence is expected. This process also implies the importance of shallow cumulus clouds and boundary-layer interaction in MJO dynamics (e.g., Johnson et al., 1999; Kikuchi and Takayabu, 2004; Lin et al.,

2004). Shallow convection can bring moisture out of the boundary layer and thereby moisten the free troposphere, leading to MSE buildup. It is hypothesized that the time for the MJO convective center to move from the Indian Ocean to the western Pacific depends on how fast the boundary-layer frictional convergence moistens the atmosphere in the western Pacific (Maloney and Hartmann, 1998). Apart from tropical wave dynamics, thermodynamic processes such as low-level moistening, surface turbulent fluxes and radiative fluxes are also important for the growth and propagation of the MJO (e.g., Sobel et al., 2014; Johnson et al., 2015; Tseng et al., 2015). However, the latter processes are not necessarily independent of wave dynamics. In the following, we briefly discuss three major current theories to account for many aspects of the MJO. They include the coupled Kelvin–Rossby wave theory, moisture mode theory, and unified dynamic moisture mode theory.

2.5.1 Convectively Coupled Kelvin–Rossby Wave Theory

Wang and Li (1994) used a 2.5-layer model that considers both the condensational heating in free atmosphere and boundary-layer sea surface temperature (SST) forcing. Using this model, they were able to simulate the eastward propagation of the MJO convection. The simulated boundary-layer convergence leads the convection, which is consistent with observations (Fig. 2.3). In their model, the perturbation condensation heating in the troposphere is proportional to the vertically integrated moisture convergence. The heating induces a Kelvin wave response to the east of the MJO convection and a Rossby wave response to the west of the convection (Fig. 2.6), thus forming a Kelvin–Rossby wave horizontal structure (e. g., Gill, 1980; Wheeler and Kiladis, 1999).

   To the west of the convection, two cyclonic Rossby wave gyres appear in the lower troposphere on both side of the equator with equatorial westerlies, accompanied by two anticyclonic Rossby wave gyres in the upper troposphere (e.g., Rui and Wang, 1990; Hsu and Li, 2012). As a Kelvin wave response to the east of the convection, the low-level low-pressure anomaly with equatorial easterly induces boundary-layer convergence and shifts the heating to the east of the MJO convection, leading to an eastward heating tendency. Because of this free atmospheric wave–convection–boundary-layer convergence interaction, the MJO envelope moves eastward (Wang and Li, 1994). This mechanism is known as the convectively coupled Kelvin–Rossby wave theory. The term convectively coupled means moist deep convection is associated with these waves (Wang and Rui, 1990; Wang and Li, 1994; Wheeler and Kiladis, 1999; Wang et al., 2016; Wang and Chen, 2017).

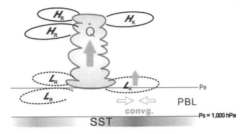

Fig. 2.6 Schematic diagram showing a Kelvin–Rossby wave couplet in response to the MJO convection with heating (Q). Also shown are the boundary-layer convergence (convg.) induced by atmospheric wave dynamics and SST. Dashed ellipses with L_R and L_K denote low-pressure anomalies (L) associated with Rossby and Kelvin waves response, respectively. Solid ellipses with H_R and H_K denote the high-pressure anomalies (H) associated with Rossby and Kelvin waves response in the upper troposphere. Red and blue shading denote positive and negative SST anomalies, respectively. Solid green arrows denote anomalous ascending motion. Ps and Pe denote pressures at the bottom and top of the atmospheric boundary layer, respectively. A black and white version of this figure will appear in some formats. For the color version, refer to the plate section.
Source: Hsu and Li, 2012; ©American Meteorological Society. Used with permission.

2.5.2 The Moisture-Mode Theory

The moisture-mode theory relies on atmospheric moisture (or MSE) and its feedback with convection (Hsu and Li, 2012; Sobel and Maloney, 2012, 2013; Wang et al., 2017; Li et al., 2020; Hu et al., 2021). The only prognostic variable is the total column water vapor or MSE. Sobel and Maloney (2012, 2013) proposed a moisture model where processes related to modulation of synoptic-eddy drying, advection of background zonal moisture gradient, and frictional convergence are parameterized as sources of MSE. Only when the gross moist stability is negative, eastward-propagating modes can be unstable because of cloud–radiation feedback or moist instability. The negative gross moist stability allows the vertical advection to maintain the MSE anomalies and strengthen the MJO (Sobel et al., 2014).

The modern view of the moisture-mode theory can be separated into two types (Hu et al., 2021). Based on a moisture budget diagnosis for a region 130°E–150°E, 0°–10°S when the MJO convection is in the eastern Indian Ocean, Hsu and Li (2012) suggest that the largest contribution to the ISO moisture tendency comes from the vertical moisture advection term, which is five times larger than its horizontal counterpart. Furthermore, the atmosphere tends to become more (less) potentially unstable to the east (west) of the convection, as noted by Hendon

(1988) and Hsu and Li (2012). For the atmosphere to be convectively unstable, $\frac{\partial \theta e}{\partial z} < 0$. New convection forms east of the existing MJO center only when the local atmosphere becomes convectively unstable and sufficient lifting to saturation occurs. Low-level moistening sets up an unstable stratification of the atmosphere and triggers new convection east of the existing MJO center for the convection to propagate eastward (Hsu and Li, 2012). The zonal asymmetry in moisture perturbation in the boundary layer as gleaned from observation is aligned in such a way that the positive specific humidity anomaly in the boundary layer leads the convection while the negative humidity lags the convection (Fig. 2.3). This can be viewed as the first type of moisture-mode theory.

Now the question arises as to what is the source of moisture to the east of the convection. Low-level convergence is postulated as a possible source for moistening the boundary layer ahead of the convection. While the low-level convergence occurs to the east of the MJO convection, surface evaporation, however, tends to decrease there (Hsu and Li, 2012). Because the background surface wind in the southeastern Indian Ocean in the boreal winter is westerly, the MJO easterly anomalies ahead of the convection erode the surface wind speed, which result in a suppressed surface evaporation, or latent heat flux (LHF). On the other hand, stronger wind speed and LHF are found to the west of MJO convection where westerly anomalies are in phase with the background wind. A weaker (stronger) LHF also induces a warmer (cold) SST anomaly to the east (west) of convection. The warm SST anomaly favors boundary-layer convergence via hydrostatic effect on sea-level pressure (Lindzen and Nigam, 1987). In the meantime, the low pressure atop the boundary layer associated with the Kelvin wave response to the east of convective heating induces convergence in the boundary layer (Fig. 2.6). Hsu and Li (2012) found that the atmospheric wave dynamics outweigh the surface air–sea process and contribute mainly to the boundary-layer convergence and moisture asymmetry, and is responsible for the eastward MJO propagation.

The next question is why negative specific humidity is observed to the west of the convection (Fig. 2.3). Because this is also the region where low-level westerlies prevail, atmospheric Ekman divergence is induced in the boundary layer so that the air above the boundary layer has to sink to replace the air that leaves. The dryness may also be due to the intrusion of the drier off-equatorial air (dry subtropical air) resulting from the twin cyclonic Rossby gyres to the west of the convection. As a result, drying becomes a common feature immediately following the active phase of the MJO (e.g., Hendon and Liebmann, 1990). During the Cooperative Indian Ocean Experiment on Intraseasonal Variability/Dynamics of the Madden–Julian Oscillation field program in late 2011, the source of the dry air was found to come from the subtropical Arabian Sea and the Indian subcontinent (Sobel et al., 2014).

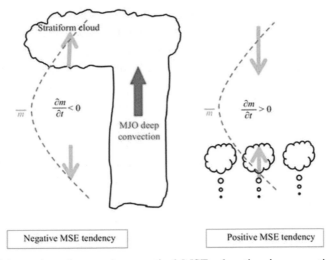

Fig. 2.7 Schematics of anomalous vertical MSE advection in generating a zonal asymmetric MSE tendency. The small clouds to the east of MJO convection denote shallow cumulus convection. The dashed lines denote the mean MSE vertical profile and arrows denote the vertical velocity anomalies.
Source: Wang et al., 2017; ©www.creativecommons.org/licenses/by/4.0/.

The moisture-mode theory also can be understood in terms of the time tendency of column-integrated MSE relative to the MJO convection center (Fig. 2.7), and this is known as the second type of the theory. Specifically, the MSE tendency increases (decreases) to the east (west) of the convection. In the column-integrated MSE budget, the MSE tendency is balanced by the sum of four terms: vertical MSE advection, horizontal MSE advection, sum of surface latent heat and sensible heat fluxes; and sum of shortwave and longwave heating rates. Li et al. (2020) noted that a positive (negative) MSE tendency anomaly occurs to the east (west) of maximum MSE anomalies, favoring an eastward propagation of the MSE maximum. The vertical and horizontal MSE advection terms make the largest and positive contribution to the MSE tendency and lead to MJO eastward propagation, while the heat flux and radiation terms contribute to westward and smaller phase speed. As a result, this combination slows down the eastward movement. Note that the zonal asymmetry of the moisture anomaly is concentrated in the boundary layer, whereas the zonal asymmetry of the column-integrated MSE tendency anomaly is mainly contributed by the free-atmosphere processes.

A question arises as to what physical processes underpin the modulation of vertical and horizontal MSE advection? As shown in Figs. 2.7 and 2.3c, to the east of convection, downward (upward) anomaly prevails in the upper (lower) troposphere (Li et al., 2020). Because the mean MSE is at minimum in the middle troposphere, the vertical profile of the vertical motion anomaly promotes an increase in the column-integrated moist entropy in advance of convection. Conversely, upward (downward) anomaly occurs in the upper (lower) troposphere at rear (or west) of the convection, which results in negative vertical MSE advection (Fig. 2.7). Through a numerical experiment with different prescribed heating profiles, Wang et al. (2017) suggested the role of the stratiform heating to the west of the convection in generating zonally asymmetric vertical motion anomalies in the upper troposphere. The zonal difference in the vertical velocity anomalies in the lower troposphere is mainly caused by the boundary-layer process (Wang and Li, 1994; Hsu and Li, 2012). The horizontal MSE advection is dominated by the meridional MSE advection in the lower troposphere due to negative diabatic heating anomalies associated with the anomalous descent of the vertically overturning circulation to the far east of convection (Kim et al., 2014; Wang et al., 2017). Diabatic heating refers to heating due to latent heat release or radiation. In the presence of the convective background, a descending anomaly causes a negative heating anomaly by inhibiting background precipitation/ convection. The negative heating anomalies then cause poleward flows via an anticyclonic Rossby gyre and result in a positive MSE advection because the MSE is maximum near the equator. As a result, the horizontal MSE advection also positively contributes to the MSE tendency in advance of the MJO convection (Maloney, 2009; Sobel et al., 2014; Wang et al., 2017).

2.5.3 The Frictional Coupled Dynamic Moisture-Mode Theory

While the coupled wave theory emphasizes the interaction between convection and equatorial wave dynamics/boundary-layer frictional effects, the moisture feedback process is neglected. The early view of the moisture-mode theory embraces the moisture–convection feedback but not the wave dynamics/boundary-layer effects and their interaction with convective heating. By integrating these two theories, Wang et al. (2016) and Wang and Chen (2017) postulate a unified dynamic-moisture-mode theory to explain a myriad of MJO phenomenon, including a coupled Kelvin–Rossby wave structure, slow eastward propagation ($\sim$5 m s^{-1}) over warm pool, zonal planetary-scale circulation, boundary-layer moisture convergence leading major convection, and amplification/decay over warm/cold SST regions. For example, observations show that MJO propagate eastward slowly over the Indo-Pacific warm pool but at much faster speed over the cold ocean in the Western Hemisphere. According to Wang et al. (2016), when MJO moves over the warm pool, local strong convective heating reduces the effective static stability. This reduces the convectively coupled eastward-propagating Kelvin wave speed to about 5 m s^{-1} over the warm pool, in agreement with observations.

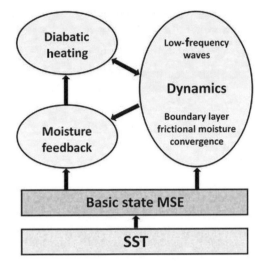

Fig. 2.8 Schematics of the frictional coupled dynamic moisture mode involving the interaction among convection, equatorial wave dynamics/boundary layer frictional moisture convergence, and the moisture feedback.
Source: Wang et al., 2016; ©www.creativecommons.org/licenses/by/4.0/.

Central to the trio-interaction theory of Wang et al. (2016) involves convective heating, moisture feedback, and equatorial waves/boundary-layer dynamics (Fig. 2.8). Under this theory, convective heating interacts with the boundary-layer frictional convergence and moisture feedback. The latter is determined by the moisture convergence induced by waves and boundary-layer processes, and lower-tropospheric MSE, which is mainly controlled by the background SST state. Using the Betts–Miller convective heating parameterization in their model, Wang et al. (2016) and Wang and Chen (2017) found that the boundary-layer moisture convergence provides the coupling among convection and Kelvin–Rossby waves, and selects eastward propagation. In the meantime, boundary-layer frictional convergence also can pump MSE upward and increase convective instability to the east of the major convection center, resulting an eastward MJO propagation (Hsu and Li, 2012). Therefore, "the frictional convergence feedback acts like an engine that drives the wave dynamics feedback and moisture feedback to generate the unstable dynamic moisture mode" (Wang et al., 2016).

2.6 The Role of the Maritime Continent in Perturbing the Eastward MJO Propagation

It has been observed that the MJO convection tends to amplify and becomes stationary over the Indian Ocean and the Western Pacific where SSTs are warm and the moisture content is high (Hsu and Lee, 2005; Wu and Hsu, 2009). During

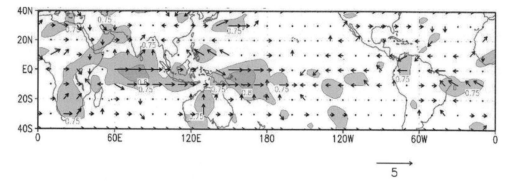

Fig. 2.9 Propagation tendency of the 30–60-day filtered OLR anomaly from November to April during 1979/80 to 1992/93. The vector denotes the speed and direction of an OLR anomaly propagating from the base point to the location where the lagged correlation reaches maximum five days later. The scale of the vector is labelled in the lower-right as 5 m s^{-1}. Shading indicates correlation coefficients greater than 0.75.

Source: Hsu and Lee, 2005; ©American Meteorological Society. Used with permission.

enhancement in these regions, the MJO convection remains stationary over the MC for certain periods before continuing its eastward journey. In other words, the eastward propagation is not a smooth progression but may consist of disjointed processes. Figure 2.9 displays the propagation tendency of the filtered OLR at each grid (Hsu and Lee, 2005). The vectors are based on five-day lagged one-point correlation map for each grid point. It represents the direction of an OLR anomaly propagating from the base point to the location where the lagged correlation is largest five days later. The eastward propagation occurs mainly in two regions, the Indian Ocean and western Pacific, where propagation is fast. The eastward propagation is weak in the MC, which seems to separate the Indian Ocean from the western Pacific. Thus, the deep convection originating from the Indian Ocean does not continuously propagate eastward through the western Pacific. Interestingly, westward propagation is noted in the eastern equatorial Pacific and tropical Atlantic. This peculiar feature explains why most model simulations of the MJO work poorly in the MC region and inevitably result in large medium and extended range weather prediction errors (e.g., Peatman et al., 2014).

The MC is comprised of narrow and steep mountains with sharp terrain gradients stretching from the islands of Sumatra, Borneo, Sulawesi, and eastward to New Guinea. High mountain tops well above 3,000 m are not unusual in some islands. For instance, the highest mountain at 4,884 m is found on the island of New Guinea, which extends across 1,600 km northwest to southeast between the equator and 10°S. What effect does the complex topography and land–sea contrast in the MC have on the MJO propagation?

In a modeling study, the blocking effect of the topography on the low-level eastward-propagating Kelvin waves slows down MJO propagation (Innes and Slingo, 2006). The deep diabatic heating anomalies, preceded by the near-surface moisture convergence and shallow heating, tends to occur on the windward side of topography in the MC, tropical South America, and tropical east Africa (Hsu and Lee, 2005). Diabatic heating is calculated as the residue from the thermodynamic equation.

For example, at day −15, a positive heating anomaly appears in the Indian Ocean and intensifies in the following 10 days (Fig. 2.10). This is followed by centers of two new deep-heating anomalies to the east of mountainous Borneo and New Guinea at day 0 over the MC (Fig. 2.10), while the anomaly over the Indian Ocean disappears. Here lag 0 corresponds to the strongest convection over the MC. The western anomaly seems to lead the eastern counterpart by about five days. Moreover, low-level moisture convergence and shallow heating anomalies start to develop on the windward side of Borneo and New Guinea at day −10 and −5, respectively, when the easterly anomaly prevails. Hsu and Lee (2005) suggest that the lifting and frictional effects of the topography and land–sea contrast induce the near-surface moisture convergence anomaly, which in turn triggers the deep heating anomaly. Subsequently, the old heating anomaly to the west of landmass weakens and the new heating anomaly east of the topography develops because of the eastward shift of the major moisture convergence center to the east of mountains. The existence of the low-level easterly anomaly is one of the important preconditions for the development of deep heating anomaly (Fig. 2.10). A strong easterly can be maintained between a heating–cooling pair with the heating anomaly to the west and the cooling anomaly (i.e., suppressed convection) in the east. The easterly then induces boundary-layer convergence and rising motion. If the easterly anomaly prevails over the tropical mountain ranges between the heating–cooling pair, the near-surface moisture convergence and shallow heating anomaly occurs on the eastern and windward side of the mountains due to lifting and frictional convergence. This setting also preconditions the lower troposphere for the later development of the new deep heating anomaly.

Using high-resolution (0.25° × 0.25°) Tropical Rainfall Measuring Mission precipitation data and reanalysis products, Wu and Hsu (2009) were able to show the blocking effect of the mountainous islands of the MC on the incoming MJO convection. Particularly noteworthy are the southward detour of the eastward-propagating MJO around Sumatra and New Guinea, and the sudden shift of deep convection from one island to another. The topographic effect causes the low-level flow and convection to split and skirt around the major islands, creating quasi-stationary features in the region. As a result, the MJO convection is stalled and weakened when the system passes through the MC.

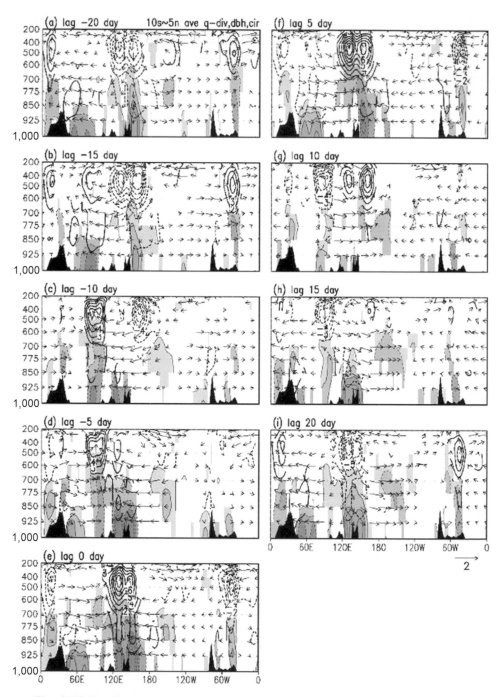

Fig. 2.10 Zonal cross sections between 10°S and 5°N of the lagged regression coefficients between the first principal component of the 200-hPa velocity potential and the diabatic heating (heavy contours), zonal wind, pressure velocity (multiplied by 100), and moisture divergence (shaded and light contours) at

Tseng et al. (2017) used a new version of the atmospheric general circulation model coupled with an 1-D ocean model (ECHAM5-SIT) that can realistically simulate the major features of the MJO. They concluded that orography and land–sea contrast in the MC affect the strength of the eastward-propagating signals, lead to stronger southward detour of the MJO convection and larger low-level moisture convergence due to frictional and lifting effects, as well as a distorted Kelvin–Rossby wave structure. Moreover, the combination of mountains and islands in the MC strengthens the climatological westerly flows in the eastern Indian Ocean (EIO) and the MC, and provides higher moisture content over the MC, possibly caused by the enhanced low-level convergence and deep convection. Consequently, the zonal moisture gradient between the EIO and MC is enhanced, which is instrumental to an eastward propagation of the MJO signal from the EIO to the MC.

2.7 Mechanisms for the Northward Propagation

The tropical ISO exhibits a pronounced seasonality in its propagation direction. While its eastward-propagating mode weakens substantially during the boreal summer, the northward propagation in the Indian monsoon region and north-westward movement in the western North Pacific, East Asia, and the South China Sea become evident. The mean speed of northward propagation is about 1 m s^{-1} (Li and Hsu, 2018). To differentiate the difference from its boreal winter phenomenon, the ISO during the boreal summer is also referred to as the boreal summer intraseasonal oscillation (BSISO) by Wang and Xie (1997), Lee et al. (2013), and others. To reveal the dynamic and thermodynamic features associated with the ISO, Fig. 2.11 displays a meridional-height cross section of 20–80-day filtered variables based on the composite of the most significant northward propagating cases during the boreal summer along the Indian sector (Jiang et al., 2004). Positive (negative) values in the *x*-axis denote the latitudes (in degrees) to the north (south) of the ISO convection center at 0. The maximum ascending motion occurs near 400 hPa and coincides with the convection center 0 in panel a.

Caption for fig. 2.10 (*cont.*). days (a) –20, (b) –15, (c) –10, (d) –5, (e) 0, (f) 5, (g) 10, (h) 15, and (i) 20. Solid and dashed lines denote positive and negative values, respectively. Contour intervals are $1 \times 10^{-6} \text{ K s}^{-1}$ and $10^{-9} \text{ g Kg}^{-1} \text{ s}^{-1}$ for the diabatic heating and moisture divergence, respectively. Length of the reference arrow is 2 m s^{-1}. Regression coefficients have been multiplied by one standard deviation of the principal component and only those that are significant at the 5% level are plotted.
Source: Hsu and Lee, 2005; ©American Meteorological Society. Used with permission.

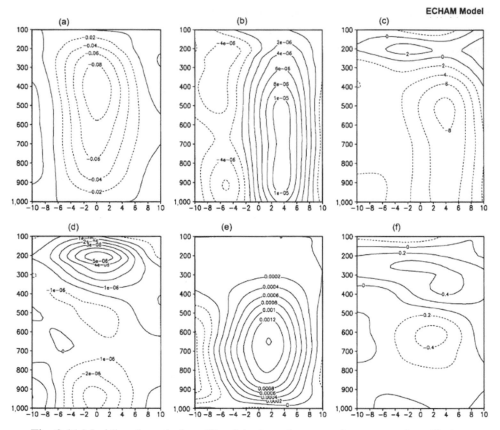

Fig. 2.11 Meridional-vertical profile of the boreal summer intraseasonal oscillation from the ECHAM model: (a) vertical velocity (hPa s^{-1}), (b) vorticity (s^{-1}), (c) geopotential height (dam), (d) divergence (s^{-1}), (e) specific humidity (kg kg^{-1}), and (f) temperature (K). The horizontal axis is the distance (°lat) with respect to the convection center; positive (negative) values mean to the north (south) of the convection center. The vertical axis is the pressure (hPa).

Source: Jiang et al., 2004; ©American Meteorological Society. Used with permission.

Associated with this vertical motion is the upper-level divergence and low-level convergence in panel d. Another noteworthy feature is the positive vorticity center with an equivalent barotropic structure located 400 km north of the convection center, and a negative vorticity to the south of the convection center in panel b. For the specific humidity, its maximum anomaly shifts 150 km to the north of the convection center in the lower level.

Based on the meridional-vertical structures of those variables, two mechanisms are proposed to explain the northward propagation. The first one is known as the vertical wind shear mechanism as motivated by the observed vertical profile of an equivalent barotropic vorticity field to the north of the convection. The second is the moisture–convection feedback mechanism, in which low-level specific humidity leads convection. For the first mechanism, the vorticity equation for the barotropic component can be simplified to a titling term only (Jiang et al., 2004):

$$\frac{\partial \xi_+}{\partial t} \propto \frac{\partial \bar{u}}{\partial z} \frac{\partial w'}{\partial y},$$ (2.1)

where the first term on the right is the vertical shear of the basic-state zonal flow and w' denotes the ISO perturbation vertical velocity in the mid-troposphere. The subscript + denotes the barotropic mode. The vertical structure of the mean flow in the boreal summer is characterized by a strong easterly shear in the Indian monsoon region. To the north of the convection center, upward motion decreases with the latitude (y). This vertical motion field twists the mean-flow horizontal vorticity and generates barotropic vorticity with a positive (or upward) vertical component north of the convection center in the presence of the easterly vertical shear. The generation of the positive barotropic vorticity may lead to the development of the barotropic divergence (D_+) in the free atmosphere as (Jiang et al., 2004)

$$\frac{\partial D_+}{\partial t} \propto f_0 \xi_+,$$ (2.2)

where f_0 is the reference latitude (12°N). Because $f_0 > 0$ in the northern hemisphere, the free-atmosphere divergence to the north of the convection center, together with the Ekman pumping effect, leads to the further boundary-layer convergence to satisfy the continuity equation. The boundary-layer moisture convergence thus favors the development of convective heating to the north of the convection and then promotes northward propagation. In short, the positive barotropic vorticity in the free atmosphere induces convergence in the boundary layer, which destabilizes the air and triggers new convection to the north of the original convection center.

For the moisture–convection feedback mechanism, the basis is grounded on the fact that specific humidity leads convection (Fig. 2.11) so that the meridional asymmetry in perturbation moisture may cause the northward shift of the convective heating, leading to northward ISO convection propagation. Within this mechanism, two processes rooted on moisture advection are proposed. The first emphasizes the role of the mean summer monsoonal flow and the second focuses on the mean meridional moisture gradient.

2.7.1 Moisture Advection by the Mean Low-Level Flow

The asymmetry of the specific humidity with respect to convection can be primarily expressed as the sum of two terms: moisture advection by the mean northward flow

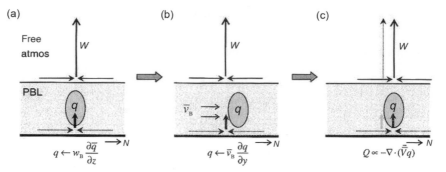

Fig. 2.12 Schematics of moisture advection by the mean Indian summer monsoon. (a) The specific humidity perturbation (q) by Ekman pumping, (b) the moisture advection by the mean northward wind in the planetary boundary layer, and (c) the northward shift of moisture convergence and convective heating.
Source: Jiang et al., 2004; ©American Meteorological Society. Used with permission.

in the boundary layer and advection of the mean specific humidity by the perturbation vertical velocity in the boundary layer as follows (Jiang et al., 2004):

$$\frac{\partial q}{\partial t} \propto -\bar{V}_B \frac{\partial q}{\partial y} - W_B \frac{\partial \bar{q}}{\partial z},\tag{2.3}$$

where $\bar{V}_B$ is the meridional component of the mean flow in the boundary layer, W_B is vertical motion at the top of the boundary layer, and $\bar{q}$ is the mean moisture. Consider a case of strong ISO convection with low-level convergence and upper-level divergence. The convergence at low level will induce upward motion in the boundary layer, bringing moisture from the surface to a certain height (Fig. 2.12). The observed mean summer flow over the East Indian Ocean has a prevailing northward component, therefore $\bar{V}_B$ is positive. The advection by the mean southerly flow shifts the anomalous humidity center to the north of the convection. Because the convective heating depends on boundary-layer moisture convergence, the shifted moisture center will lead to the northward displacement of convective heating and thus the northward movement of the convection.

2.7.2 Moisture Advection Due to the Mean Meridional Moisture Gradient

This hypothesis considers the northward propagation through the advection effect by the ISO wind in the presence of the mean meridional specific humidity gradient. Climatologically, the maximum moisture is located near 20°N over the north Indian Ocean in the boreal summer. To the south of this maximum, the meridional gradient of the mean specific humidity is positive. Considering the advection by the perturbation meridional wind, the anomalous moisture advection is

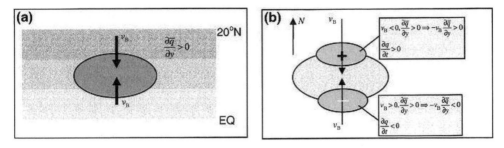

Fig. 2.13 Schematics of moisture advection by the mean specific humidity gradient and the meridional component of the wind in the boundary layer. (a) The mean meridional humidity field is advected by perturbation wind on both sides of the convection center. (b) Positive (negative) moisture perturbation to the north (south) of the convection center leads to northward movement.

Source: Jiang et al., 2004; ©American Meteorological Society. Used with permission.

$$\frac{\partial q}{\partial t} \propto -V_{\mathrm{B}} \frac{\partial \bar{q}}{\partial y}. \tag{2.4}$$

In response to the ISO heating, the perturbation wind has a southward (northward) component to the north (south) of the convection center (Fig. 2.13). As a result of the advective process, the perturbation moisture attains a meridional asymmetry, with a positive (negative) specific humidity anomaly center to the north (south) of the convection center. The asymmetry moisture distribution further leads to the northward shift of the convective heating and the ISO convection.

2.8 The Quasi-Biweekly (QBW) Oscillation

It should be noted that there is another kind of ISO that operates at a time scale of 10–20 days in the Indian Ocean, South Asia, western North Pacific, East Asia, and elsewhere (e.g., Krishnamurti and Bhalme, 1976; Yasunari, 1979; Sikka and Gadgil, 1980; Webster, 1986; Kikuchi and Wang, 2009; Chen and Sui, 2010). Over the Indian sector for each of the boreal summer during 1973–1977, Sikka and Gadgil (1980) noted a succession of orderly and northward propagating maximum cloudiness (i.e., rain) starts near the equator and ends in the foothills of the Himalayas with an average lifetime of 15 days. When the maximum cloudiness zone is over central India, it corresponds to an active Indian monsoon with a strong upper-level tropical easterly jet. The break monsoon occurs when the precipitation belt has moved northward to the foothills of the Himalayas and subsidence is dominant over India with a weak easterly jet. According to Annamalai and Slingo (2001), this quasi-biweekly (QBW) mode accounts for about a fourth of the subseasonal monsoon variability while the MJO makes up two-thirds of the variability.

Kikuchi and Wang (2009) studied the life cycle of the QBW oscillation during the boreal summer (JJA) and austral summer (DJF) from a global perspective. Using an outgoing longwave radiation (OLR) anomaly with a 12–20-day bandpass filter, the initiation, propagation, and demise of the QBW variation is analyzed by identifying QBW deep convective events through a tracking method. In short, this method is based on the value of OLR anomaly at a given time (<-10 W m^{-2}), its spatial extent ($>15°$ longitude), duration of the event (>5 days), and minimum OLR anomaly at its mature stage (<-20 W m^{-2}).

During the boreal summer, most QBW activity is initiated in the South Asia–western Pacific region, encompassing the equatorial and subtropical monsoon areas (Fig. 2.14). Minor initial locations of QBW events are found in the Caribbean Sea, Uruguay, and South Pacific. In the Asian monsoon region, QBW convection first occurs in the equatorial western Pacific (Fig. 2.14b) and moves northwestward to the Philippine Sea before branching out westward into the South China Sea and northeastward to the west of the Midway Island near the date line (Fig. 2.14c). For the westward track, QBW convection merges with the events originated from the Indian Ocean. The convective events initiated from the western Indian Ocean tend to move eastward, followed by northward propagation to the Arabian Sea or the Bay of Bengal. In the South Pacific, the QBW activity is also originated in the equatorial western Pacific, move poleward and eastward through a long journey and dissipate in the southeastern Pacific. Another track which is rather short occurs in the Caribbean Sea and moves westward through Mexico to the eastern Pacific.

During the austral summer, QBW activity is mainly initiated in the Southern Hemisphere (Fig. 2.15) and tracks eastward, except in Australia and the western North Pacific where convective events tend to move northwestward to the Philippine Sea. The trajectory of QBW convection in Australia is unusual, which will be discussed later. Eastward movement is evident in the eastern North Pacific, North Atlantic–western Africa, South America, South Africa–south Indian Ocean, and the Indian Ocean regions. Another notable track emerges from the Coral Sea, propagates eastward to the dateline and dissipates in the south central Pacific.

Besides the tracking method, Kikuchi and Wang (2009) applied the extended empirical orthogonal function (EEOF) analysis to the OLR anomaly from ± 8 days with a four-day interval. Conventional empirical orthogonal analysis gives a measure of covariance (or correlation) of the entire domain in terms of eigenvalues and their associated eigenvectors. By extension, EEOF analysis combines spatial/temporal covariance structure to produce eigenmodes that are space/time dependent (e.g., Weare and Nasstrom, 1982; Lau and Chan, 1986). As a result, this method can indicate propagating features of an entire field such as Rossby wave trains as interpreted from a consecutive time series of convective anomalies. For example, focusing on the Asian monsoon region (Fig. 2.16), a convective anomaly emerges

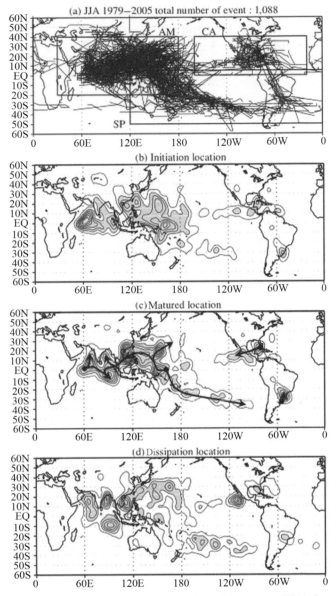

Fig. 2.14 The quasi-biweekly events in the boreal summer (JJA) for the period 1979–2005. (a) Tracks of each individual event, (b) initial, (c) mature, and (d) termination location. Contour intervals are 0.2 and values greater than 0.4 are shaded. Arrows in (c) denote the propagation paths.

Source: Kikuchi and Wang, 2009; ©American Meteorological Society. Used with permission.

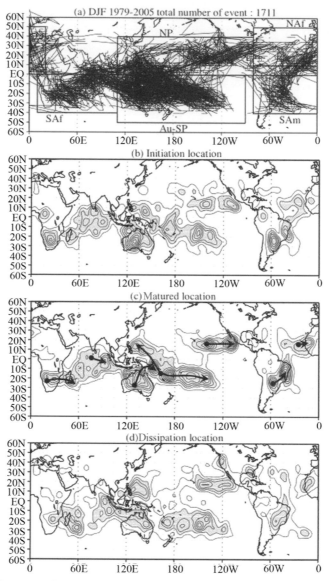

Fig. 2.15 Same as Fig. 2.14 but for the austral summer (DJF).
Source: Kikuchi and Wang, 2009; ©American Meteorological Society. Used with permission.

over the equatorial Pacific near the date line on day –4 and moves westward to 150°E four days later (Kikuchi and Wang, 2009). It then tracks northwestward to the Philippines Sea by day 4 and a Rossby wave response becomes clear in both hemispheres, with a stronger signal in the northern branch. The QBW event continues its westward movement to the South China Sea and reaches its mature stage by day 8. The phase speed of the event is 6 m s^{-1} and the zonal wavelength is approximately 70° longitude. Note the similar patterns of both the convective anomaly and wind anomaly at day –8 and day 8, a result of the bandpass filtering when the average life cycle of the QBW events is set to be 16 days. Another noteworthy feature is that the anomaly appears to initiate from the equatorial Western Indian Ocean on day –4 and then moves both northward to the Arabian Sea and southward to the central South Indian Ocean on day 0.

As shown in Figs. 2.14 and 2.15, the QBW activity can be classified into a westward and eastward moving modes. The westward mode is found in the Asian monsoon, and Central American regions during the boreal summer, and in the western North Pacific during the austral summer. This behavior resembles equatorial Rossby (ER) waves modified by the monsoon mean flow (Kikuchi and Wang, 2009). Figure 2.17 displays a spectral analysis of the OLR data with frequency (or period) as the ordinate and zonal wavenumber as the abscissa for five equatorial waves as they reside in equatorial regions and possess wavelike features (Frank and Roundy, 2006; Schreck et al., 2012). Values are normalized by dividing by the background spectrum. Tropical depression (TD)-type disturbances are commonly referred to as easterly waves (Frank and Roundy, 2006).

The eastward-propagating MJO and Kelvin waves are also evident. Frank and Roundy (2006) showed that westward-propagating ER and mixed Rossby-gravity (MRG) waves are most active during the typhoon season in the western North Pacific. They also illustrated a classic ER wave pattern in the Northern Hemisphere, in which a strong cyclonic gyre is located about one-quarter wavelength to the northwest of the genesis center. This one-quarter phase lag between the gyre circulation and convection is also noted in Chen and Chou (2014). Thus, the ER waves are conducive to genesis development by inducing rotation and local low-level vorticity. In addition, the ER waves may modulate the background vertical wind shear, causing easterly shear anomalies at the genesis locations (Frank and Roundy, 2006). The easterly shear favors trapping Rossby waves in the lower troposphere and provides a favorable condition for unstable Rossby waves (Xie and Wang, 1996). Interestingly, the easterly shear occurs in the monsoon region from the northern Indian Ocean to the western North Pacific during the boreal summer.

Based on 10–20-day filter data for eight summers (JASO) in the western North Pacific, Chen and Sui (2010) noted that the QBW event features an alternating low-level cyclonic and anticyclonic vorticity anomalies in a wave train, which is oriented with a southeast–northwest pattern, and a wavelength of approximately 3,500 km. The wave train appears to originate from the equatorial region and tracks westward to reach 150°E before propagating northwestward toward the south of Japan. Some features of the QBW (e.g., phase speed and group velocity)

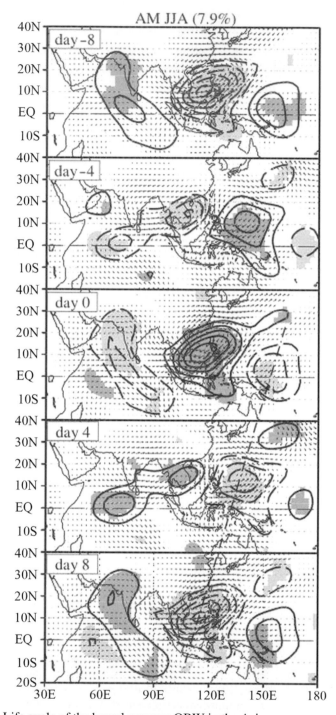

Fig. 2.16 Life cycle of the boreal summer QBW in the Asian monsoon region from the EEOF analysis with five time intervals separated by 4-day intervals. The

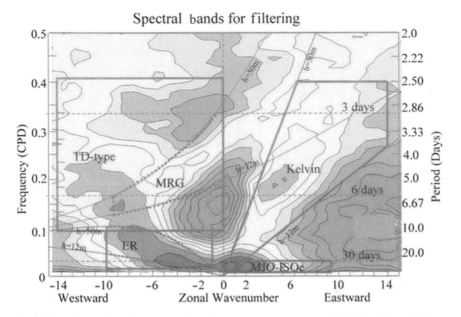

Fig. 2.17 Normalized wavenumber-frequency spectrum of OLR adapted from Wheeler and Kiladis (1999). Thick black lines encompass the regime of filter bands. Thin dashed lines denote shallow-water dispersion curves in a dry, motionless atmosphere for the equivalent depths indicated by labels. A black and white version of this figure will appear in some formats. For the color version, refer to the plate section.
Source: Frank and Roundy, 2006; ©American Meteorological Society. Used with permission.

resemble equatorially trapped Rossby mode. Given a large variance accounted for by ER waves, Chen and Sui (2010) further suggest that these waves might contribute to the origin of QBW oscillations in the tropical western North Pacific.

During the austral summer, the QBW modes in many regions of the world are related to extratropical Rossby wave trains, for example, over Australia and South Pacific (Fig. 2.15). The train in Australia appears to initiate from its southern portion, propagates straight across the continent, and then dissipates in its northern coast. Extratropical Rossby waves can penetrate to lower latitudes where upper tropospheric winds are westerlies (Webster and Holton, 1982). The so-called

Caption for fig. 2.16 (*cont.*). 12–20-day bandpass-filtered OLR anomalies (contours) are used and contour interval is 0.01. Lag-regression are applied to obtain 850-hPa wind fields (vectors) and TRMM 3B42 precipitation (shading). The thick dash-dotted line is drawn to illustrate the propagation of a convective anomaly.
Source: Kikuchi and Wang, 2009; ©American Meteorological Society. Used with permission.

"westerly ducts" of the tropical upper troposphere occur in Australia and extend eastward to south Pacific and become a preferred location for incoming Rossby waves (Kikuchi and Wang, 2009). The signature of Rossby wave train is more evident through the EEOF analysis in Kikuchi and Wang (2009) (not shown).

2.9 Summary

This chapter discusses the intraseasonal oscillation (ISO), which is comprised of the Madden–Julian Oscillation (MJO) and quasi-biweekly (QBW) oscillation. For MJO, the focus is on its physical description, periodicity, initiation process, propagation, and role of the Indonesian maritime continent in the intervention of the propagation. For the QBW oscillation, initiation, propagation, and demise are presented from a global perspective.

   The current prevailing hypothesis regarding the periodicity and initiation of the MJO is the recharge–discharge paradigm of the MSE. That is, the MJO period is primarily determined by the slow recharge time for column MSE before the formation of the deep convection in the Indian Ocean and the subsequent discharge by vertical motions during and after MJO convection. Observations show that the MJO convection is initiated in the western Indian Ocean. This process appears to be a direct Rossby wave response to antecedent, downstream negative heating anomalies (i.e., suppressed MJO convection) over the eastern Indian Ocean. The low-level easterly anomalies over the Indian Ocean resulting from anticyclonic Rossby wave gyres cause moisture convergence against the east African highland, which contributes to the initiation of the deep MJO convection.

   For the eastward propagation of MJO deep convection, three theories are proposed. The first is the convectively coupled Kelvin–Rossby wave theory. The convective heating anomalies in the troposphere induces a Kelvin (Rossby) wave response to the east (west) of the convection. As a Kelvin wave response to the east of convection, equatorial low-level easterlies induce boundary-layer convergence and shifts the heating to the east of the MJO convection. Because of the free atmospheric wave–convection–boundary–layer convergence interaction, the MJO envelope tends to move eastward. The second theory, moisture-mode theory, invokes the moisture and convection feedback, wave dynamics and boundary-layer interaction, and can be separated into two types. Under the first type, low-level moistening sets up an unstable atmospheric layer and triggers new convection to the east of the existing MJO center, causing the convection to propagate eastward. The boundary-layer convergence is a possible source for moistening the low-level atmosphere ahead of the convection. The second type of the moisture-mode theory emphasizes the column MSE tendency asymmetry in the eastward propagation theories. The third theory deviates somewhat from the second and emphasizes the trio interaction of convective heating, moisture feedback, and equatorial waves/ boundary-layer dynamics. This is called the frictional coupled dynamic moisture-mode theory because it emphasizes the frictional convergence feedback that drives the wave dynamics and moisture feedback.

During the boreal winter, the eastward propagation of MJO convection is fast over the equatorial Indian Ocean and western Pacific but becomes stalled over the Indonesian maritime continent (MC), which encompasses many islands with steep mountains. In particular, a southward detour of the convection around Sumatra and New Guinea and the sudden shift in convection from one island to another are evident. Studies show that topographic blocking generates extra lifting and sinking processes in such a way that the convection remains quasi-stationary near the major topography when the MJO envelope passes through the MC. Orography and land–sea contrast in the MC tend to cause discontinuity of the eastward-propagating MJO convection, presenting a challenge in weather prediction.

In the boreal summer, the northward propagation of ISO convection over the Indian monsoon region, the western North Pacific, and East Asian becomes evident. This phenomenon is also referred to as the boreal summer instraseasonl oscillation (BSISO). Two mechanisms are suggested to account for this northward propagation. The first is known as the vertical wind shear mechanism as stimulated by the observed vertical profile of an equivalent barotropic vorticity structure to the north of the convection. In the presence of the mean easterly shear in the monsoon region, the vorticity twisting term results in a positive (or upward) vorticity and a free-atmosphere barotropic divergence to the north of the BSISO convection, which favors boundary-layer convergence and convective heating. This promotes northward propagation of convection. The second is the moisture–convection feedback mechanism. Under this mechanism, two processes linked to moisture advection are proposed. The first emphasizes the role of the mean summer low-level northward monsoon flow and the second focuses on the mean meridional moisture gradient.

For the QBW oscillation, its activity is initiated in the equatorial western Indian and western Pacific regions during the boreal summer. In the western Pacific, the QBW anomaly first tracks northwestward through the Philippines Sea and then bifurcates into two branches: one westward to the South China Sea and another northeastward to the Midway. The westward track branch merges with events from the Indian Ocean. The westward mode of the QBW anomaly in the western North Pacific resembles equatorial Rossby waves modified by the monsoon mean flows. Over the Indian Ocean, the northward propagation of the QBW convection over the Arabian Sea and the Bay of Bengal is also evident. In the South Pacific, the QWB activity is characterized by a long southeastward journey from the equatorial western Pacific to the southeast Pacific. There is also an indication of the QBW events over the Central America from the Gulf of Mexico to the eastern North Pacific. During the austral summer, QBW activity is mainly confined to the Southern Hemisphere, and tracks eastward, with the exception over Australia and the western North Pacific. The QBW modes appear to be related to the extratropical Rossby wave trains.

Exercises

2.1 Discuss and illustrate graphically the zonal/vertical structure of horizonal winds, convergence/divergence, vertical motion, and humidity associated with a major MJO convection over the Indonesian Maritime Continent.

2.2 (a) Discuss and draw a schematic diagram of the atmospheric wave structure at both the lower and upper troposphere associated with a major MJO convection. (b) Explain how the wave structure you depicted from (a) could account for the eastward shift of the MJO envelope.

2.3 What processes are important for the initiation of the MJO convection in the Indian Ocean?

2.4 How might the moisture advection by the mean low-level monsoonal flows during the boreal summer explain the northward propagation of the intraseasonal oscillation?

2.5 Describe the initiation and propagation of the quasi-biweekly oscillation as represented by the outgoing longwave radiation anomalies in the South Asia–western Pacific region during the boreal summer.

References

Annamalai, H., and J. M. Slingo, 2001: Active-break cycles: Diagnosis of the intraseasonal variability of the Asian summer monsoon. *Clim. Dyn.* **18**, 85–102.

Blade, I., and D. L. Hartmann, 1993: Tropical intraseasonal oscillations in a simple non-linear model. *J. Atmos. Sci.*, **50**, 2922–2939.

Charney, J. G., and A. Eliassen, 1964: On the growth of hurricane depression. *J. Atmos. Sci.*, **21**, 68–75.

Chen, G., and C. Chou, 2014: Joint contribution of multiple equatorial waves to tropical cyclogenesis over the western North Pacific. *Mon. Wea. Rev.*, **142**, 79–93.

Chen, G., and C.-H. Sui, 2010: Characteristics and origin of quasi-biweekly oscillation over the western North Pacific during boreal summer. *J. Geophys. Res.*, **115**, D14113.

Frank, W. M., and P. E. Roundy, 2006: The role of tropical waves in tropical cyclogenesis. *Mon. Wea. Rev.*, **134**, 2397–2417.

Gill, A. E., 1980: Some simple solutions for heat-induced tropical circulation. *Quart. J. Roy. Meteorol. Soc.*, **106**, 447–462.

Hendon, H. H., 1988: A simple model of the 40-50 day oscillation. *J. Atmos. Sci.*, **45**, 569–584.

Hendon, H. H., and B. Liebmann, 1990: The intraseasonal (30-50 day) oscillation of the Australian summer monsoon. *J. Atmos. Sci.*, **47**, 2909–2923.

Hendon, H. H., and M. L. Salby, 1994: The life cycle of the Madden–Julian Oscillation. *J. Atmos. Sci.*, **51**, 2225–2237.

Higgins, W., and W. Shi, 2001: Intercomparison of the principal modes of interannual and intraseasonal variability of the North American monsoon systems. *J. Climate*, **14**, 403–417.

Hsu, H.-H., and M-.Y. Lee, 2005: Topographic effects on the eastward propagation and initiation of the Madden–Julian oscillation. *J. Climate*, **18**, 795–809.

Hsu, H.-H., B. Hoskins, and F.-F. Jin, 1990: The 1985/86 intraseasonal oscillation and the role of the extratropics. *J. Atmos. Sci.*, **47**, 823–839.

Hsu, P.-C., and T. Li, 2012: Role of the boundary layer moisture asymmetry in causing the eastward propagation of the Madden–Julian oscillation. *J. Climate*, **25**, 4914–4931.

Hu, F., T. Li, J. Gao, and L. Hao, 2021: Reexamining the moisture mode theories of the Madden–Julian Oscillation based on observational analyses. *J. Climate*, **34**, 839–853.

Hung, C.-S., and C.-H. Sui, 2018: A diagnostic study of the evolution of the MJO from Indian Ocean to maritime continent: Wave dynamics versus advective moistening processes. *J. Climate*, **31**, 4095–4615.

Hung, C.-W., and M. Yanai, 2004: Factors contributing to the onset of the Australian summer monsoon. *Quart. J. Roy. Meteorol. Soc.*, **130**, 739–758.

Inness, P. M., and J. M. Slingo, 2006: The interaction of the Madden–Julian oscillation with the maritime continent in a GCM. *Quart. J. Roy. Meteorol. Soc.*, **132**, 1645–1667.

Jiang, X., T. Li, and B. Wang, 2004: Structures and mechanisms of the northward propagating boreal summer intraseasonal oscillation. *J. Climate*, **17**, 1022–1039.

Johnson, R. H., et al., 1999: Trimodal characteristics of tropical convection. *J. Climate*, **12**, 2397–2418.

Johnson, R. H., P. E. Ciesielski, J. H. Ruppert, and M. Kasumata, 2015: Sounding-based thermodynamic budgets for DYNAMO. *J. Atmos. Sci.*, **72**, 598–622.

Jones, C., 2000: Occurrence of extreme precipitation events in California and relationships with the Madden–Julian oscillation. *J. Climate*, **13**, 3576–3587.

Julia, C., 2012: Assessing the influence of the MJO on strong precipitation events in subtropical, semi-arid north-central Chile (30°S). *J. Climate*, **25**, 7003–7013.

Kemball-Cook, S. R., and B. C. Weare, 2001: The onset of convection in the Madden–Julian oscillation. *J. Climate*, **14**, 780–791.

Kikuchi, K., and Y. N. Takayabu, 2004: The development of organized convection associated with the MJO during TOGA COARE IOP: Trimodel characteristics. *Geophys. Res. Lett.*, **31**: L10101.

Kikuchi, K., and B. Wang, 2009: Global perspective of the quasi-biweekly oscillation. *J. Climate*, **22**, 340–358.

Kiladis, G. N., et al., 2014: A comparison of OLR and circulation-based indices for tracking the MJO. *Mon. Wea. Rev.*, **142**, 1697–1715.

Kim, D., J. Kug, and A. H. Sobel, 2014: Propagating versus nonpropagating Madden–Julian Oscillation events. *J. Climate*, **27**, 111–125.

Krishnamurti, T. N., and H. N. Bhalme, 1976: Oscillations of a monsoon system. Part I. Observational aspects. *J. Atmos. Sci.*, **33**, 1937–1954.

Lau, K. M., and P. H. Chan, 1986: Aspects of the 40-50 day oscillation during the northern summer as revealed from outgoing longwave radiation. *Mon. Wea. Rev.*, **114**, 1354–1367.

Lau, K. M., and L. Peng, 1987: Origin of low-frequency (intraseasonal) oscillations in the tropical atmosphere. Part I: Basic theory. *J. Atmos. Sci.*, **44**, 950–972.

Lee, J., and Coauthors, 2013: Real-time multivariate indices for the boreal summer intraseasonal oscillation over the Asian summer monsoon region. *Clim. Dyn.*, **40**, 493–509.

Li, T., 2014: Recent advance in understanding the dynamics of the Madden–Julian Oscillation. *J. Meteorol. Res.*, **28**, 1–33.

Li, T., and P.-C. Hsu, 2018: *Fundamentals of Tropical Climate Dynamics*. Springer.

Li, T., J. Ling, and P.-C. Hsu, 2020: Madden–Julian Oscillation: Its discovery, dynamics, and impact on East Asia. *J. Meteorol. Res.*, **34**, 20–42.

Li, T., C. B. Zhao, P.-C. Hsu, and T. Nasuno, 2015: MJO initiation processes over the tropical Indian Ocean during DYNAMO/CINDY2011. *J. Climate*, **28**, 2121–2135.

Li, T., and Coauthors, 2018: A paper on the tropical intraseasonal oscillation published in 1963 in a Chinese journal. *Bull. Amer. Meteorol. Soc.*, **99**, 1765–1779.

Lin, J., B. Mapes, M. Zhang, and M. Newman, 2004: Stratiform precipitation, vertical heating profiles, and the Madden–Julian Oscillation. *J. Atmos. Sci.*, **61**, 296–309.

Lindzen, R. S., 1974: Wave-CISK in tropics. *J. Atmos. Sci.*, **31**, 156–179.

Lindzen, R. S., and S. Nigam, 1987: On the role of sea surface temperature gradients in forcing low-level winds and convergence in the tropics. *J. Atmos. Sci.*, **44**, 2418–2436.

Madden, R. A., and P. R. Julian, 1971: Detection of a 40-50 day oscillation in the zonal wind in the tropical Pacific. *J. Atmos. Sci.*, **28**, 702–708.

Madden, R. A., and P. R. Julian, 1972: Description of global-scale circulation cells in the tropics with a 40-50 day period. *J. Atmos. Sci.*, **29**, 3138–3158.

Maloney, E. D., 2009: The moist static energy of a composite tropical intraseasonal oscillation in a climate model. *J. Climate*, **22**, 711–729.

Maloney, E. D., and D. L. Hartmann, 1998: Frictional moisture convergence in a composite life cycle of the Madden–Julian Oscillation. *J. Climate*, **11**, 2387–2403.

Nakazawa, T., 1988: Tropical super clusters within intraseasonal variations over the western Pacific. *J. Meteorol. Soc. Japan*, **66**, 823–839.

Ooyama, K., 1964: A dynamic model for the study of tropical cyclone development. *Geofis. Int.*, **4**, 187–198.

Peatman, S., A. J. Matthews, and D. P. Stevens, 2014: Propagation of the Madden–Julian Oscillation through the Maritime Continent and scale interaction with the diurnal cycle of precipitation. *Quart. J. Roy. Meteorol.*, **140**, 814–825.

Rui, H., and B. Wang, 1990: Development characteristics and dynamic structure of tropical intraseasonal convection anomalies. *J. Atmos. Sci.*, **47**, 33–61.

Schreck, C. J., J. Molinari, and A. Aiyyer, 2012: A global view of equatorial waves and tropical cyclogenesis. *Mon. Wea. Rev.*, **140**, 774–788.

Sikka, D. R., and S. Gadgil, 1980: On the maximum cloud zone and the ITCZ over Indian longitudes during the southwest monsoon. *Mon. Wea. Rev.*, **108**, 1840–1853.

Sobel, A., and E. Maloney, 2012: An idealized semi-empirical framework for modeling the Madden–Julian Oscillation. *J. Atmos. Sci.*, **69**, 1691–1705.

Sobel, A., and E. Maloney, 2013: Moisture modes and the eastward propagation of the MJO. *J. Atmos. Sci.*, **70**, 187–192.

Sobel, A., S. Wang, and D. Kim, 2014: Moist static energy budget of the MJO during DYNAMO. *J. Atmos. Sci.*, **71**, 4276–4291.

de Souza, E. B., and T. Ambrizzi, 2006: Modulation of the intraseasonal rainfall over tropical Brazil by the Madden–Julian oscillation. *Int. J. Climatol.*, **26**, 1759–1776.

Sperber, K. R., 2003: Propagation and the vertical structure of the Madden–Julian oscillation. *Mon. Wea. Rev.*, **131**, 3018–3037.

Tseng, K. C., C.-H. Sui, and T. Li, 2015: Moistening processes for Madden–Julian oscillations during DYNAMO/CINDY. *J. Climate*, **28**, 3041–3058.

Tseng, W.-L. and Coauthors, 2017: Effects of orography and land-sea contrast on the Madden–Julian oscillation in the maritime continent: A numerical study using ECHAM5-SIT. *J. Climate*, **30**, 9725–9741.

Waliser, D. E., 2006: Intraseasonal variability. In *The Asian Monsoon*, B. Wang, Ed., Springer, 203–257, Berlin.

Wang, B., 1988: Dynamics of tropical low-frequency waves: An analysis of the moist Kelvin wave. *J. Atmos. Sci.*, **45**, 2051–2065.

Wang, B., and G. Chen, 2017: A general theoretical framework for understanding essential dynamics of Madden–Julian Oscillation. *Clim. Dyn.*, **49**, 2309–2328.

Wang, B., and T. Li, 1994: Convective interaction with boundary-layer dynamics in the development of a tropical intraseasonal system. *J. Atmos. Sci.*, **51**, 1386–1400.

Wang, B., and H. Rui, 1990: Dynamics of the coupled moist Kelvin–Rossby wave on an equatorial plane. *J. Atmos. Sci.*, **47**, 397–413.

Wang, B., and X. Xie, 1997: A model for the boreal summer intraseasonal oscillation. *J. Atmos. Sci.*, **54**, 72–86.

Wang, B., F. Liu, and G. Chen, 2016: A trio-interaction theory for Madden–Julian Oscillation. *Geosci. Lett.*, **3**, 34, doi 10.1186/s40562-016-0066-z.

Wang, L., T. Li, E. Maloney, and B. Wang, 2017: Fundamental causes of propagating and nonpropagating MJOs in MJOTF/GASS models. *J. Climate*, **30**, 3743–3769.

Weare, B. C., and J. S. Nasstrom, 1982: Examples of extended empirical orthogonal function analysis. *Mon. Wea. Rev.*, **110**, 481–485.

Webster, P. J., 1986: Variable and interactive monsoon. In *Monsoons*, J. Fein and P. Stephens, Eds., Wiley Interscience, 269–330, New York.

Webster, P. J., and J. R. Holton, 1982: Cross-equatorial response to middle-latitude forcing in a zonally varying basic state. *J. Atmos. Sci.*, **39**, 722–733.

Wheeler, M., and G. N. Kiladis, 1999: Convectively coupled equatorial waves: Analysis of clouds and temperature in the wavenumber-frequency domain. *J. Atmos. Sci.*, **56**, 374–399.

Wu, C.-H., and H.-H. Hsu, 2009: Topographic influence on the MJO in the maritime continent. *J. Climate*, **22**, 5433–5448.

Xie, X. S., and B. Wang, 1996: Low-frequency equatorial waves in vertically sheared zonal flow. Part II: Unstable waves. *J. Atmos. Sci.*, **53**, 3589–3605.

Xie, Y. B., S. J. Chen, I.-L. Zhang, and Y.-L. Hung, 1963: A preliminary statistic and synoptic study about the basic currents over southeastern Asia and the initiation of typhoons. *Acta Meteor. Sinica*, **33**, 206–217 (in Chinese).

Yasunari, T., 1979: Cloudiness fluctuation associated with the northern hemisphere summer monsoon. *J. Meteorol. Soc. Japan*, **57**, 227–242.

Zhang, C., 2005: Madden-Julian oscillation. *Rev. Geophys.*, **43**, RG2003.

Zhao, C., T. Li, and T. Zhao, 2013: Precursor signals and processes associated with MJO initiation over the tropical Indian Ocean. *J. Climate* **26**, 291–307.

3

Climate Variability. Part II: Interannual to Interdecadal Variability

3.1 Interannual Variability

3.1.1 Introduction

It is well known that the El Niño–Southern Oscillation (ENSO) is the most prominent mode of interannual climate variability in the tropics. The ENSO is a coupled, tropical ocean–atmosphere system that fluctuates on a time scale of two to seven years in the Pacific (Philander, 1990). The ENSO extremes are labeled as either a warm or cold phase, yet its amplitude varies across a continuum with essentially Gaussian statistics (Trenberth, 1997). Characterizing the warm (cold) ENSO phase is the presence of the anomalously warm (cold) sea surface temperatures (SSTs) in the eastern and/or central equatorial Pacific known as the El Niño (La Niña) event.

The Southern Oscillation (SO) is the atmospheric component of the ENSO and describes the large-scale surface pressure seesaw between the Pacific and the Indian Oceans. Pressure variations near the eastern south Pacific tend to vary inversely with those in the Indonesia. Owing to the appearance of this standing wavelike diploe pattern in the zonal direction across two southern oceans, the term SO was coined. Fluctuations in the SO have been monitored in terms of an index consisting of the normalized monthly mean sea-level pressure difference between Tahiti and Darwin (Chen, 1982; Chu and Katz, 1985). Rasmusson and Carpenter (1982) first used the term ENSO to describe the ocean–atmosphere interaction and the relationship between oceanic El Niño conditions and the negative phase of the atmospheric SO phenomenon (i.e., pressures become anomalously low in the eastern Pacific and anomalously high in the tropical western Pacific).

El Niño is the Spanish term for the Christ Child, which has been used for centuries by mariners and fishermen in South America to define the annual occurrence of warm, southward-flowing oceanic coastal waters off Ecuador and Peru around the Christmas time. Although the warming in SSTs off the Peru occurs every year as a regular annual event, there are strong large-scale warming events across the eastern to central equatorial Pacific every two to seven years (e.g., Rasmusson and Carpenter, 1982). Now only these strong events are referred to as El Niño. During an El Niño event, anomalously warm SSTs and its attendant deep convective heating in the tropical Pacific force an upper-level stationary wave. Such a wave provides a link between the tropical forcing and extratropical response, causing climate anomalies such as drought, excessive rainfall and/or flooding, temperature extremes, and wildland fires over North America and other parts of the world (e.g., Hastenrath, 1985; Ropelewski and Halpert, 1987; Glantz et al., 1991; Hoerling and Kumar, 1997; Cayan et al., 1999; Guo et al., 2017). The opposite condition of El Niño is termed as La Niña, which means the Girl Child, describing the large-scale cooling across the equatorial Pacific. La Niña also has a significant effect on the global atmospheric and oceanic circulation and climate.

As the air moves over the ocean, the friction between moving air and underlying water creates stress and transfers momentum from the air to the ocean surface, setting water in motion. Wind stress drives the ocean current and stress is a force per unit area applied tangentially to the water's surface. Stress can be expressed in the form $\tau = \rho C_d |\vec{V}| \vec{V}$, where ρ is the air density, C_d the drag coefficient, and $\vec{V}$ the horizontal wind vector. Wind stress at the ocean surface not only causes horizontal movement of water but also leads to vertical motion near the ocean surface. When the wind stress leads to a divergence of surface water, deeper water below rises to take its place (i.e., upwelling). Conversely, when there is a convergence of the surface water, sinking occurs (i.e., downwelling). Upwelling of subsurface water and sinking of surface water occur throughout the oceans. Nutrients induced by upwelling generate high phytoplankton biomass, most of which are eaten by zooplankton and fish. Many of the world's great fisheries are in coastal upwelling regions such as Peru, Ecuador, the west coast of North America, and off northwest Africa. During a major El Niño event, upwelling of cold and nutrient water is suppressed and being replaced by warm, nutrient-poor waters. As a result, large numbers of fish and marine plants die. For example, the El Niño of 1972–1973 reduced the annual Peruvian anchovy catch by more than 40%. Because much of this fish is converted into fishmeal, the world's fishmeal production was greatly reduced in 1972.

Figure 3.1 displays the climatological monthly mean SSTs and low-level winds over the tropical Pacific. A warm pool in the western Pacific ($\geq 28°C$) and a cold tongue ($<26°C$) and a strong frontal zone extending from the equator to 5°N in the eastern Pacific are evident. In January, a slack zonal SST gradient and weak warming occurs in the eastern equatorial Pacific. By April, the 26° isotherm contour is very close to the South American coast, suggesting a maximum warming occurs off the coast. With the approach of austral winter (July), enhanced

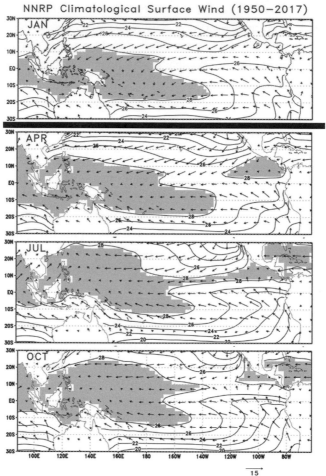

Fig. 3.1 Long-term monthly mean sea surface temperature (SST) and surface wind vectors for the tropical Pacific during 1950–2017 from the Extended Reconstructed Sea Surface Temperature (ERSST v4) for January (JAN), April (APR), July (JUL), and October (OCT). Contour interval is 2°C. Shading indicates SST greater than 28°C. Winds are in m s^{-1}.

southeasterlies increase the vertical mixing in the upper ocean, causing SSTs in the eastern Pacific to drop appreciably from April to July. Consequently, the axis of the equatorial cold tongue extends farther westward to 135°W. By October, the cold tongue is fully developed with its axis expanding westward to about 140°W and this pattern remains until November. Therefore, a westward expansion of the cold tongue throughout an annual cycle is noted. In contrast, SST variations in the western Pacific warm pool region are comparatively small from one season to another.

In addition to the zonal asymmetry, the SST pattern in the eastern Pacific exhibits a meridional asymmetry with warmer (cooler) temperature to the north (south) of the equator. The equatorial cold tongue and its attendant frontal zone show a strong seasonality and is strongest in September/October and weakest in March/April. Water in the eastern equatorial Pacific is cold because southeast winds along the Peruvian coast cause coastal upwelling. The Peru Current carries the cooler water northward and then westward from the South American coast. In addition, the steady easterly trades straddling on both side of the equator cause Ekman transport of surface water away from the equator towards both poles, inducing upwelling of cold water and generating a cold tongue. The shallow thermocline in the eastern Pacific allows cold water to be upwelled more efficiently.

It is also interesting to note that the cold tongue region does not lie exactly on the equator. It is actually located just south of the equator and its location is primarily determined by the orientation of the western coastline of South America. This is better seen in July and October (Fig. 3.1). In the case of an eastern boundary, southerly winds crossing the equator would force a westward Ekman transport south of the equator and an eastward Ekman transport north of the equator. This process leads to an upwelling south and downwelling north of the equator, thus cooling south and warming north of the equator, enhancing the meridional asymmetry in SSTs with respect to the equator. Because the west coast of South America tilts in a northwest to southeast direction in low latitudes and southeast winds prevail, this geographic setting and the prevailing wind direction favor warmer (cooler) SSTs to the north (south) of the equator in the eastern Pacific, and a meridional SST asymmetry across the equator is evident (Xie, 1998).

The tropical western Pacific is a warm pool region because of several contributing factors. The sun crosses the equator twice a year and solar radiation is nearly directly overhead and most intense throughout the year at low latitudes. The winds are light in the tropical western Pacific so evaporative cooling is kept at minimum. Also there is no cold ocean current and no major upwelling. Moreover, the thermocline is relatively deep (~200 m) in the western Pacific so the deep pool of warm water remains in the upper layer of the ocean which inhibits large SST fluctuations throughout the year. All those ingredients combined to make the tropical western Pacific a warm pool and keep SST changes from one season to another rather small.

Figure 3.2 shows the SST and surface winds over the tropical Pacific for the very strong El Niño year of 1997–1998 and the very strong La Niña year of 1998–1999. Note the extensive warming across the tropical Pacific during November–April of 1997–1998 and the absence of warming over the central and eastern equatorial Pacific for the same six months of 1998–1999. The surface wind

NOAA ERSST.V4 and NNRP Surface Wind
El Nino (Nov.–Apr., 1997–1998)

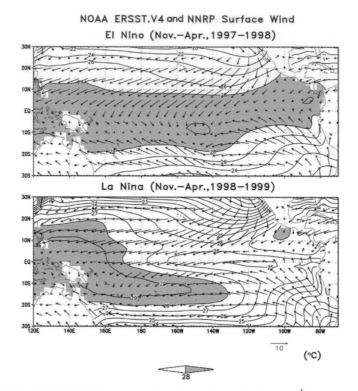

La Nina (Nov.–Apr., 1998–1999)

Fig. 3.2 (Top) SST (°C in contours) and surface winds (m s^{-1} in vectors) from November 1997 to April 1998, representing a strong El Niño event and (bottom) also for SST and surface winds, except from November 1998 to April 1999 for a strong La Niña event. Regions of SST greater than 28°C are shaded.
Source: Extended reconstructed Sea Surface Temperature (ERSST) v4 from International Comprehensive Ocean-Atmosphere Dataset (ICOADS) on a 2°grid. Winds are derived from the NCEP/NCAR Reanalysis I on a 2.5°grid.

over the western/central equatorial Pacific also has changed dramatically from the strong easterlies during the La Niña to weak westerlies during the El Niño episode. Also note the strong signal of the equatorial cold tongue extending from the South American coast to the central Pacific during the La Niña year. Figure 3.3 displays the precipitation and surface wind for these two climatic extremes. Precipitation corresponds rather well with warm SSTs. For the El Niño episode, the region of maximum precipitation belt has shifted eastward to central and eastern equatorial Pacific where the latter is also known as the equatorial dry zone. For the La Niña episode, the region of maximum precipitation is more limited and confined in the tropical western Pacific.

Each El Niño has its own characteristic onset and demise time, duration, evolution pattern, magnitude and location of maximum warming, propagation direction, and so on. Although each episode behaves differently, there is a tendency

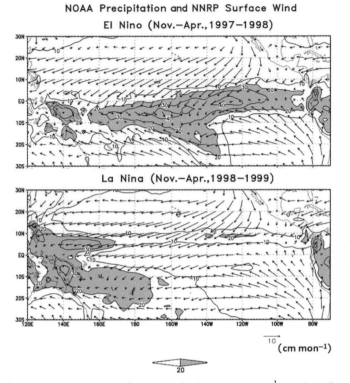

NOAA Precipitation and NNRP Surface Wind
El Nino (Nov.–Apr., 1997–1998)

La Nina (Nov.–Apr., 1998–1999)

Fig. 3.3 Same as Fig. 3.2 but for precipitation (cm mon^{-1}) and surface winds. Regions where monthly precipitation greater than 20 cm mon^{-1} are shaded.
Source: Precipitation data from NOAA's PRECipitation REConstruction Dataset (PREC) on a 2.5°grid. Winds are derived from the NCEP/NCAR Reanalysis I on a 2.5°grid. ©www.psl.noaa.gov/data/gridded/data.prec.html

for the maximum amplitude of major events to occur near the end of the calendar year so it is phase-locked to the annual cycle. Typically, an El Niño event runs for one year or so. It starts to develop from the boreal spring, increases in strength in the following summer and autumn, peaks in the winter, and then decays in the subsequent spring. The El Niño and La Niña events are also known to change their phase preferentially around March–May when surface winds are weak, sea-surface temperatures in the equatorial cold tongue are relatively warm, and the east–west sea-surface temperature gradient along the equatorial Pacific is slack (Fig. 3.1). This transition season is the time when atmosphere and ocean coupling is weakest. This phenomenon is known as the "spring predictability barrier" of ENSO when most climate models have a hard time of making forecasts accurately (Webster and Yang, 1992). The season in this book refers to the Northern Hemisphere. Consequently, the skill of ENSO prediction for the coming summer is very low when the forecast is made in the antecedent winter or earlier. El Niño or La Niña events usually decay after their spring season.

ENSO is a naturally occurring phenomenon with a significant global impact on society. During El Niño events, warm pools of seawater shift eastward to the central equatorial Pacific. The warm water, together with latent heat released during condensation, fuels the atmosphere with enormous warmth and moisture. This leads to storms, rainfall, and changing local/regional/global atmospheric circulation patterns in a distant part of the world, known as the ENSO teleconnections. Regionally, the enhanced tropical convection results in a strong ascending motion, which leads to prominent Hadley-type circulation in the central Pacific between the equator and the subtropics (e.g., Chu, 1995). Globally, these changes in atmospheric circulation can have effects thousands of miles away from the source region by rearranging the jet streams and areas of rising and sinking motions in the atmosphere. In short, the diabatic heating associated with tropical precipitation not only drives a local response in the atmospheric circulation but can also induce a remote response through the excitation of a stationary Rossby wave train to influence midlatitude climate (Hoskins and Karoly, 1981). One of such an example is the Pacific North American (PNA) teleconnection pattern (Wallace and Gutzler, 1981). As a result, droughts, heavy rainfall and flooding, wildland fire, and shifts in tropical storm patterns are natural hazards common to the imprints of ENSO (e.g., Glantz, 2001).

Through atmospheric teleconnections, which are linkages of climate anomalies at some distance away from the heat source region in the central or eastern equatorial Pacific, ENSO influences extratropical climate patterns over North America and Asia. For example, steered by an altering jet stream from the Pacific, California and Gulf coast of the USA see more midlatitude storms and rainfall during an El Niño winter, while a warm and dry winter occurs over the northern tier of the USA, southern Alaska, and western Canada (e.g., Ropelewski and Halpert, 1987). The very strong El Niño of 1982–1983 was responsible for the loss of nearly 2,000 lives and displacement of hundreds of thousands from their homes. Droughts and bush fires in Australia, South Africa, Indonesia, Philippines, India, and South America caused these losses. In the meantime, there were floods in the USA, South America, and Cuba. More tropical cyclones (TCs) than usual affected Hawaii and Tahiti, which were not hurricane-prone regions (e.g., Sadler, 1983; Chu and Wang, 1997; Chu, 2004). The 1997–1998 El Niño was among the strongest events in history. In fact, the 1982–1983 and 1997–1998 El Niño are the two strongest events in the twentieth century, and the 2015–2016 is the strongest by far in the twentieth-first century. Because of droughts, floods, famine, and disease, the death toll associated with the 1997–1998 El Niño was more than 20,000 resulting from this event around the world. ENSO also influences marine ecosystems, agricultural production, public health, and other human activity.

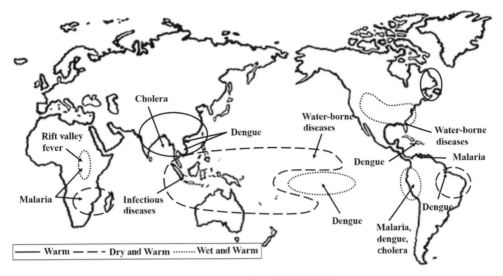

Fig. 3.4 Public health impacts of the 1997–1998 El Niño event.
Source: Adapted from NOAA OGP website.

The adverse effects of the ENSO on human health also received much attention worldwide (Glantz, 2001). When temperature and precipitation changed significantly from normal conditions, there is a greater likelihood of epidemic incidence. For example, many areas in tropical regions such as northeast Brazil, southeast Asia, and Pacific island nations experienced increased risk for dengue fever outbreaks during the 1997–1998 El Niño event due to accelerated productivity of mosquitoes (NOAA/OGP website, Fig. 3.4). Likewise, dengue fever is a problem in the Caribbean during La Niña. Malaria epidemics are another well-known infectious disease transmitted by mosquitoes and linked to ENSO. The epidemics affect Venezuela, Columbia, Ecuador, Peru, India, Pakistan, and east Africa. Rift valley fever is another disease that occurred in eastern Africa following excessive rainfall during the 1997–1998 event. This viral disease, also transmitted by mosquitoes, affects mainly cattle but can also spread to human beings. Water borne diseases are infections that are transmitted by microbes or parasites in water, and they occurred during the 1997–1998 event. They can spread to people who make skin contact with the contaminated water or eat food exposed to infected water.

3.1.2 Tropical Ocean–Atmosphere Feedbacks and Climate Variability

To facilitate the discussion in the book, it is first necessary to introduce some ocean–atmosphere feedback processes in the tropical Pacific and Atlantic. Here we present four feedback processes. In addition, the Bjerknes feedback will be presented in Section 3.1.3.

3.1.2.1 Wind-Evaporation–SST (WES) Feedback

There is a feedback that occurs between the wind-driven evaporation and SST perturbation (Xie and Philander, 1994). SST and some atmospheric variables such as wind speed govern evaporation, also known as latent heat flux, over the ocean surface, according to the bulk formula. Consider an initially asymmetric SST anomalous couplet with cold (warm) SSTs to the south (north) of the equator (Fig. 3.5). The warm SST anomalies create a low-pressure area because of the heating. This pattern induces a cross-equatorial flow and anomalously south-easterly flows to the south of the equator and anomalously southwesterly flows to the north of the equator due to the Coriolis effects. Under the background easterly trade winds, southeasterlies will be enhanced and thus lead to greater surface evaporation to the south of the equator, resulting in further cooling therein. To the north of the equator, anomalously southwesterly flows are against the mean easterlies, leading to weaker wind speeds and reduced evaporative cooling. As a result, extra cooling and warming resulting from wind and evaporation interaction would strengthen the original meridional SST perturbation, which in turn increase the meridional SST gradient and anomalous winds, and thus provides a positive feedback to enhance the interhemispheric asymmetry in both winds and SST. This is known as the wind-evaporation–SST (WES) feedback.

3.1.2.2 Footprinting Mechanism

The WES feedback may also act to propagate the subtropical ocean warming equatorward. Consider a positive SST anomaly in the northern subtropical ocean, which creates a low-pressure area around this SST perturbation because of heating. Westerly (easterly) anomalies are found to the south (north) of the low-pressure center. Against a background of easterly trades, warm (cold) SST occurs south (north) of the initial SST maximum via the WES feedback. As a result, the initial subtropical SST anomalies tend to propagate equatorward and this meridional propagation help extend the subtropical SST and wind variability to the deep tropics in the so-called

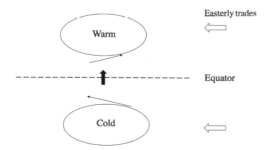

Fig. 3.5 An illustration showing the wind-evaporation–SST (WES) feedback across the equator.

seasonal "footprinting mechanism" (Vimont et al., 2003). In particular, the tropical atmosphere (e.g., zonal wind anomalies) during spring and summer is forced by the subtropical or midlatitude SST anomalies of the preceding winter.

3.1.2.3 SST and Cloud Feedback

The presence of clouds may also alter the zonal SST asymmetry along the equatorial Pacific (Li and Philander, 1996; Li, 1997). In the eastern Pacific, a capping inversion is formed due to the large-scale subsidence associated with the Pacific Walker circulation. The inversion acts as a lid for the vertical development of deep clouds and thus favors the development of boundary-layer clouds (e.g., marine stratus clouds). A positive feedback exists between the SST and the marine stratus clouds in the subtropics. When the SST is cold, the static stability of the lower atmosphere is larger, which leads to stronger subsidence inversion. Strong inversion can prevent moist boundary layer from mixing with dry air above and keep the low layer moist, thus providing more low-level cloud amounts. An increase in the cloud decks shields the ocean from shortwave radiation and promotes the initial cooling. This positive feedback mechanism enhances the cooler SSTs of the eastern Pacific and contributes to the zonal asymmetry of SSTs. Therefore, low cloud cover is negatively correlated with the underlying SST in areas where the climatological mean SST is low. However, in regions where the SST is high such as the western Pacific warm pool, cloud and SST feedback is complicated. This is because the warm ocean enhances deep convection, producing cumulonimbus clouds with deep convective clouds that block the solar radiation from reaching the surface and thus cool the ocean. However, those clouds also can warm the ocean surface by trapping longwave radiation below the cloud base. This effect offsets the negative shortwave radiation feedback.

3.1.2.4 SST and Water Vapor Feedback

According to the Clausius–Clapeyron equation, the saturation water vapor pressure in the atmosphere increases exponentially with temperature at a rate of $\sim 7\%$ K^{-1}. If the relative humidity remains constant, atmospheric water vapor concentration will increase with increasing temperatures at almost the same rate. Because water vapor absorbs infrared radiation and is an important greenhouse gas, higher concentration of atmospheric water vapor will result in higher SSTs by increasing downward longwave radiation (Wang et al., 2017). As SST increases, more water vapor from the ocean will evaporate into the overlying atmosphere, and a positive feedback mechanism is established. This SST–water vapor feedback enhances the hydrological cycle and likely leads to the intensification of heavy precipitation and a significant increase in the total wet-day rainfall in a warmer world as seen from observations and climate model simulations. The strengthening of the hydrological cycle also alters water resources both in quantity and quantity (IPCC AR5, 2014).

3.1.3 Pioneers of the ENSO Studies and Bjerknes Feedback

Bjerknes (1966, 1969) first developed a conceptual model relating the zonal pressure seesaw between the extreme ends of the great Pacific basin (i.e., the SO) to the Walker circulation and SST anomalies in the equatorial Pacific. The Walker circulation, named by Bjerknes (1969), is a zonal atmospheric circulation cell along the equatorial Pacific with rising air over the warm and lower pressure over the Indonesian region and sinking air and higher pressure over the cold eastern Pacific (Fig. 3.6 top). The vertical air movement of this thermally direct circulation cell is connected by easterlies in the lower troposphere and westerlies in the upper troposphere. The surface air flows from cold air with higher pressures in the east toward warm air with lower pressures in the west. The zonal pressure gradient in the atmosphere caused by the zonal SST contrast drives an equatorial zonal asymmetric Walker circulation, which enhances the surface easterlies across the equatorial Pacific and thus supports the cold tongue.

   Bjerknes suggested that during an El Niño episode (or a negative phase of the SO), there was a weakening of the zonal SST gradient across the Pacific basin, brought about by warming of the equatorial eastern Pacific, and that the trade wind field across the Pacific slackened, in response to ocean warming (Fig. 3.6 bottom). Weakened trades imply a reduction in the zonal sea-level pressure gradient. As a result, the surface air flows down the zonal pressure gradient towards the area of anomalously lower pressure in the east, resulting in weakened easterly trade winds. The weakened trades also reduce the upwelling of the cold subsurface water in the eastern equatorial Pacific, which in turn further reduce the zonal SST gradient. Therefore, a positive feedback exists between the tropical ocean and atmosphere as surface winds drive ocean currents, and these currents redistribute surface thermal gradients that affect wind fields through hydrostatic effects on sea-level pressure (Lindzen and Nigam, 1987). The strength of the zonal SST gradient controls the strength of the winds. In short, strong (weak) SST gradients lead to strong (weak) easterly trades which then result in strong (weak) SST gradients, and this is known as the Bjerknes feedback. This positive ocean–atmosphere feedback in the tropical Pacific would lead to a never-ending climate state. As such, the Bjerknes' positive feedback mechanism describes the rapid growth but cannot explain the dynamics that brought a warm event to the end. Moreover, the simultaneous slackening of trade winds and warming of the entire Pacific Ocean are not consistent with observations.

   It was Wyrtki (1975) who provided the observational basis for modifications to Bjerknes hypothesis for ENSO, by adding the thermocline adjustment in the

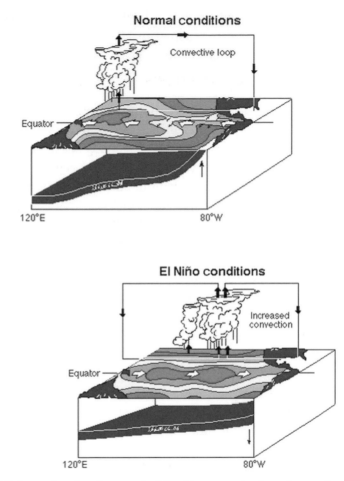

Fig. 3.6 Schematics showing tropical Pacific atmospheric and oceanic conditions during the normal conditions (top) and ENSO conditions (bottom). For the normal conditions, the Pacific Walker circulation is marked by surface easterly winds, ascending motion, and strong convection in the far western Pacific, and upper-level westerlies and sinking motion in the eastern Pacific. A steep thermocline across the Pacific is seen. For the El Niño case, the Walker circulation is split into two small cells, joined by rising motion and convection in the central equatorial Pacific with anomalously westerly winds near the surface. A rather flat thermocline is evident.

Source: www.pmel.noaa.gov/elnino/what-is-el-nino.

tropical Pacific Ocean, and this is also sometimes known as the Bjerknes–Wyrtki feedback. Wyrtki found that in the year before a warm episode occurred, trade winds strengthened, increasing the zonal gradient of sea level across the Pacific by building up the water in the western Pacific. In a simple way, the sea level can be viewed as a measure of the depth of the warm-water layer above the thermocline. The thermocline is a zone separating the warm, well-mixed surface layer from the cold waters below, in which temperature decreases sharply with depth. In the tropical Pacific, the thermocline is shallow in the eastern Pacific and relatively deep in the western Pacific. The tropical ocean heat content is positively related to the thermocline depth, with a higher heat content being proportional to deeper thermocline depth and vice versa.

Any prolonged relaxation of central Pacific trade winds would then lead to a dynamical sequence in which the accumulated western Pacific warm water (i.e., heat content) would not be maintained under a reduced zonal sea-level gradient, and thus would travel eastward as a wavelike body of upper-layer oceanic water mass (i.e., oceanic Kelvin waves), deepening the thermocline and raising the sea level and SST of the eastern equatorial Pacific (Wyrtki, 1975). This important advance based on the observational approach combined with the subsequent numerical model of Zebiak and Cane (1987) provide the foundation of what is later known as the "recharge–discharge" theory (Jin, 1997). The other prevailing theory is the "delayed oscillator" mechanism. There are two other mechanisms to explain the oscillatory nature of ENSO: the western Pacific oscillator (Weisberg and Wang, 1997; Wang, 2001) and the advective–reflective oscillator (Picaut et al., 1997). The western Pacific oscillator emphasizes changes in equatorial winds in the western Pacific for the negative feedback of the coupled atmosphere–ocean processes, while the advective–reflective oscillator emphasizes the anomalous zonal advection associated with wave reflection at the eastern and western boundary. Here, we will focus on the delayed oscillator and recharge oscillator because these two are popular theories to account for the oscillatory nature of ENSO phase transition. Sarachik and Cane (2010) contain very comprehensive and theoretical treatments of the ENSO dynamics. Li and Hsu (2018) present fundamental dynamics of tropical climate relevant to the ENSO phenomenon.

3.1.4 Delayed Oscillator Theory

Suarez and Schopf (1988) and Battisti and Hirst (1989) were the first to use linear equatorial ocean dynamics to explain the turnabout of ENSO from a warm to a cold state. If the SST in the eastern equatorial Pacific is anomalously warm then the patch of wind anomalies induced by the warm SST in the central equatorial Pacific is westerly. The delayed oscillator relies on two properties of the upper ocean. One is that information propagates rapidly eastward in the form of oceanic Kelvin waves and more slowly westward in the form of oceanic Rossby waves. The other is that the thermocline is significantly deeper in the western Pacific than that in the eastern Pacific. Consequently, SSTs are very sensitive to thermocline displacement in the eastern Pacific, but in-situ SST and thermocline-depth changes are not necessarily in phase.

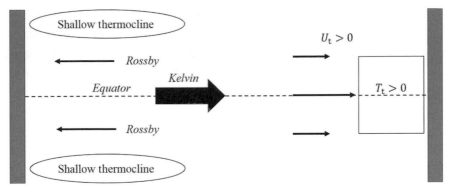

Fig. 3.7 Schematic showing the equatorial oceanic Kelvin waves excited by westerly wind bursts and off-equatorial Rossby waves in the western Pacific. Note the delayed deepening of the thermocline and rise of sea surface temperatures in the eastern Pacific (T_t).

One process that may affect SST changes in the eastern Pacific is known as westerley wind bursts (Luther et al., 1983; McPhaden et al., 1988; Harrison and Giese, 1991). Based on daily ship data in the equatorial western Pacific during 1958–1987, Chu et al. (1991) showed that westerlies indeed increase their frequency of occurence during an ENSO cycle. The westerly anomalies not only advect warm water eastward but also send downwelling Kelvin waves eastward, which will suppress thermocline in the east (Fig. 3.7). Oceanic Kelvin waves are typically 5–10 cm high (i.e., a slight rise in sea level), hundreds of kilometers wide, and a few degrees warmer than surrounding waters. Their phase speed is two to three m s^{-1}, and they cross the Pacific in about two to three months. In deepening the thermocline in the equatorial eastern Pacific, cold upwelled water becomes less efficient. Thus, SST increases and sea level rises in the eastern Pacific. At the eastern boundary, the equatorial Kelvin waves split into northward and southward moving coastal Kelvin waves and partially reflected as oceanic Rossby waves.

The patch of westerly anomalies in the central Pacific also excite near-equatorial upwelling Rossby signals which propagate westward to the west of the patch. McCreary (1983) originally demonstrated how oceanic Rossby waves might be involved in generating the low-frequency oscillatory nature of ENSO. These Rossby signals are characterized by shallower thermocline anomalies due to forcing by anomalous cyclonic wind stress curl associated with westerly anomalies. These Rossby waves take longer time to reach the western boundary of the vast Pacific basin, where they reflect back towards the east in the equatorial wave guide as "upwelling" or "negative" Kelvin signals, which then progressively raise the thermocline depth and reduce the initial warming in the eastern Pacific. As a result, the thermocline rises at the east and the original warm SST now becomes cold, and the ocean–atmosphere system is then about to begin the cold phase of the ENSO cycle. This delayed oscillator mechanism provides the negative feedback necessary to terminate a warm phase and leads to relatively cooler SSTs, reduced sea level, and a shallow thermocline in the east with a reversed conditions in the west, which is a manifestation of a cold phase of ENSO. The positive feedback of Bjerknes gives the pure growth but not the ENSO oscillation. The equatorial ocean waves offer a mechanism to reverse the phase of perturbations of the thermocline depth. By invoking both the Bjerknes feedback and equatorial ocean wave dynamics, a self-sustaining oscillation between the atmosphere and ocean with a period like ENSO can be generated. In short, the delayed oscillator mechanism can account for the oscillating nature of multi-year time scale of ENSO and requires the ocean basin with an east–west dimension of greater than 13,000 km. This implies that ENSO could not occur in the Indian or the Atlantic Ocean based on this theory.

Mathematically, the delayed oscillator model may be understood from a single equation with a delay term and involves both positive and negative feedbacks (Wang, 2001):

$$\frac{dT}{dt} = AT - BT(t - \eta) - \varepsilon T^3,$$

where T is the SST anomaly in the eastern equatorial Pacific, and A, B, and η are model constants. Conceptually, the first term on the right-hand side of the equation represents the positive feedback by local atmosphere–ocean processes in the eastern equatorial Pacific. The second term provides the negative feedback by free Rossby waves generated in the eastern Pacific, propagated and reflected back from the western boundary, and returned as Kelvin waves to reverse the thermocline depth anomalies in the eastern Pacific. The last term denotes damping in the model system.

3.1.5 Recharge–Discharge Mechanism

Based on observations, Wyrtki (1975, 1985) noted that the oceanic heat content or the volume of warm water in the equatorial Pacific builds up prior to a warm event (Fig. 3.8). Here the heat content is represented by the integrated water volume above the 20°C isotherm between 5°N and 5°S, and 120°E and 80°W. The buildup of warm water volume and its subsequent discharge are the key ingredients in the successful simulation of ENSO-like behavior of the equatorial Pacific SSTs and low-level winds (Cane and Zebiak, 1985; Zebiak and Cane, 1987). Invoking the relationships between ocean heat content, SSTs and zonal wind stress, Jin (1997) proposed a "recharge–discharge" oscillator theory of ENSO. It is the recharge–discharge process that makes the coupled ocean–atmosphere system in the tropical Pacific to oscillate on interannual time scales.

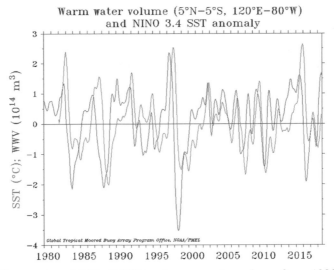

Fig. 3.8 Time series of Niño-3 SST and warm water volume from 1980 to 2017. A black and white version of this figure will appear in some formats. For the color version, refer to the plate section.
Source: NOAA/PMEL.

The discharge–recharge paradigm of Jin (1997) is illustrated in a schematic (Meinen and McPhaden, 2000). Given warm SST anomalies in the eastern equatorial Pacific (panel I in Fig. 3.9), the westerly wind anomaly in the central Pacific responds to the zonal temperature and pressure gradients and thermocline become anomalously deep in the eastern Pacific (i.e., below the dashed line) and shallow in the western Pacific (above the dashed line). The thermocline depth is highly related to the upper ocean heat content. A deeper thermocline corresponds to higher heat content and vice versa. The discharge of the warm water to higher latitudes is caused by the divergence of upper-ocean poleward mass transports resulting from the westerly wind anomaly near the equator (i.e., the Sverdrup transport induced by wind stress curl). In simple terms, the Sverdrup balance expresses a relationship between the wind stress near the ocean surface and the vertically integrated meridional mass transport of ocean water. Here the word "discharge" implies that the heat is being released or discharged meridionally from the tropics. This leads to a shoaling of zonal mean thermocline, which causes cooling in the eastern Pacific and the east–west thermocline slope becomes rather flat (II). The east–west thermocline tilt is in a quasi-equilibrium balance with the zonal wind stress. Note that at this transition phase, zonal wind anomalies in the central Pacific and SST anomalies in the eastern Pacific are near normal.

The shoaling of thermocline allows cold water from the deeper ocean to be pumped efficiently to the upper ocean by climatological upwelling, and the surface

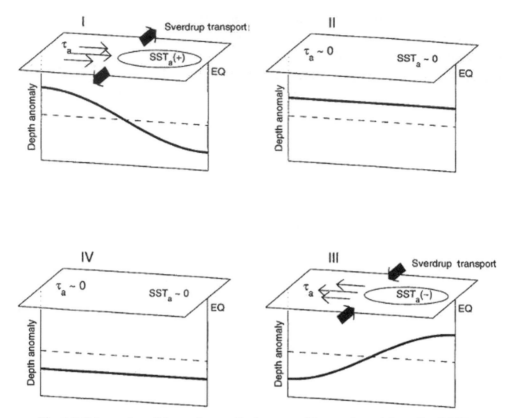

Fig. 3.9 Schematics of the recharge–discharge oscillator, adapted from Jin (1997). The model considers SST anomalies in the eastern equatorial Pacific, zonal wind stress anomalies in the central equatorial Pacific, and thermocline depth anomalies. Panels I and III correspond to the El Niño and La Niña oscillations, respectively. All quantities are anomalies from the climatological mean conditions. Arrows indicate anomalous zonal wind stress, τ_a, and thick arrows denote the corresponding Sverdrup transports. SSTa are sea surface temperature anomalies. Depth anomaly in the ocean refers to the time mean along the equator. Dashed line is for zero anomaly. Panels II and IV are the transition phases.

Source: Meinen and McPhaden, 2000; ©American Meteorological Society. Used with permission.

water in the eastern Pacific subsequently becomes colder. This cooling favors the development of easterly wind anomaly in the central Pacific via the Bjerknes–Wyrtki feedback, corresponding to a La Niña event (III). The enhanced trade winds, in response to cold SST anomalies, raise the thermocline depth in the east and deepen it in the west (III). Subsequently, the relatively warm water volume from the off-equatorial regions is transported meridionally via the Sverdrup relation to the cold eastern equatorial Pacific (a recharge process). The zonal mean thermocline deepens resulting from the recharging of equatorial heat content and sets the stage for the next warm event (IV). Also note that the zonal mean thermocline variation lags behind the central Pacific zonal wind anomalies or the eastern Pacific SST anomalies (from I to II or from III to IV). This discharge of the equatorial heat content in the warm phase and recharge at the cold phase provide a necessary negative feedback to account for the phase transition from an El Niño to a La Niña event and vice versa.

The recharge–discharge theory is also consistent with actual observations, as shown by Capotondi et al. (2015) and Timmermann et al. (2018). Before the peak of an El Niño event, as indicated by January 1 of the mature El Niño year, Fig. 3.10a, the equatorial thermocline depth reaches its maximum depth in the Niño-3 region (~120°W), which is an indication of enough warm water already built up in the equatorial belt (a recharge state). The positive depth anomalies along the equator then decrease rapidly and changed to negative anomalies after the peak event. That is, heat is being "discharged" eastward and poleward because of the mass exchange between the equatorial and off-equatorial regions via ocean dynamics adjustment (Cane and Zebiak, 1985; Jin, 1997). For regions farther to the west in the central equatorial Pacific (175°W), as represented by the Niño-4 (Fig. 3.10b), the thermocline evolution bears a similar pattern to the Niño-3, although its amplitude is much weaker. In the warm pool region (165°E), the thermocline changes are very small and there is no sign reversal of thermocline anomalies after the peak event, suggestive of absence of warm water discharge (Fig. 3.10c).

3.1.6 Eastern and Central Pacific El Niño

In the composites of Rasmusson and Carpenter (1982), an initial warming occurred off of the South American coast and then propagated westward to the equatorial central Pacific. The SST anomalies are largest near the coast and are accompanied by weak and opposite polarity in western equatorial Pacific. This canonical ENSO evolution, however, is not observed in some other ENSO years such as the 1977–1978 and 2004–2005 events. In fact, a different type of El Niño with the largest warming in the central equatorial Pacific has emerged (Larkin and Harrison, 2002; Ashok et al., 2007; Kao and Yu, 2009; Kug et al., 2010; Capotondi et al., 2015; Timmermann et al., 2018). Based on a self-organizing map analysis and statistical distinguishability test, Johnson (2013) found nine different patterns that characterize the tropical Pacific SST anomaly fields for the last 60 years. These nine patterns represent different flavors of ENSO but they can be broadly classified into two types: Central Pacific (CP), Modoki (a Japanese word meaning similar but different), dateline El Niño or warm pool El Niño and Eastern Pacific (EP) or canonical El Niño events. CP events have its largest sea surface temperature

Thermocline depth evolution

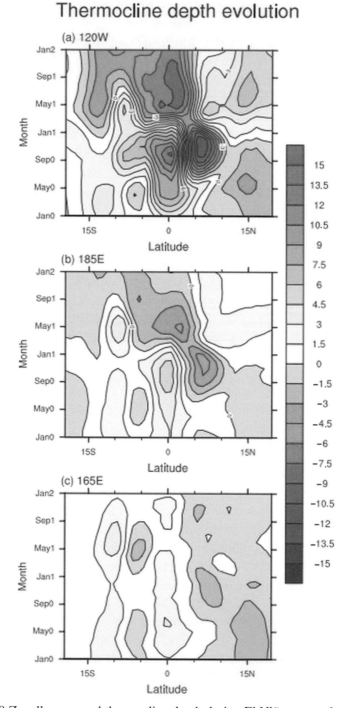

Fig. 3.10 Zonally averaged thermocline depth during El Niño events from year 0, year 1 (event peak), to year 2 for the Niño-3 region (5°N–5°S, 150°W–90°W) Caption for fig. 3.10 (*cont.*). as indicated by 120W at top (a), the Niño-4 region

anomalies (SSTA) centered in the central equatorial Pacific and EP has its largest SST amplitudes in the equatorial eastern Pacific. The propagating feature of the SSTA associated with CP type of ENSO is less clear, compared to the EP type (Kao and Yu, 2009). Ashok et al. (2007), Lee and McPhaden (2010), and Lu et al. (2020) found that CP El Niño events tended to occur more frequently than EP events in the twenty-first century. A new method of separating CP from EP El Niño events is recently presented in Lu et al. (2020).

The EP type of ENSO is associated with the basin-wide thermocline and surface wind variations. This is in contrast to the CP type, which is characterized by regional variations of ocean temperatures and winds (Kao and Yu, 2009). In particular, the thermocline variations appear to exert less influence on the generation of CP ENSO type and zonal advection and air–sea heat fluxes become more important for the central Pacific SSTA (Kao and Yu, 2009; Kug et al., 2009; Newman et al., 2011; Timmermann et al., 2018). However, the equatorial Pacific thermocline depth at the peak phase of a CP event may determine whether a warming, cooling, or neutral state in the eastern Pacific will follow as the CP event decays (Yu and Kim, 2010). The composite SSTA for EP El Niño (a) and EP La Niña (b) are shown in Fig. 3.11. The label 0 refers to the peak-intensity month and −3 (+3) means three months before (after) the peak warming. For both El Niño and La Niña of EP type, the initial SSTA occur off the South American coast along the equator but the magnitude of warming is much stronger than that of the cooling. The SSTA appear to propagate westward with time.

In comparison, for the two climatic extremes of CP type, the largest SST anomalies are found over the central equatorial Pacific and there is little indication for the direction of SST propagation (Fig. 3.11). For the CP type of ENSO, SST cooling in the central Pacific during La Niña events is a little stronger than SST warming during El Niño events in the same region. Therefore, there is an amplitude asymmetry between the warm and cold events in the central Pacific, which may result from nonlinearities in the ENSO dynamics (Takahashi et al., 2011; Choi et al., 2013; Dommenget et al., 2013). Wang et al. (2019) unraveled the "strong basin-wide (SBW)" type or the extreme El Niño, distinguished by their extraordinary intensity (maximum SSTA > 2.5°C) and ubiquitous Pacific

(5°N–5°S, 160°E–150°W) as indicated by 185E in (b), and a region displaced 20° to the west of the Niño-4 region in the western Pacific in (c). The depth of the 15°C (m) is used as a proxy for thermocline depth. A black and white version of this figure will appear in some formats. For the color version, refer to the plate section.
Source: Capotondi et al., 2015; ©American Meteorological Society. Used with permission.

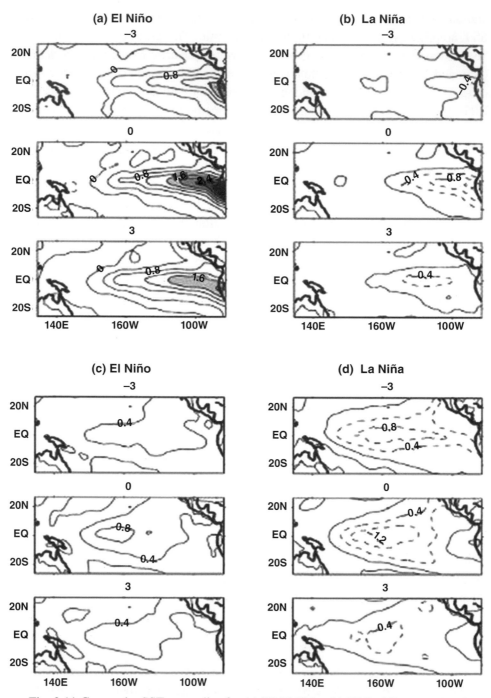

Fig. 3.11 Composite SST anomalies for (a) EP El Niño, (b) EP La Niña, (c) CP El Niño, and (d) CP La Niña. Contour intervals are 0.4°C and negative values are dashed. The lag month refers to the peak-intensity month and is labeled above each panel.

Source: Kao and Yu, 2009; ©American Meteorological Society. Used with permission.

basin-wide warming features. El Niño onset regime has changed from eastern Pacific origin to western Pacific origin with more frequent occurrence of extreme events since the 1970s. This regime change is found to arise from a background warming in the western Pacific and the associated increased zonal SST and vertical ocean temperature gradients in the equatorial Central Pacific (Wang et al., 2019).

In the eastern Pacific where thermocline feedback is dominant, warmer SSTs induce weaker easterly wind stress which leads to the deeper thermocline, resulting in warmer SSTs due to strong upwelling. This is a positive feedback process (An and Jin, 2001). In the central Pacific, the zonal advective feedback, that is, the advection of the zonal temperature gradients by the anomalous zonal currents, together with thermocline feedback, is also important for the development of SST anomalies associated with the CP El Niño (Ren and Jin, 2013; Fedorov et al., 2015; Guan and McPhaden, 2016). Specifically, positive equatorial SSTA weaken trade winds, which reduce the oceanic transport of cold water from the eastern Pacific and thus reinforce the initial warm ocean temperatures. The zonal advective is a positive feedback and effective in driving the central Pacific's SST changes because of the large mean zonal SST gradients in the central Pacific between the western Pacific warm pool and the eastern Pacific cold tongue. This is different from the EP El Niño where the thermocline–depth displacement is more influential on the in-situ SSTs. Moreover, a unique wind-driven vertically slanted subsurface temperature response due to higher oceanic baroclinic modes makes thermocline feedback less effective in the central Pacific than in the eastern Pacific (Zhao et al., 2021). Therefore, changes in local SST for the central and eastern Pacific depend on different key dynamical processes. Atmospheric forcing from the subtropical North Pacific by the footprinting mechanism described in Section 3.1.2 may also generate the CP type SSTA. The subtropical atmospheric circulation can first induce positive SSTA off Baja California during boreal winter (Vimont et al., 2003), which then spread southwestward in the following seasons through the subtropical atmosphere–ocean interaction (i.e., the Pacific Meridional Mode) and reach the tropical central Pacific to yield a Central Pacific type of El Niño (Kim et al., 2012; Paek et al., 2017). The Pacific Meridional Mode and the Atlantic Meridional Mode will be described in Section 3.3.

3.1.7 ENSO Forecasts and Recent Improvements

Because ENSO has an immense impact on global climate and the society worldwide, many researchers, institutes, and national meteorological centers around the world have actively performed real-time ENSO forecasting. The Climate Prediction Center of the National Centers for Environmental Prediction of NOAA, together with the International Research Institute (IRI) for Climate and

Society, has been providing an ENSO diagnostic discussion on a regular basis, which describes the current status of ENSO conditions and their long-lead forecasts derived from a suite of climate models. ENSO predictions have been displayed each month on a graph known as the "ENSO prediction plume."

In general, dynamical or statistical models or a combination of both models are used in the forecasting enterprise. For dynamical models, the degree of complexity varies from simple, linear shallow-water equations for both ocean and atmosphere, to intermediate coupled ocean–atmosphere models, to hybrid coupled models (e.g., statistical atmosphere and comprehensive ocean circulation), to fully coupled ocean–atmosphere–land–ice models with multiple vertical layers. In comparison to dynamical models, statistical models are simple and use less computer time and storage space. Most statistical models are based on regression techniques and will be described later.

Latif et al. (1994, 1998) provided a comprehensive study of ENSO prediction performance based on some dynamical and statistical models. They concluded that all models performed better than the simple persistence forecast for typical ENSO indices on lead times of 6–12 months. Based on a collection of 15 statistical and dynamical climate models, Barnston et al. (1999) assessed the predictive skill of SST forecasts during the extraordinarily strong 1997–1998 El Niño event and the 1998 La Niña onset. Neither the group of dynamical or statistical models performs significantly better than the other. Many models were able to forecast some degrees of the warming one to two seasons prior to the onset in boreal spring 1997, but none of any models predicted the strength until the warming became very strong in late spring.

By considering the persistence of initial conditions, trend, and climatology of past ENSO events, Knaff and Landsea (1997) developed a multiple regression model to forecast Niño-3 and Niño-3.4 region SSTA. They called it an ENSO-CLIPER model. The idea of the ENSO-CLIPER model is derived from the tropical cyclone community in which a simple CLIPER (i.e., Climatology plus Persistence) scheme has long been used as a benchmark against other more sophisticated models for storm-track prediction. Landsea and Knaff (2000) compared forecast skills of their ENSO-CLIPER model with other dynamical and statistical models for the very strong 1997–1998 El Niño event. They noted that at short lead time (up to eight months ahead), the ENSO-CLIPER has the smallest root-mean-squared error (RMSE) among all models tested. This result is rather intriguing because forecasts made by the ENSO-CLIPER are considered as a baseline reference against those from the more complex dynamical models.

Subsequently, Jin et al. (2008) examined ENSO prediction in retrospective forecasts from 10 different coupled atmosphere–ocean models for the period of 1980–2001 (22 years). Retrospective forecasts, also known as reforecasts or hindcasts, are used to calibrate subsequent real-time forecasts, generated by the same model (Saha et al., 2006) or calibrated against the reanalysis which can be approximately regarded as "observations." A simple method to calibrate forecasts against reanalysis is used. Let's say the variable of interest is the January temperature where a Gaussian distribution can be reasonably assumed. The raw model forecast is plotted along with the reanalysis in the form of a statistical distribution. This allows the mean forecast error in the model (e.g., cold bias) to be corrected. Similarly, if the spread of the real-time operational forecasts is too small, it can be inflated by scaling the standard deviation of the raw forecasts to the reanalysis. Consequently, the model's forecasts are transformed to a distribution that is consistent with the observed mean state and the observed interannual variability. It is through this calibration process that raw operational forecasts from a dynamical climate model are transformed to retrospective forecasts. In Jin et al. (2008), almost all coupled models have problems in simulating the mean state and mean annual cycle of SST in the Niño-3.4 region. The growing phase of both warm and cold ENSO events is better predicted than the corresponding decaying phase. A stronger El Niño is better predicted than a weak one.

Although the coupled dynamical models developed in the 1980s and 1990s have limited success in ENSO prediction (Jin et al., 2008), the newer version of these models made remarkably improvement. The forecast skill of dynamical models, evaluated during 2002–2011, exceeds that of statistical models (Barnston et al., 2012) who attributed this difference to the improvement of the observing and the sophisticated data assimilation systems for model initialization, higher spatial resolution, the improved model physics, and better understanding of the coupled atmosphere–ocean processes in the tropics important for ENSO development, maturity, demise, and transition (Guilyardi et al., 2009). The coupled dynamical models are run frequently using new information in the atmosphere–ocean system on a shorter time span (hours, days, to a few weeks). These qualities, through complex data assimilation schemes, allow dynamical models to ingest more recent observations and see better real-time changes in the ENSO evolution. By comparison, most statistical models, which were developed 30 years ago, have remained unchanged structure-wise. They are built on monthly or seasonal data and receive inputs that are relatively older in comparison to dynamical models. That is, they are unable to update their forecasts on a submonthly or even shorter time scale as new information becomes available. These statistical models also do not include the subsurface ocean heat content in the tropical Pacific as potential predictors (Barnston et al., 2012). As a result, dynamical ENSO prediction models in recent years appear to outperform their statistical counterparts, particularly during the boreal spring.

The predictive skill of the most recent 2015–2016 El Niño event, which was among the top three in the historical records (the other two major El Niño events are 1982–1983 and 1997–1998), was studied extensively based on 15 dynamical models and 10 statistical models (L'Heureux et al., 2017). They noted that the prediction for this most recent event was successful in comparison to the low ENSO predictability in the previous decades (Latif et al., 1994, 1998; Jin et al., 2008; Barnston et al., 2012). For example, back in May 2015, an average from a subset of dynamical models predicted in excess of 2°C for the following fall/winter seasons. This agrees rather well with the subsequent Niño-3.4 index which peaked in December 2015 with a value of 2.6°C. Moreover, the temporal anomaly correlation (AC) between forecasts and observations within a 26-month sliding window is as high as 0.9 for lead up to 4 months based on the dynamical multimodel ensemble (MME) average for the study period DJF 2013/2014 to FMA 2016 (Fig. 3.12). This is better than the past IRI/CPC plume history (gray lines). For the same AC of 0.9, the statistical models only go out to lead two. In terms of the RMSE, the dynamical ensemble has the smallest errors for leads up to

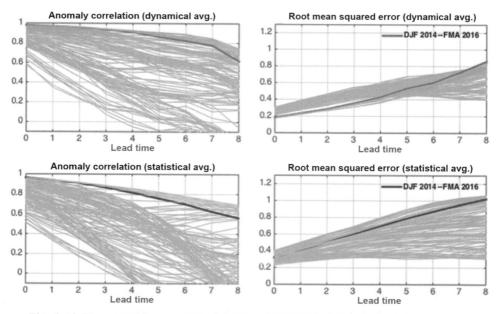

Fig. 3.12 The anomaly correlation (left) and RMSE (right) between observations and multimodel ensemble averages of the dynamical (top) and statistical (bottom) forecasts of the Niño-3.4 index. The thick red and blue lines denote the skill for the period between DJF 2013/14 to FMA 2016. The gray lines are the skills of past windows of 26 consecutive overlapping seasons and each slides by one season. The thin red and blue lines correspond to windows that overlap with the DJF 2013/2014–FMA 2016 period.
Source: L'Heureux et al., 2017; ©American Meteorological Society. Used with permission.

four as seen in the model plume while the statistical model errors are not so small at all lead.

Given the numerous models available in the seasonal forecasting community, it is tempting to combine forecasts from various models because each of them has strength and weakness in capturing different aspects in ENSO prediction. Indeed, a common approach to synthesize the forecast information is to take a simple average from multi model outputs. This approach is generally found to provide a better result than that from a single best model run. However, the simple average approach places an equal weight on each model, without being able to discriminate the good ones from the bad ones. To overcome this problem, Raftery et al. (2005) suggested that the Bayesian model averaging (BMA) can be used to assess the performance of the individual models and assign greater weights to better performing models. As a result, this method can be used as a basis for selecting models and hence to improve forecasts from multi models. In addition, BMA provides probabilistic forecasts that describe the uncertainty of the prediction. This differs from the deterministic approach which gives a definite information for an event occurrence. In the following, an example of how to improve ENSO prediction using the BMA is illustrated (Zhang et al., 2019).

The data used are the three-month running mean seasonal SST forecasts in the Niño-3.4 region from 1982 to 2010 from four climate models managed by the Climate Prediction Center (CPC) of the National Centers for Environmental Prediction of NOAA. Three statistical models and one dynamical model with lead times from one to seven months are used. The three statistical models are the Constructed Analogues (CA) (van den Dool, 1994), Canonical Correlation Analysis (CCA) (Barnston et al., 1994; He and Barnston, 1996) and Markov Model (MKV) (Xue and Leetmaa, 2000). For statistical models, historical hindcasts are used. The dynamical model used is the fully coupled Climate Forecast System Version 2 (CFSv2), which provides a vast array of products for subseasonal and seasonal forecasting with an extensive set of retrospective forecasts (Saha et al., 2014). At the operational center, retrospective forecasts perform the forecasts again using information that was not available at the time of original forecasts. That information might be new observations, updated data assimilation system, or forecasts from other models. A cold bias in the CFSv2 forecasts before 1998 was first corrected. For convenience, the term "forecasts" mean retrospective forecasts or hindcasts here. The observation data are available on the CPC's website. Details on statistical methods of the BMA can be found in Raftery et al. (2005) and Zhang et al. (2019).

To distinguish the difference in predictive skills between dynamical and statistical models, the BMA is first applied to the purely statistical models and then the combination of statistical and dynamical models. Here the BMA deterministic forecasts are first used, which are the weighted average from a suite of models subject to the constraint that the sum of the weights much be unity. The BMA probabilistic forecasts will be presented next in the attributes diagram. The BMA is developed for each season independently and the weight for each model is generally not the same. In some seasons, the weight is mainly distributed on one model while a particular model may not receive any weight.

Figure 3.13a shows the summary of the RMSE for statistical-dynamical BMA model in different target seasons and lead times. The RMSE has a negative orientation so the smaller the value the better is the forecast. In general, RMSE are larger at longer leads. A typical issue of many of the models in ENSO predictions is the poor performance when forecasts go through boreal spring (e.g., Webster and Yang, 1992; Kirtman and Min, 2009). This so-called spring predictability barrier is also reflected in this figure. For example, the forecast for July–August–September (JAS) made in the preceding January (lead 5 month) has a RMSE of 0.6–0.7. This value is relatively high compared to the target season of January–February–March (JFM), also at five month's lead. The boreal spring is a transition season when the tropical ocean–atmosphere interaction in the Pacific is usually the weakest and the ENSO signal is relatively not well defined. Figure 3.13b is the difference in RMSE between the statistical-dynamical BMA and the statistical BMA models. For most of the target seasons and leads, the values are negative which indicate that the overall forecast performance of the former is better than the latter.

It is also of interest to compare the BMA forecasts with that of the multi model ensemble average (MMEA) forecasts, which assumes an equal weight for each model. Results of the differences between BMA and MMEA methods in terms of RMSE are shown in Fig. 3.14. The majority of RMSE are negative (about 67% of all the lead months and target seasons), indicating that the BMA forecasts outperform MMEA forecasts for most of the time. For some seasons such as DJF to FMA, MMEA is better than BMA but the differences of RMSE between these two methods are very small (~0.00–0.05). Besides providing a deterministic forecast, BMA also considers the uncertainty of each model's forecasts and used this uncertainty to construct a predictive distribution. In other words, one of the main advantages of the BMA method is that it provides a forecast distribution that can be used to probabilistic analysis and prediction.

The simple forecast performance measure such as the RMSE is a convenient and quick view but a comprehensive understanding of forecast quality can be achieved through a graphical format such as reliability or attributes diagram. The attributes diagram is an elaboration of the reliability diagram that includes the calibration function, refinement distribution, and reference lines related to the algebraic decomposition of the Brier score and the Brier skill score (Wilks, 2011).

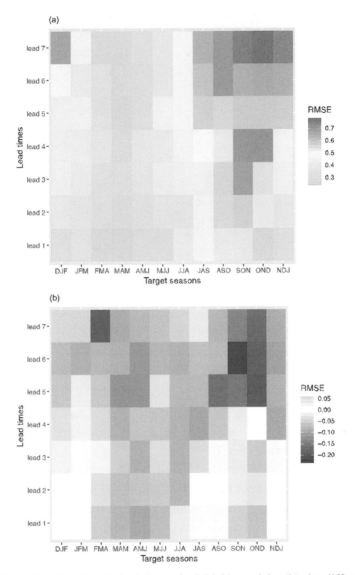

Fig. 3.13 (a) RMSE of statistical-dynamical BMA models, (b) the difference of RMSE between the statistical-dynamical models and statistical BMA models for various leads (months) and target seasons. A black and white version of this figure will appear in some formats. For the color version, refer to the plate section.
Source: Zhang et al., 2019; ©www.creativecommons.org/licenses/by/4.0/. Used with permission from Springer Nature.

To assess the overall statistical-dynamical BMA forecast reliability, the attributes diagram (Hsu and Murphy, 1986) for forecast probabilities of seasonal Ocean Niño Index (ONI) larger than 0.5 are displayed (Fig. 3.15). The CPC uses the ONI values in excess of +0.5°C (less than –0.5°C) for at least five overlapping seasons

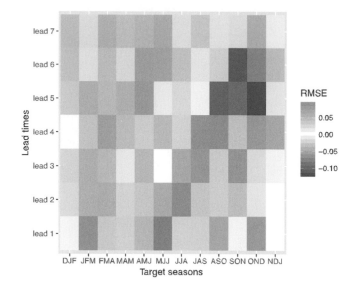

Fig. 3.14 Difference of RMSE between the BMA and multi model ensemble average (MMEA). A black and white version of this figure will appear in some formats. For the color version, refer to the plate section.
Source: Zhang et al., 2019; ©www.creativecommons.org/licenses/by/4.0/. Used with permission from Springer Nature.

(three-month average) as criteria of El Niño (La Niña) episodes. Here, the attributes diagrams are made by pooling all target seasons, lead times, and years together (more than 2,000 events).

In Fig. 3.15, the points show observed relative frequency of ONI > 0.5, conditional on each of the $i = 11$ ($i = 0.1, 0.2, 0.3, \ldots, 1$) possible forecast bins. It is noticeable that the points are very close to the perfect forecast line (45° line) which indicates that the overall statistical-dynamical BMA probability forecasts are consistent with the observed ONI > 0.5 frequency and El Niño forecasts are reliable. In this case, no points fall on the no-resolution line, and the forecast exhibit a substantial degree of resolution. The numbers next to the points express the relative frequency with which the event has been predicted (over the reference period and at all events) with different levels of probability.

When forecasts are frequently much different from the climatological value of the predictand, such forecasts are said to have sharpness. Sharpness is the tendency to forecast extreme values, rather than values clustered around the mean, e.g., a forecast of climatology has no sharpness. In Fig. 3.15, there is a tendency for forecast probabilities to be near zero (0.31) or at low forecast probabilities (e.g., 0.19). Therefore, the forecasts in this case exhibit sharpness. Points in the shaded area bounded by the lines of "no skill" and the overall sample climatology (i.e., the

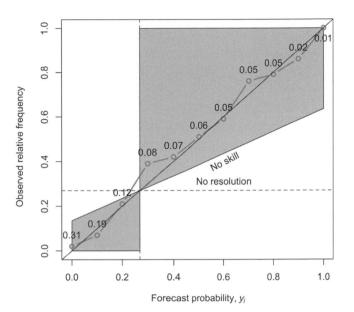

Fig. 3.15 Attributes diagram for forecasts probability of seasonal ONI larger than 0.5. This was made by pooling all target seasons, lead months, and years together with the 0.1 forecast bin width. The relative frequency of each of the forecast values is shown in the numbers. The 45° solid line indicates perfect reliability. The no-resolution line is plotted at the same level of the sample climatological probability. The no-skill line is halfway between the perfect reliability and no-resolution lines.

Source: Zhang et al., 2019; ©www.creativecommons.org/licenses/by/4.0/. Used with permission from Springer Nature.

vertical solid line) indicate the positive contribution to forecast skill (i.e., Brier skill). In this case, all the points are located inside the shaded area.

The attributes diagram for the probability of ONI < –0.5 is shown in Fig. 3.16. In this diagram, the reliability curve lies mainly above the 45° line especially for higher forecast probabilities. This indicates that the statistical-dynamical BMA model slightly under-forecasts the probability of ONI < –0.5 for higher forecast probabilities (forecast probabilities too low). Nevertheless, the overall probability forecast still contributes positively to the prediction skill. The probability forecast of ONI < –0.5 also exhibit sharpness. If an ONI greater than 0.5 can be simply regarded as an El Niño event and a value smaller than –0.5 as a La Niña event, El Niño forecasts seem to be more reliable than La Niña. Overall, the statistical-dynamical BMA presents a reliable probability forecast of ONI. Therefore, BMA can be applied not only to improve the forecast skill, but also provide a reliable probability forecast for ONI. Besides BMA, machine learning approaches also appear promising for improving the conventional ENSO prediction methods (Ham et al., 2019).

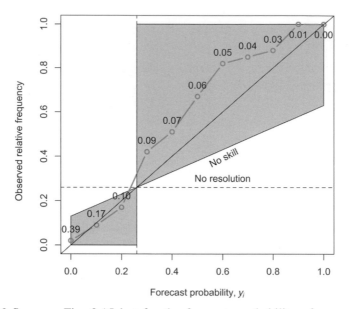

Fig. 3.16 Same as Fig. 3.15 but for the forecasts probability of seasonal ONI smaller than –0.5.

Source: Zhang et al., 2019; ©www.creativecommons.org/licenses/by/4.0/. Used with permission from Springer Nature.

3.1.8 North Atlantic Oscillation (NAO)

The North Atlantic Oscillation (NAO) is a major climate pattern in the North Atlantic, Europe, and North America. If reflects fluctuations in the surface pressure between the Icelandic low and the Azores high, and the NAO index is usually defined through changes in atmospheric pressures between the center of these two weather systems (Fig. 3.17). The NAO index also can be obtained by projecting the NAO loading pattern to the daily 500-hPa height field over the Northern Hemisphere. The loading pattern refers to the first mode of a rotated empirical orthogonal function analysis using monthly standardized 500-hPa height anomalies over 0–90°N (Climate Prediction Center, NOAA/NCEP). The positive phase of the NAO is indicative of a deeper than normal Icelandic low and stronger than normal subtropical high over the North Atlantic Ocean and the eastern USA. Conversely, a negative phase of the NAO reveals the weakening of both weather systems. The NAO is related to the Arctic Oscillation (AO) or northern annular mode (NAM). There is a clear interseasonal and interannual variation in the NAO. By presenting the meridional pressure gradient between the Icelandic low and the Azores high, the NAO index provides a measure of the strength and direction of surface westerly winds and the preferred location of midlatitude storm tracks across the North Atlantic and the downstream Europe.

(a)

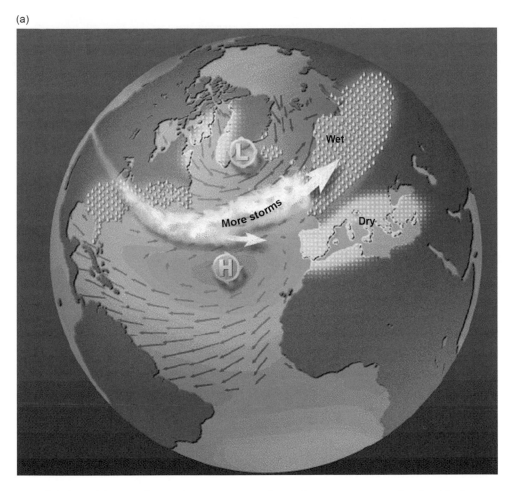

Fig. 3.17 Changes in surface pressure and weather patterns associated with the
(a) positive phase of the North Atlantic Oscillation (NAO) and (b) negative phase
of the NAO. A black and white version of this figure will appear in some formats.
For the color version, refer to the plate section.
Source: www.ldeo.columbia.edu/res/pi/NAO/.

The NAO has a substantial effect on climate in Europe and the eastern seaboard
and southeastern USA, particularly during the winter. When the NAO is in a
positive phase, the enhanced meridional pressure gradient drives strong westerly
winds blowing across the North Atlantic. The warm Gulf Stream and North
Atlantic Current flow stronger than usual, which transport more heat and moisture
between the North Atlantic and the surrounding land masses. On average, their
paths are pushed further north. In the atmosphere, the jet stream and storm tracks
are on a more northerly course. Changes in ocean and atmosphere systems
bring moist air and storms into northern Europe. As a result, winters in northern

(b)

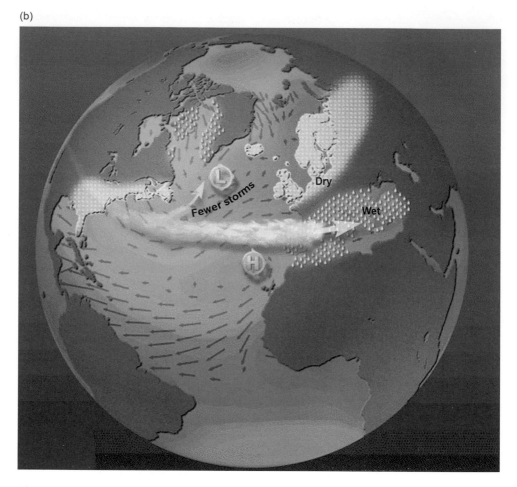

Fig. 3.17 (*cont.*)

Europe are wet and mild during the positive NAO phase. The eastern USA also tends to be wet and mild in winters. For the negative NAO phase, the reduced pressure gradient leads to suppressed westerlies and weaker and fewer winter storms across the North Atlantic. The jet stream is also weaker and variable, allowing winter storms to develop in a far south location. Thus, southern Europe and the Mediterranean Sea experience more storms while northern Europe suffer cold and dry winter. The weaker jet stream during the negative NAO phase also allows cold air outbreaks to surge southward, causing the southeastern USA to be cold and dry in winters.

Unlike ENSO, the NAO is an internally generated atmospheric mode. It affects the North Atlantic Ocean by inducing changes in surface winds, thus altering the momentum, heat and freshwater fluxes at the air–sea interface. These changes affect ocean circulation and have effects on marine and ecosystems, including the distribution and population of fish and shellfish, the production of zooplankton, among others (Hurrell et al., 2001). Note that the spatial pattern of NAO in the summertime is a bit different from winter (Folland et al., 2009). The shift in the NAO phase also influences the direction of tropical storm paths across the North Atlantic because those storms tend to track around the periphery of the Azores high. When the NAO is in the positive phase, tropical cyclones tend to track closer to the east coast of the USA while during the negative phase of the NAO, storms tend to recurve before reaching the east coast of the USA. Murakami et al. (2016) suggested that the summer NAO is a key element for the improved predictions of frequency of landfalling TCs over the US.

3.1.9 Pacific Meridional Mode and Atlantic Meridional Mode

The discovery of an interhemispheric SST mode in the tropical Atlantic was first noted by Hastenrath and Heller (1977) and since then has gained a considerable attention over the last 20 years (Nobre and Shukla, 1996; Chang et al., 1997; Xie and Tanimoto, 1998; Servain et al., 1999; Chiang and Vimont, 2004). Hastenrath and Heller (1977) showed that the juxtaposition of a warm anomaly in the tropical North Atlantic and cold anomaly in the South Atlantic is conducive to a thermally direct meridional circulation with subsidence and low rainfall over northeast Brazil, a northward shift of the Inter-tropical Convergence Zone (ITCZ) towards the warmer North Atlantic, and rainfall change in the Sahel (Lamb, 1978). This meridional SST gradient drives southerly winds across the equator and the Coriolis force bends the boundary layer flows towards the east to the north of the equator.

More recently, using the maximum covariance analysis (MCA) to both the SST and wind vectors during March–May, Chiang and Vimont (2004) and Chang et al. (2007) were able to demonstrate the coupled variability of the subtropical North Atlantic warming, cold SST in the south Atlantic, and low-level winds in the tropical Atlantic, as shown in Fig. 3.18b. This SST dipole pattern with strong southerly flows over the anomalous meridional SST gradients, together with tropical convection embedded in the ITCZ, is dubbed as the Atlantic Meridional Mode (AMM) and appears to operate at both interannual and decadal time scales. Figure 3.18d shows the corresponding regression map of the AMM SST expansion coefficient onto satellite rainfall data. Because the climatological mean ITCZ in the central/western Atlantic reaches its southernmost position (near the equator) in boreal spring and moves northward until about August/September (Fig. 3.19), an anomalous northward displacement of the ITCZ during March–May is implied when the AMM is in the positive phase as in Fig. 3.18d. It should be

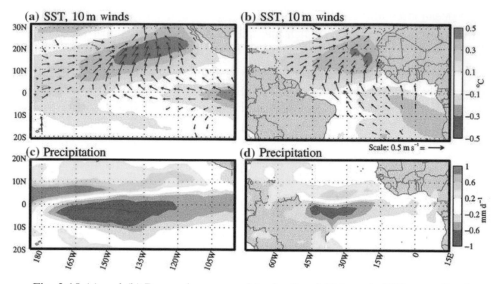

Fig. 3.18 (a) and (b) Regression maps of the leading MCA mode SST normalized expansion coefficient on SST and 10-m wind vectors in the Pacific (left) and Atlantic (right). Same in (c) and (d) but for precipitation (mm d^{-1}). A black and white version of this figure will appear in some formats. For the color version, refer to the plate section.
Source: Chiang and Vimont, 2004; ©American Meteorological Society. Used with permission.

noted that prior to the MCA, the seasonal cycle is removed, data are detrended, a three-month running mean is applied, and the ENSO influence is removed from the original input data of SST and surface wind for each grid point to extract the AMM.

The boreal spring is the season when the AMM is most pronounced but the pattern revealed in Fig. 3.18b also persists in other seasons. Under the mean easterly trades, the interaction of SST, wind, and latent heat flux through the WES feedback, as explained in Section 3.1.2, accounts for the existence of this interhemispheric SST dipole mode in the tropical Atlantic. Based on climate model experiments, the AMM in the tropics is conjectured to influence the extratropical portion of the NAO (Okumura et al., 2001) and also modulates the Atlantic hurricane activity (Kossin and Vimont, 2007; Vimont and Kossin, 2007; Patricola et al., 2014). The AMM is likely forced by the NAO or ENSO through trade-wind variations (Xie and Tanimoto, 1998; Czaja et al., 2002). As described in the preceding subsection, the NAO is the dominant wintertime climate mode with a large-scale pressure seesaw between the subtropical (Azores) high and the subpolar low in the North Atlantic. Chiang and Vimont (2004) showed that the northern mid-latitude circulation anomalies in boreal winter induced by NAO alter the trade

NOAA Extended Reconstructed V4 Climatological SSTs
NNRP Climatological Surface Wind (1950–2017)

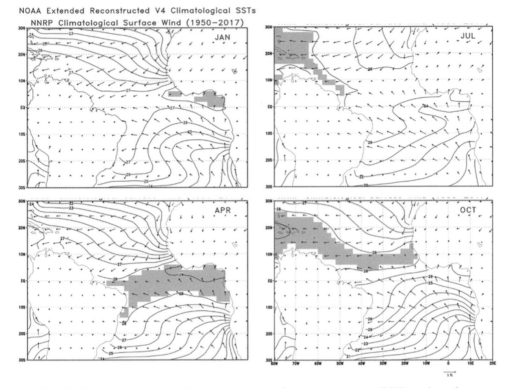

Fig. 3.19 Long-term monthly mean sea surface temperature (SST) and surface wind vectors for the tropical Atlantic during 1950–2017 from the Extended Reconstructed Sea Surface Temperature (ERSST v4) for January (JAN), April (APR), July (JUL), and October (OCT). Contours are in °C. Shading indicates SST greater than 28°C. Winds are in m s^{-1}.

wind in the northern tropical Atlantic, and that in turn leads to anomalous SST maximizing in boreal spring (Fig. 3.19b). Therefore, the AMM and NAO may interact with each other and generate feedback processes.

Although the AMM is largely independent of ENSO (Chiang and Vimont, 2004), it is possible that Pacific ENSO can still excite AMM variability through atmospheric teleconnection patterns (Covey and Hastenrath, 1978; Enfield and Mayer, 1997). Further studies are needed to demonstrate their physical relationships and also in connection with the NAO. In short, the anomalous trade winds in the North Atlantic subtropics drive the SST anomalies through modulation of latent heat flux, giving rise to the dipole SST pattern, the meridional SST gradient, and the preferred location of enhanced tropical convection (Chiang and Vimont, 2004).

The MCA is intended to find pairs of linear combinations of two sets of vector data such that their covariances are maximized, subject to the constraint that the left and right vectors are orthonormal. This method is similar to the more commonly used singular value decomposition (Bretherton et al., 1992). Applied to the AMM studies, this method is used to extract the cross-covariance matrix between wind and SST over the domain of the tropical Atlantic (32°N–21°S and 74°W to the West African coastline) to find the leading mode of coupled variability. The AMM time series of expansion coefficients are calculated by projecting SST or the 10-m wind data onto the spatial structure resulting from the MCA so there are two indices, one for SST and the other for wind.

Akin to the spatial pattern and the principal mechanisms of the AMM, there is a counterpart in the Pacific, named the Pacific Meridional Mode (PMM; Chiang and Vimont, 2004). The PMM is characterized by an anomalous warming in the subtropical eastern North Pacific and cooling over the equatorial cold tongue regions with strong anomalous southerlies and southwesterlies toward the warm lobe (Fig. 3.18a). For the rainfall regression map (Fig. 3.18c), a slight meridional displacement of the ITCZ in the Pacific is also clear, from its climatological mean position around 5°N in the eastern Pacific to a broader region to the north of this mean latitude. An intensification of rainfall in the mean ITCZ latitudes from 150°W to the dateline and a broad zonal band of less rainfall in the south Pacific is also noted. For both the Pacific and Atlantic, the SST expansion coefficient peaks in boreal spring, following the corresponding wind peak in late winter/early spring. Possibly due to persistence, the PMM and AMM signals are still strong during the hurricane season following the boreal spring (Hu and Huang, 2006). Through the cross correlation analysis between the SST and wind expansion coefficients, the wind appears to lead the SST by a month in both basins, suggesting that the trade wind drives SST variability. Because the cold lobe of the PMM is collocated in the Pacific equatorial cold tongue area where ENSO signal is strong, there is a concern whether the PMM is a part of ENSO. It is noted that the ENSO influence from the SST and wind data was already linearly removed in Fig. 3.18a by subtracting the linear regression fit to the cold tongue index averaged between 6°N and 6°S, and 180° and 90°W (Chiang and Vimont, 2004). By further just selecting the ENSO neutral years, Chiang and Vimont (2004) redid the MCA. They found the new PMM pattern resembles that shown in Fig. 3.18a when all years (1948–2001) were used, suggesting that the PMM owes its existence regardless of ENSO variability.

It is, however, likely that the PMM is forced externally by some major climate modes such as ENSO or the North Pacific Oscillation (NPO, Rogers, 1981) through variations in the North Pacific subtropical trades with some WES feedback in the deep tropics (Chiang and Vimont, 2004; Chang et al., 2007). Just like the AMM in the Atlantic, the strength of northeast trade winds in the north Pacific is modulated by the large-scale semi-permanent subtropical high-pressure system associated with fluctuations in the NPO. An analogous meridional mode is also found in the South Pacific (Zhang et al., 2014).

The positive PMM is also thought to trigger some El Niño developing events (Chang et al., 2007; Zhang et al., 2009), especially Central Pacific El Niño events (Stuecker, 2018) in which the maximum positive SSTA are located in the tropical Central Pacific rather than in the tropical eastern Pacific. Chang et al. (2007) reported that the time series of the expansion coefficient of PMM averaged from January to May highly correlates with the time series of the following November–December–January averaged cold tongue index ($r = 0.68$), which is an SST based index commonly used to gauge the ENSO variability. This indicates that most of the El Niño events are preceded by positive PMM events. However, it is revealed that the PMM effect on ENSO is nonlinear (Thomas and Vimont, 2016), indicating that El Niño events are more easily initialized by PMM than La Niña events. Stuecker (2018) also indicates that Central Pacific El Niño triggers positive PMM events through excitement of Aleutian low. This indicates that Central Pacific El Niño and PMM cannot be considered two independent dynamical processes.

As a major climate mode in the Pacific, the PMM also has been shown to exert its influence on tropical cyclone activity in the western North Pacific, eastern North Pacific, and the central North Pacific (Zhang et al., 2016; Murakami et al., 2017). Discussions of these studies will be presented in Chapter 4. The monthly time series of PMM and AMM from 1948 are calculated by Dr. D. J. Vimont and are available from the website at www.aos.wisc.edu/~dvimont/MModes/Data.html.

3.2 Decadal to Interdecadal Variability

3.2.1 Pacific Decadal Oscillation

The Alaskan and the Pacific Northwest fishermen have long reported good and bad salmon catches on a decadal time scale. The alternating fishery production has subsequently been connected to the climate fluctuations in the Pacific basin. In the late twentieth century, the low-frequency variation of the North Pacific system was revealed (e.g., Trenberth and Hurrell, 1994; Mantua et al., 1997; Minobe, 1997; Zhang et al., 1997) and was named the Pacific Decadal Oscillation (PDO) (Mantua et al., 1997).

The PDO has been defined as the leading empirical orthogonal function (EOF) of the North Pacific (20°N–70°N) monthly averaged SST anomalies, after removing the annual cycle and the global mean SSTs. The spatial pattern of the PDO and the corresponding time series of the leading principal component, defined as the PDO index, are shown in Fig. 3.20. Long stretches of positive and negative values of the index alternate against each other. Positive (negative) values

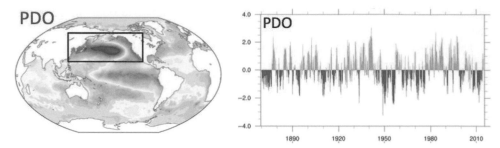

Fig. 3.20 Spatial pattern of the Pacific Decadal Oscillation (PDO) (left) derived from SSTs and time series of the PDO (right) from 1870 to 2014. A black and white version of this figure will appear in some formats. For the color version, refer to the plate section.
Source: www.climatedataguide.ucar.edu/.

of the PDO index correspond to negative (positive) SST anomalies in the central and western North Pacific and positive (negative) SST anomalies along the west coast of North America all the way to the Gulf of Alaska. In addition, SST anomalies along the tropical central and eastern Pacific are positively correlated with the PDO index. The spatial patterns associated with the positive (negative) PDO phase mimic the warm (cold) ENSO episode. However, PDO's footprint is more evident in the midlatitude North Pacific, in contrast to ENSO's tropical signal. Moreover, the PDO climatic signals, unlike those associated with ENSO which operates on the interannual time scales, wax and wane approximately every 20–30 years from one polarity to another (Fig. 3.20).

The similar spatial pattern of ENSO and PDO implies that the ENSO may modulate the PDO. Using a bivariate first-order autoregressive model, Newman et al. (2003) demonstrated that the PDO is driven by ENSO, North Pacific sea surface temperature anomalies, and random atmospheric forcing. Therefore, the PDO and ENSO are not independent of each other but are inherently related. Early PDO studies suggested that the observed SST patterns seen in Fig. 3.20 result from a physical mode that involves a positive feedback between ocean and atmosphere in the North Pacific, giving rise to the required period of decadal variability (Latif and Barnett, 1994, 1996). Liu (2012) suggested that stochastic forcing is the major mechanism for driving interdecadal variability and that the time scale of such variability is determined mainly by Rossby wave propagation in the extra-tropics. More recent studies, however, indicate that the PDO SST pattern is not generated by a single phenomenon, but instead is a result of several different physical processes, involving remote tropical forcing and local atmosphere–ocean interactions, that act on different time scales (Newman et al., 2016; Liu and Lorenzo, 2018). Specifically, the Aleutian low, teleconnections from the tropics, and midlatitude ocean dynamics are three key players in driving the PDO variability.

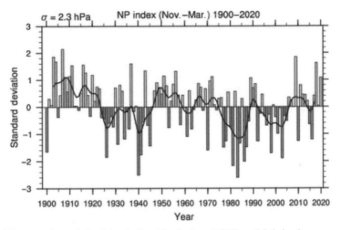

Fig. 3.21 Time series of the North Pacific Index (NPI), which is the area-weighted sea-level pressure over the region 30°N–65°N, 160°E–140°W.
Source: www.climatedataguide.ucar.edu/climate-data/.

In the following, we will discuss the aforementioned three processes. The strength of the Aleutian low, a semi-permanent low-pressure system near the Aleutian Islands, can be represented by the North Pacific Index (NPI), which is the area-weighted sea-level pressure over the region 30°N–65°N, 160°E–140°W (www.climatedataguide.ucar.edu/climate-data/). Negative (positive) NPI values indicate that the Aleutian low is stronger (weaker) than normal (Fig. 3.21). The NPI is characterized by extended periods, two to three decades in duration, of positive (e.g., 1900–1924, 1947–1976) and negative (e.g., 1925–1946, 1977–2003) values. Phase transitions of the NPI occur in 1925, 1947, and 1977 (e.g., Deser et al., 2004). Also note that the interdecadal variations in the NPI run approximately inverse to the PDO series, suggesting the important role of the Aleutian Low to the PDO variations. A stronger Aleutian cyclone, corresponding to a positive PDO phase, favors enhanced westerly wind speeds and colder air temperature over midlatitude North Pacific oceans. These effects cool the underlying ocean via ocean mixing and through loss of sensible and latent heat fluxes to the atmosphere west and south of the cyclone (Fig. 3.20). Anomalously warm waters associated with southerlies prevail in the extreme eastern North Pacific. Southerly anomalies on the eastern flank of the cyclone increase onshore Ekman transport, with the effect of suppressing coastal upwelling. This effect, together with northward advection of warm and moist water by southerly anomalies, results in anomalously warming along the west coast of North America. The reduced upwelling and anomalously warming limit the nutrient supply for fish and thus result in a decreased fish population and catch along the west coast of North America. Cross correlations between the NPI and PDO reveals that the NPI leads the PDO by several months during the cold season (Newman et al., 2016). This implies that atmospheric variations, driven by the fluctuations of the Aleutian low, leads changes in the North Pacific SST represented by the PDO (Trenberth and Hurrell, 1994). The changes in the Aleutian low are related to unpredictable weather noise and also to remote forcing from interannual and decadal tropical variability via "atmospheric bridge."

From the teleconnections perspective, the variability of the extratropical atmosphere–ocean system is shown to be linked to ENSO-induced SST anomalies in the tropical Pacific (Alexander, 1990, 1992; Lau and Nath, 1994, 1996). That is, the atmospheric "bridge" extends from the tropical Pacific to the midlatitude northern oceans during the northern winter through atmospheric Rossby waves by tropical convection and the associated divergent outflows in regions where vorticity gradient is strong (Hoskins and Karoly, 1981; Sardeshmukh and Hoskins, 1988).

Figure 3.22 displays a schematic of the atmospheric bridge, a term first introduced by Lau and Nath (1994) (Alexander et al., 2002). Because ENSO shifts tropical precipitation and wind patterns in the Pacific, which then force changes in the surface heat (shortwave and longwave radiation, latent and sensible), momentum and freshwater fluxes in the North Pacific and North Atlantic through atmospheric teleconnections (Trenberth et al., 1998). Thus, during ENSO events, an atmospheric bridge extends from the equatorial Pacific to higher latitudes. As a result, the Aleutian low is enhanced (weakened) during the mature phase of ENSO and the associated changes in wind mixing, surface heat fluxes, and the Ekman transport lead to a positive (negative) PDO SST pattern SSTA pattern. As mentioned previously, the Aleutian low alone is a major driver for the PDO, it is expected that the atmospheric bridge resulting from ENSO can lead to a PDO-like

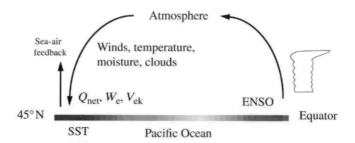

Fig. 3.22 Schematic of the atmospheric bridge between the tropical and North Pacific. The bridge results from changes in the Hadley and Walker circulation, Rossby waves, and interaction between the quasi-stationary flow and storm tracks during ENSO. Q_{net} refers to the net surface heat flux, and W is the entrainment rate into the mixed layer from below.
Source: Alexander et al., 2002; ©American Meteorological Society. Used with permission.

SST pattern. Indeed, oceanic cooling is observed over the central North Pacific and the east coast of North America during a warm ENSO events (e.g., Lau and Nath, 1996; Alexander et al., 2002). On the other hand, anomalous warming is found over the subtropical North Atlantic and the west coast of North America. This anomalous SST pattern is reminiscent of the PDO pattern (Fig. 3.20). Using the time series of the leading EOF mode over the tropical Pacific SSTA as an ENSO index, Newman et al. (2016) showed that ENSO leads PDO throughout the year, although there is a seasonal dependence on the lagged correlation between these two indices. Besides ENSO, low-frequency anomalies in the tropical Pacific may cause PDO variability (e.g., Nakamura et al., 1997; Deser et al., 2004). Based on GCM experiments, Alexander et al. (2002) estimated that tropical Pacific decadal variations may contribute to about ¼ to ½ of the PDO related variability via atmospheric bridge.

In midlatitudes, SST anomalies are caused primarily by variations in surface energy fluxes, wind-driven horizontal advection and diffusion, and vertical motions involving entrainment of water through the ocean mixed layer (e.g., Frankignoul and Reynolds, 1983; Haney, 1985). While the atmosphere tends to drive the ocean by one to two months over most of the midlatitudes, the ocean also feeds back gradually to the atmosphere (e.g., Davis, 1976; Kushnir and Lau, 1992). It is through the midlatitude air–sea interaction and seasonal displacement of the ocean mixed layer depth, SST anomalies are maintained from one winter to the next while the intervening summer shows little or no sign of persistence (Namias and Born, 1970, 1974; Alexander and Deser, 1995). Because of strong surface wind stress, SST anomalies transport and mix downward into the deep mixed layer in winter. When the mixed layer shoals in spring, heat anomalies at depth could remain intact under the seasonal thermocline in the summer. The anomalies are then mixed back toward the surface when the mixed layer deepens again in the following fall and winter as wind-induced mixing enhances again. It is through this process and large heat capacity of the ocean that surface temperature anomalies can propagate downward and store for several seasons and return to influence surface in subsequent seasons. As such, the decorrelation time scale of midlatitude SST anomalies in successive winters is greater than one year. It is termed "reemergence mechanism" by Alexander and Deser (1995). By this mechanism, interannual to interdecadal PDO SST variability is enhanced (Newman et al., 2003; Schneider and Cornuelle, 2005).

Decadal PDO variability may also be related to oceanic Rossby waves excited by anomalous wind stress curl in the central North Pacific associated with fluctuations in the Aleutian low (e.g., Schneider et al., 2002; Qiu and Chen, 2005; Taguchi et al., 2007). These slow-moving ocean waves, travelling westward at a wave phase speed of a few centimeter per second, take about 3–10 years to cross the basin and result in persistent SST anomalies in the western Pacific, particularly in the Kuroshio and Oyashio frontal zones. While Kuroshio is a warm current, Oyashio is a cold ocean current. Both currents are part of large ocean gyres: the subpolar gyre and the massive subtropical gyre. They are separated by a sharp meridional SST gradient called the subarctic frontal zone. The Kuroshio Current is the northward-flowing waters of the western boundary current. When it leaves the coast of Japan, it flows eastward into the Pacific as a free jet, known as the Kuroshio Extension, which separates the warm subtropical waters from the cold waters to the north. In the western Pacific, large SST variations associated with the PDO occur in the Kuroshio Extension and Oyashio frontal zones.

When the Aleutian low is strong, the SST in the central and North Pacific cools. These negative SST anomalies tend to extend westward via propagation of Rossby waves, subducted into the mixed-water region between Japan and 150°E, and 36°N–42°N (Qiu and Chen, 2010), and then lead to subsequent cooling in the western Pacific. This delayed cooling tends to reduce heat and moisture fluxes into the atmosphere. Figure 3.23 displays that SSTA in the mixed-water region between the Kuroshio Extension and Oyashio Current exhibit decadal variations and resemble the PDO time series. A strong Aleutian low also implies that it expands and shifts equatorward. In response, the gyre circulation and the border between

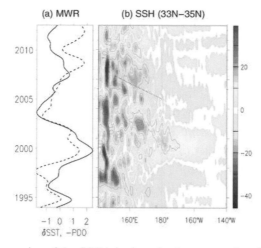

Fig. 3.23 (a) Time series of the SSTA in the mixed-water region (MWR, solid) and the PDO index (with sign reversed, dashed). The MWR spans from the coast of Japan to 150°E, and 36°N–42°N. Both indices are normalized by their respective standard deviations. (b) Satellite-derived sea surface height anomalies (cm), averaged over 33°N–35°N. The dotted line marks a slow westward phase speed of the ocean process. A black and white version of this figure will appear in some formats. For the color version, refer to the plate section.
Source: Newman et al., 2016; ©American Meteorological Society. Used with permission.

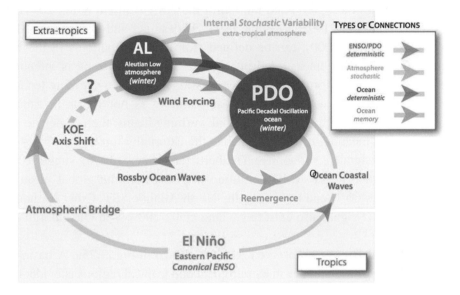

Fig. 3.24 Schematic showing processes involved in the PDO. A black and white version of this figure will appear in some formats. For the color version, refer to the plate section.
Source: Newman et al., 2016; © American Meteorological Society. Used with permission.

two ocean gyres shift southward. Through ocean gyre dynamics (e.g., recirculation of gyres, heat transports), an initial SST anomaly could enhance the decadal PDO variability. While in most of the North Pacific, SST variability is driven primarily by atmospheric forcing, in this area (Kuroshio Extension, Oyashio frontal zones, and the mixed-water region in between), persistent warm (cool) SST anomalies tend to enhance (reduce) heat and moisture fluxes into the atmosphere.

It should be noted that the PDO does not represent a single phenomenon, but rather a combination of several different physical processes that operate in the tropics and extratropics at different time scales. (Fig. 3.24). Therefore, the PDO should not be thought of as an independent phenomenon such as ENSO and complex interactions among many different processes render predicting the turnaround of the PDO phase extremely difficult. Because of this difficulty, using the PDO as a single predictor for forecasting changes in other variables such as temperature and precipitation is a challenge.

3.2.2 Atlantic Multidecadal Oscillation

The North Atlantic Ocean appears to alternate between warm and cold phases over a roughly 70-yr cycle called the Atlantic Multidecadal Oscillation (AMO) and has been linked to long-term hurricane activity over the North Atlantic (e.g., Goldenberg et al., 2001) and rainfall in the USA (Enfield et al., 2001). The AMO (Delworth et al., 1993; Kushnir, 1994; Schlesinger and Ramankutty, 1994; Delworth and Mann, 2000), can be defined from the SST patterns in the North Atlantic after removing linear trend in the data. This linear trend is intended to represent the externally forced, greenhouse-gas-induced global warming tendency. However, because the tendency of global warming is not linear in time, for example as complicated by the recent global warming hiatus, there is debate about what is the most meaningful way to separate the externally forced variability from the internally generated variability. Trenberth and Shea (2006) suggested one simple way to modify the AMO definition is to subtract the global mean SST averaged between 60°N and 60°S from the North Atlantic SST. Other methods for defining the AMO signal also exist (e.g., Ting et al., 2009; Van Oldenborgh et al., 2009).

The spatial pattern of the observed AMO is shown in Fig. 3.25a. A basin-wide warming with large magnitudes in extratropical and tropical regions is evident. The cold phase of the AMO prevailed in the periods 1900–1930 and 1965–1995, while its warm phase dominated during 1930–1965 and after 1995 (Fig. 3.25b). Based on observations and a suite of climate models, Ting et al. (2009) suggested that the observed, long-term North Atlantic SST variability is characterized by a forced global warming trend and a multidecadal oscillation that most likely results from internal variability.

Now the question is what are the physical causes of the AMO? Because of the long memory and slow change in ocean, the AMO has been traditionally attributed to oceanic processes such as changes in the Atlantic meridional overturning circulation. This circulation consists of a northward flow of warm, salty water in the upper layer of the North Atlantic (e.g., the Gulf Stream), the southward flow of colder water in the deep layer, and sinking of dense seawater in high latitudes and gradual rising of less dense water in temperate and lower latitudes that are part of the thermohaline circulation. "Thermo" refers to temperature and "haline" refers to salt content. When heat is transferred to water, its molecules move faster and the increased space between fast-moving molecules decreases the density. Therefore, warm seawater expands and becomes less dense than the surrounding cooler water. Likewise, when salt is dissolved in water, it adds to the mass of the water and makes the water denser than freshwater. Changes in salinity are brought about by addition or removal of freshwater, mainly through evaporation and precipitation. For example, salinity is increased by evaporation and decreased by precipitation (rain and snow) at the ocean surface. In polar regions, freezing and melting of sea ice also change its salt content.

However, similar multidecadal fluctuations have been found without ocean circulation variability (Clement et al., 2015). Thus, stochastic forcing from the

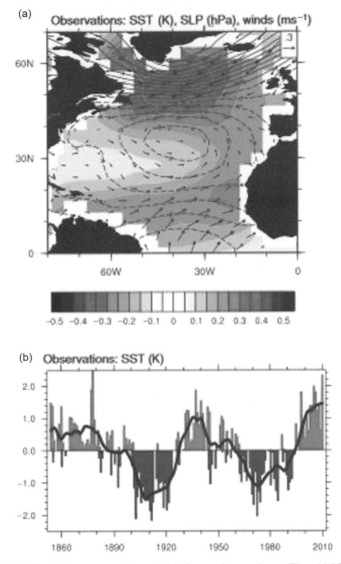

Fig. 3.25 (a) Spatial pattern of the AMO from observations. The AMO index is defined as the average SST for 0°–60°N and 80°W –0°, after detrending to isolate the natural variability. In (a), the AMO index is regressed to SST (shaded), SLP (contours), and surface winds (vectors). (b) Time series of annual mean anomalies of the standardized AMO index (colored bars) and the 10-yr running average of the index (black line). A black and white version of this figure will appear in some formats. For the color version, refer to the plate section.

Source: Clement et al., 2015; ©Science. Used with permission.

atmospheric circulation in midlatitudes is also suggested as the driver of the AMO (Clement et al., 2015). Yuan et al. (2016) hypothesized that low clouds and dust (i.e., aerosols) may be responsible for amplifying the tropical portion of the AMO through radiative effects. Moreover, Brown et al. (2016) postulated that the mechanisms for the tropical and extratropical North Atlantic AMO variability are different. For the extratropical North Atlantic, the AMO-like SST variability is due to oceanic heat convergence associated with the Atlantic meridional overturning circulation but the tropical portion of the North Atlantic requires cloud feedback. Decadal variations in surface heat fluxes are also found to be associated with the AMO-like SST variability in the North Atlantic (O'Reilly et al., 2016).

3.3 Summary

This chapter begins by introducing the El Niño and La Niña phenomena, background climatology of surface winds, SST, and precipitation in the tropical Pacific. The impact of El Niño and La Niña on worldwide rainfall variations, tropical cyclones, and public health are also presented. Tropical atmosphere–ocean feedbacks and climate variability are then discussed. These include wind-evaporation–SST feedback, footprinting mechanism, SST–cloud feedback, and SST–water vapor feedback in the Pacific equatorial cold tongue and warm pool regions, and SST–water vapor feedback. This is followed by the early studies of ENSO and Bjerknes positive feedback, which describes how a strong zonal SST gradient enhances the strong easterly trade that amplifies the initial SST gradient. Two widely accepted theories to account for the oscillatory nature of ENSO phase transition are discussed: the delayed oscillator mechanism and recharge–discharge paradigm. The former relies on ocean wave propagation and changes in the thermocline displacement for the turnabout of ENSO from one state to another. The recharge–discharge oscillator theory emphasizes the zonal mean thermocline variation and meridional upper-ocean mass transport resulting from the surface wind anomaly. The discharge of the equatorial oceanic heat content in the warm phase and recharge at the cold phase provides the fundamental negative feedback for the phase transition of extreme ENSO events.

Studies also show that there are at least two types of El Niño (e.g., Eastern Pacific and Central Pacific). While the Eastern Pacific (EP) El Niño has its largest SST amplitudes in the eastern equatorial Pacific and the initial warming propagates westward, the SST anomalies associated with the Central Pacific (CP) type of El Niño are largest in the central equatorial Pacific and the propagating feature is less clear. Physical processes for the generation of these two types appear to be different. The basin-wide thermocline displacement and surface wind variations, as discussed in the previous section, are responsible for the EP type. In contrast, the zonal advection and regional air–sea heat fluxes are postulated to account for the CP type of El Niño. Moreover, atmospheric forcing from the subtropical North Pacific via the footprinting mechanism may also lead to CP events.

Because ENSO has an immense impact on the global climate, it is of interest to know the methodology used for its prediction and predictive results. Dynamical, statistical, or combination of both methods are used in the ENSO forecasting enterprise. Dynamical models use mathematics and the laws of physics to describe the state of climate, and the ocean component is sometimes linked to the atmosphere and/or land and ice component to be a fully coupled atmosphere–ocean model in many national meteorological centers across the world. Statistical forecasting models are generally based on regression techniques. Some researchers use a hybrid approach by exploring the statistical relationships between the predictand (e.g., Niño-3.4 SST) and the predictors provided by dynamical models. The current coupled dynamical models show significant improvement relative to the older versions. For example, the very strong 2015–2016 El Niño event was well predicted six months ago by the NCEP/CPC. To enhance the ENSO forecast skill, the Bayesian model averaging (BMA) is shown through the attributes diagram. The BMA method also can provide probabilistic forecasts that describe the uncertainty of the prediction.

The North Atlantic Oscillation (NAO) is a climate mode that operates in the North Atlantic on the interannual time scale. It reflects fluctuations in the surface pressure between the Icelandic low and the Azores high. The NAO has impact on climate in Europe and eastern and southeastern USA, particularly during winter. Variations in the NAO also influence the direction of tropical storms path across the North Atlantic. For example, when the NAO is in the positive phase, tropical cyclones are closer to the east coast of the USA, while storms tend to recurve before reaching the USA during the negative phase of the NAO.

The Atlantic Meridional Mode (AMM) is another climate mode operating on an interannual time scale, which is marked by the coupled variability of the subtropical North Atlantic warming, cold SST in the South Atlantic, and low-level northward directed wind in the tropical Atlantic. The interaction of SST, wind and latent heat fluxes through the WES feedback accounts for the existence of the interhemispheric SST dipole mode in the tropical Atlantic. The AMM is linked to the Atlantic hurricane activity. A similar counterpart to the AMM is the Pacific Meridional Mode (PMM) in the eastern Pacific. Studies suggest that the PMM can excite CP El Niño events. However, CP El Niño may also trigger positive PMM events through excitement of Aleutian low, with the implication that CP El Niño and PMM cannot be considered two independent physical processes.

On a longer time scale, the Pacific Decadal Oscillation (PDO) and Atlantic Multidecadal Oscillation stand out as the major climate modes in the Pacific and

Atlantic, respectively. For the PDO, the Aleutian low, teleconnections from the tropics, and midlatitude ocean dynamics are postulated as three key players in driving the decadal variability. Studies suggest that the PDO should not be considered as single phenomenon, but arises from several different physical processes that span both the tropics and extratropics at different time scales. Because of the complex interaction among many processes, predicting the phase shift of the PDO is extremely difficult. The AMO affects SST of the North Atlantic and thus the climate of the surrounding continents (North America, Europe). The AMO is traditionally thought to be driven by changes in the Atlantic meridional overturning circulation in the ocean. However, stochastic forcing from the atmospheric circulation in midlatitudes is also suggested as the driving force of the AMO. Other mechanisms such as cloud feedback and decadal variations in surface heat fluxes also contribute to the multidecadal variability in the North Atlantic.

Exercises

3.1 Why is there an equatorial cold tongue in the eastern equatorial Pacific and warm pool in the tropical western Pacific?

3.2 Describe the wind-evaporation–SST (WES) feedback process.

3.3 (a) What is the Bjerknes' feedback? (b) How can it be used to explain the ENSO cycle?

3.4 What is the recharge–discharge mechanism and how can it be used to explain the ENSO evolution?

3.5 How does the El Niño of the central Pacific type differ from that of the eastern Pacific? Comment on the following aspects: the anomalous sea surface temperature pattern and its propagation; subsurface ocean temperature evolution; precipitation; and any decadal changes.

3.6 What is the Atlantic Meridional Mode?

3.7 (a) Describe the major characteristics of the Pacific Decadal Oscillation (PDO) and its possible climate impacts (e.g., temperature, precipitation, biological systems?) (b) What is the role of the Aleutian low and atmospheric bridge in driving the PDO variations?

References

Alexander, M. A., 1990: Simulation of the response of the North Pacific Ocean to the anomalous atmospheric circulation associated with El Niño. *Clim. Dyn.*, **5**, 53–65.
Alexander, M. A., 1992: Midlatitude atmosphere-ocean interaction during El Niño. Part I: The North Pacific Ocean. *J. Climate*, **5**, 944–958.

Alexdander, M. A., and C. Deser, 1995: A mechanism for the recurrence of wintertime midlatitude SST anomalies. *J. Phys. Oceanogr.*, **25**, 122–137.

Alexander, M. A., and Coauthors, 2002: The atmospheric bridge: The influence of ENSO teleconnection on air–sea interaction over the global oceans. *J. Climate*, **15**, 2205–2231.

An, S.-I., and F.-F. Jin, 2001: Collective role of thermocline and zonal advective feedbacks in the ENSO mode. *J. Climate*, **14**, 3421–3432.

Ashok, K., and Coauthors, 2007: El Niño Modoki and its possible teleconnection. *J. Geophys. Res.*, **112**, C11007.

Barnston, A. G., and Coauthors, 1994: Long-lead seasonal forecasts: Where do we stand? *Bull. Amer. Meteorol. Soc.*, **75**, 2097–2114.

Barnston, A. G., M. H. Glantz, and Y. He, 1999: Predictive skill of statistical and dynamical climate models in SST forecasts during the 1997–98 El Niño episode and the 1998 La Niña onset. *Bull. Amer. Meteorol. Soc.*, **80**, 217–243.

Barnston, A. G., and Coauthors, 2012: Skill of real-time seasonal ENSO model predictions during 2002–2011. *Bull. Amer. Meteorol. Soc.*, **93**, 631–651.

Battisti, D. S., and A. C. Hirst, 1989: Interannual variability in a tropical atmosphere-ocean model: Influence of the basic state, ocean geometry and nonlinearity. *J. Atmos. Sci.*, **46**, 1687–1712.

Bjerknes, J., 1966: A possible response of the atmospheric Hadley circulation to equatorial anomalies of ocean temperature. *Tellus*, **18**, 820–829.

Bjerknes, J., 1969: Atmospheric teleconnections from the equatorial Pacific. *Mon. Wea. Rev.*, **97**, 163–172.

Bretherton, C., C. Smith, and J. M. Wallace, 1992: An intercomparison of methods for finding coupled patterns in climate data. *J. Climate*, **5**, 541–560.

Brown, P. T., M. S. Lozier, R. Zhang, and W. Li, 2016: The necessity of cloud feedback for a basin-scale Atlantic multidecadal oscillation. *Geophys. Res. Lett.*, **43**, GL068303.

Cane, M. A., and S. E. Zebiak, 1985: A theory for El Niño and the Southern Oscillation. *Science*, **228**, 1085–1087.

Capotondi, A., and Coauthors, 2015: Understanding ENSO diversity. *Bull. Amer. Meteorol. Soc.*, **96**, 921–938.

Cayan, D. R., K. T. Redmond, and L. G. Riddle, 1999: ENSO and hydrologic extremes in the western United States. *J. Climate*, **12**, 2881–2893.

Chang, P., L. Ji, and H. Li, 1997: A decadal climate variation in the tropical Atlantic Ocean from thermodynamic air–sea interactions. *Nature*, **385**, 516–518.

Chang, P., and Coauthors, 2007: Pacific meridional mode and El Niño–Southern Oscillation. *Geophys. Res. Lett.*, **34**, L16608.

Chen, W. Y., 1982: Assessment of Southern Oscillation sea-level pressure indices. *Mon. Wea. Rev.*, **110**, 800–807.

Chiang, J. C. H., and D. J. Vimont, 2004: Analogous Pacific and Atlantic meridional modes of tropical atmosphere-ocean variability. *J. Climate*, **17**, 4143–4158.

Choi, K.-Y., G. A. Vecchi, and A. T. Wittenberg, 2013: ENSO transition, duration, and amplitude asymmetries: Role of the nonlinear wind stress coupling in a conceptual model. *J. Climate*, **26**, 9462–9476.

Chu, P.-S., 1995: Hawaii rainfall anomalies and El Niño. *J. Climate*, **8**, 1697–1703.

Chu, P.-S., 2004: ENSO and tropical cyclone activity. In *Hurricanes and Typhoons: Past, Present, and Future*. R. J. Murnane and K.-B. Liu, Eds. Columbia University Press, 297–332.

Chu, P.-S., and R. W. Katz, 1985: Modeling and forecasting the Southern Oscillation: A time-domain approach. *Mon. Wea. Rev.*, **113**, 1876–1888.

Chu, P.-S., and J. Wang, 1997: Tropical cyclone occurrences in the vicinity of Hawaii: Are the differences between El Niño and non-El Niño years significant? *J. Climate*, **10**, 2683–2689.

Chu, P.-S., J. Frederick, and A. J. Nash, 1991: Exploratory analysis of surface winds in the equatorial western Pacific and El Niño. *J. Climate*, **4**, 1087–1102.

Clement, A., and Coauthors, 2015: The Atlantic multidecadal oscillation with a role of ocean circulation. *Science*, **350**, 320–324.

Covey, D. L., and S. Hastenrath, 1978: The Pacific El Niño phenomenon and the Atlantic circulation. *Mon. Wea. Rev.*, **106**, 1280–1287.

Czaja, A., P. van der Vaart, and J. Marshall, 2002: A diagnostic study of the role of remote forcing in tropical Atlantic variability. *J. Climate*, **15**, 3280–3290.

Davis, R. E., 1976: Predictability of sea surface temperature and sea level pressure anomalies over the North Pacific Ocean. *J. Phys., Oceanogr.*, **6**, 249–266.

Delworth, T. L., and M. E. Mann, 2000: Observed and simulated multidecadal variability in the Northern Hemisphere. *Clim. Dyn.*, **16**, 661–676.

Delworth, T. L., S. Manabe, and R. J. Stouffer, 1993: Interdecadal variations in the thermohaline circulation in a coupled ocean-atmosphere model. *J. Climate*, **6**, 1993–2011.

Deser, C., A. S. Phillips, and J. W. Hurrell, 2004: Pacific interdecadal climate variability: Linkages between the tropics and the North Pacific during boreal winter since 1990. *J. Climate*, **17**, 3109–3124.

Dommenget, D., T. Bayr, and C. Frauen, 2013: Analysis of the non-linearity in the pattern and time evolution of El Niño Southern Oscillation. *Clim. Dyn.*, **40**, 2825–2847.

Enfield, D. B., and D. A. Mayer, 1997: Tropical Atlantic sea surface temperature variability and its relation to El Niño-Southern Oscillation. *J. Geophys. Res.*, **102**, 929–945.

Enfield, D. B., A. M. Mestas-Nuñez, and P. J. Trimble, 2001: The Atlantic multidecadal oscillation and its relation to rainfall and river flows in the continental U.S. *Geophys. Res. Lett.*, **28**, 2077–2080.

Fedorov, A. V., S. Hu, M. Lengaigne, and E. Guilyardi, 2015: The impact of westerly wind bursts and ocean initial state on the development, and diversity of El Niño events. *Clim. Dyn.*, **44**, 1381–1401.

Folland, C. K., and Coauthors, 2009: The summer North Atlantic Oscillation. Past, Present, and Future. *J. Climate*, **22**, 1082–1103.

Frankignoul, C., and R. W. Reynolds, 1983: Testing a dynamical model for midlatitude sea surface temperature anomalies. *J. Phys. Oceanogr.*, **13**, 1131–1145.

Glantz, M. H., 2001: *Currents of Change: Impacts of El Niño and La Niña on Climate and Society*, 2nd ed. Cambridge University Press.

Glantz, M. H., R. W. Katz, and N. Nicholls, 1991: *Teleconnections Linking Worldwide Climate Anomalies*. Cambridge University Press.

Goldenberg, S. B., C. Landsea, A. M. Mesias-Nuñez, and W. M. Gray, 2001: The recent increase in Atlantic hurricane activity: Causes and implication. *Science*, **293**, 474–479.

Guan, C., and M. J. McPhaden, 2016: Ocean processes affecting the twenty-first-century shift in ENSO SST Variability. *J. Climate*, **29**, 6861–6879.

Guilyardi, E., and Coauthors, 2009: Understanding El Niño in ocean-atmosphere general circulation models: Progress and challenges. *Bull. Amer. Meteorol. Soc.*, **90**, 325–340.

Guo, Y.-Y., M. Ting, Z. Wen, and D. E. Lee, 2017: Distinct patterns of tropical Pacific SST anomaly and their impacts on North American climate, *J. Climate*, **30**, 5221–5241.

Ham, Y.-G., J.-H. Kim, and J.-J. Luo, 2019: Deep learning for multi-year ENSO forecasts. *Nature*, **573**(7775), 568–572.

Haney, R. L., 1985: Midlatitude sea surface temperature anomalies: A numerical hindcast. *J. Phys. Oceanogr.*, **15**, 787–799.

Harrison, D. E., and B. S. Giese, 1991: Episodes of surface westerly winds as observed from islands in the western tropical Pacific. *J. Geophys. Res.*, **96**, 3221–3237.

Hastenrath, S., 1985: *Climate and Circulation of the Tropics*. D. Reidel Publishing Company.

Hastenrath, S., and L. Heller, 1977: Dynamics of climate hazards in northeast Brazil. *Quart. J. Roy. Meteorol. Soc.*, **103**, 77–92.

He, Y. and A. G. Barnston, 1996: Long-lead forecasts of seasonal precipitation in the tropical Pacific Islands Using CCA. *J. Climate*, **9**, 2020–2035.

Hoerling, M. P., and A. Kumar, 1997: Why do North American climate anomalies differ from one El Niño event to another? *Geophys. Res. Lett.*, **24**, 1059–1062.

Hoskins, B. J., and D. J. Karoly, 1981: The steady linear response of a spherical atmosphere to thermal and orographic forcing. *J. Atmos. Sci.*, **38**, 1179–1196.

Hsu, W.-R., and A. H. Murphy, 1986: The attributes diagram: A geometrical framework for assessing the quality of probability forecasts, *Int. J. Forecast.*, **2**, 285–293, doi:10.1016/0169-2070(86)90048-8.

Hu, Z. Z., and B. Huang, 2006: Physical processes associated with the tropical Atlantic meridional gradient. *J. Climate*, **19**, 5500–5518.

Hurrell, J., Y. Kushnir, and M. Visbeck, 2001: The North Atlantic Oscillation. *Science*, **291**, 603–605.

Intergovernmental Panel on Climate Change (IPCC), 2014: *Fifth Assessment Report*. WMO/UNEP, Geneva.

Jin, E. K., and Coauthors, 2008: Current status of ENSO prediction skill in coupled ocean-atmosphere models. *Clim. Dyn.* **31**, 647–664.

Jin, F.-F., 1997: An equatorial ocean recharge paradigm for ENSO. Part I: Conceptual mode. *J. Atmos. Sci.*, **54**, 811–829.

Johnson, N., 2013: How many ENSO flavors can we distinguish? *J. Climate*, **26**, 4816–4817.

Kao, H.-Y., and J.-Y. Yu, 2009: Contrasting Eastern-Pacific and Central-Pacific types of ENSO. *J. Climate*, **22**, 615–632.

Kim, S. T., J.-Y. Yu, A. Kumar, and H. Wang, 2012: Examination of the two types of ENSO in the NCEP CFS model and its extratropical association. *Mon. Wea. Rev.*, **140**, 1908–1923.

Kirtman, B. P., and D. Min, 2009: Multimodel ensemble ENSO prediction with CCSM and CFS. *Mon. Wea. Rev.*, **137**, 2908–2930.

Knaff, J. A., and C. W. Landsea, 1997: An El Niño-Southern Oscillation climatology and persistence (CLIPER) forecasting scheme. *Wea. Forecasting*, **12**, 633–651.

Kossin, J. P., and D. J. Vimont, 2007: A more general framework for understanding Atlantic hurricane variability and trends. *Bull. Amer. Meteorol. Soc.*, **88**, 1767–1781.

Kug, J.-S., F.-F. Jin, and S.-I. An, 2009: Two types of El Niño events: Cold tongue El Niño and warm pool El Niño. *J. Climate*, **22**, 1499–1515.

Kug, J.-S., J. Choi. S.-I. An, and A.-T. Wittenberg, 2010: Warm pool and cold tongue El Niño events as simulated by GFDL2.1 coupled GCM. *J. Climate*, **23**, 1226–1239.

Kushnir, Y., 1994: Interdecadal variations in North Atlantic sea surface temperature and associated atmospheric conditions. *J. Climate*, **7**, 141–157.

Kushnir, Y., and N.-C. Lau, 1992: The general circulation model response to a North Pacific SST anomaly: Dependence on time scale and pattern polarity. *J. Climate*, **5**, 271–283.

Lamb, P. J., 1978: Large-scale tropical Atlantic circulation patterns associated with sub-saharan weather anomalies. *Tellus*, **30**, 240–251.

Landsea, C. W., and J. A. Knaff, 2000: How much skill was there to forecasting the very strong 1997-98 El Niño? *Bull. Amer. Meteorol. Soc.*, **81**, 2107–2119.

Larkin, N. K., and D. E. Harrison, 2002: ENSO warm (El Niño) and cold (La Niña) event life cycles: Ocean surface anomaly patterns, their symmetries, asymmetries, and implications. *J. Climate*, **15**, 1118–1140.

Latif, M., and T. P. Barnett, 1994: Causes of decadal climate variability over the North Pacific and North America. *Sciences*, **266**, 634–637.

Latif, M., and T. P. Barnett, 1996: Decadal climate variability over the North Pacific and North America: Dynamics and predictability. *J. Climate*, **9**, 2407–2423.

Latif, M., and Coauthors, 1994: A review of ENSO prediction studies. *Clim. Dyn.*, **9**, 167–179.

Latif, M., and Coauthors, 1998: A review of the predictability and prediction of ENSO. *J. Geophys. Res.*, **103**(C7) 14375–14393.

Lau, N. C., and M. J. Nath, 1994: A modeling study of the relative roles of tropical and extratropical SST anomalies in the variability of the global atmosphere-ocean system. *J. Climate*, **7**, 1184–1207.

Lau, N. C., and M. J. Nath, 1996: The role of the "atmospheric bridge" in linking tropical Pacific ENSO events to extratropical SST anomalies. *J. Climate*, **9**, 2036–2057.

Lee, T., and M. J. McPhaden, 2010: Increasing intensity of El Niño in the central equatorial Pacific. *Geophys. Res. Lett.*, **37**, L14603.

L'Heureux, M. L., and Coauthors, 2017: Observing and predicting the 2015/16 El Niño. *Bull. Amer. Meteorol. Soc.*, **98**, 1363–1382.

Li, T., 1997: Air–sea interactions of relevance to the ITCZ: Analysis of coupled instabilities and experiments in a hybrid coupled GCM. *J. Atmos. Sci.*, **54**, 134–147.

Li, T., and P.-C. Hsu, 2018: *Fundamentals of Tropical Climate Dynamics*. Springer.

Li, T., and S. G. H. Philander, 1996: On the annual cycle of the eastern equatorial Pacific. *J. Climate*, **9**, 2986–2998.

Lindzen, R. S., and S. Nigam, 1987: On the role of sea surface temperature gradients in forcing low-level winds and convergence in the tropics. *J. Atmos. Sc.*, **44**, 2440–2458.

Liu, Z., 2012: Dynamics of interdecadal climate variability: A historical perspective. *J. Climate*, **25**, 1963–1995.

Liu, Z., and E. D. Lorenzo, 2018: Mechanisms and predictability of Pacific decadal variability. *Curr. Clim. Change Rep.*, **4**, 128–144, doi.org/10.1007/s40641–018–0090-5.

Lu, B., P.-S. Chu, S.-H. Kim, and C. Karamperidou, 2020: Hawaiian regional climate variability during two types of El Niño. *J. Climate*, **33**, 9929–9943.

Luther, D. S., D. E. Harrison, and R. A. Knox, 1983: Zonal winds in the equatorial Pacific and El Niño. *Sciences*, **222**, 327–330.

Mantua, N. J., et al., 1997: A Pacific interdecadal climate oscillation with impacts on salmon production. *Bull. Amer. Meteorol. Soc.*, **78**, 1069–1079.

McCreary, J. P., 1983: A model of tropical ocean-atmosphere interaction. *Mon. Wea. Rev.*, **111**, 370–387.

McPhaden, M. J., and Coauthors, 1988: The response of the equatorial Pacific Ocean to a westerly wind burst. *J. Geophys. Res.*, **C9**(92), 9464–9468.

Meinen, C. S., and M. J. McPhaden, 2000: Observations of warm water volume changes in the equatorial Pacific and their relationship to El Niño and La Niña. *J. Climate*, **13**, 3551–3559.

Minobe, S., 1997: A 50–70 year climatic oscillation over the North Pacific and North America. *Geophys. Res. Lett.*, **24**, 683–686.

Murakami, H., and Coauthors, 2016: Statistical-dynamical seasonal forecasts of North Atlantic and U.S. landfalling tropical cyclones using the high-resolution GFDL FLOR coupled model. *Mon. Wea. Rev.*, **144**, 2101–2123.

Murakami, H., and Coauthors, 2017: Dominant role of subtropical warming in extreme eastern Pacific hurricane season: 2015 and the future. *J. Climate*, **30**, 243–264.

Nakamura, H., G. Lin, and T. Yamagata, 1997: Decadal climate variability in the North Pacific during the recent decades. *Bull. Amer. Meteorol. Soc.*, **78**, 2215–2225.

Namias, J., and R. M. Born, 1970: Temporal coherence in North Pacific sea-surface temperature patterns. *J. Geophys. Res.*, **75**, 5952–5955.

Namias, J., and R. M. Born, 1974: Further studies of temporal coherence in North Pacific sea surface temperatures. *J. Geophys. Res.*, **79**, 797–798.

Newman, M., G. P. Compo, and M. Alexander, 2003: ENSO-forced variability of the Pacific decadal oscillation. *J. Climate*, **16**, 3853–3857.

Newman, M., M. A. Alexander, and J. D. Scott, 2011: An empirical model of tropical ocean dynamics. *Clim. Dyn.*, **37**, 1823–1844.

Newman, M., et al., 2016: The Pacific decadal oscillation, revisited. *J. Climate*, **29**, 4399–4427.

Nobre, P., and J. Shukla, 1996: Variations of sea surface temperature, wind stress, and rainfall over the tropical Atlantic and South America. *J. Climate*, **9**, 2464–2479.

Okumura, Y., S.-P. Xie, A. Numaguti, and Y. Tanimoto, 2001: Tropical Atlantic air-sea interaction and its influence on the NAO. *Geophys. Res. Lett.*, **28**, 1507–1510.

O'Reilly, C. H., L. M. Huber, T. Woollings, and L. Zanna, 2016: The signature of low-frequency oceanic forcing in the Atlantic multidecadal oscillation. *Geophys. Res. Lett.* **43**, 2810–2818.

Paek, H., J.-Y. Yu, and C. Qian, 2017: Why were the 2015/2016 and 1997/1998 extreme El Niños different? *Geophys. Res. Lett.*, **44**, doi:10.1002/2016GL071515.

Patricola, C. M., R. Saravanan, and P. Chang, 2014: The impact of the El Niño–Southern Oscillation and Atlantic meridional mode on seasonal Atlantic tropical cyclone activity. *J. Climate*, **27**, 5311–5328.

Philander, S. G. H., 1990: *El Niño, La Niña, and the Southern Oscillation*. Academic Press, 293pp.

Picaut, J., F. Masia, and Y. du Penhoat, 1997: An advective-reflective conceptual model for the oscillatory nature of ENSO. *Science*, **277**, 663–666.

Qiu, B., and S. Chen, 2005: Variability of the Kuroshio Extension jet, recirculation gyre, and mesoscale eddies on decadal time scales. *J. Phys. Oceanogr.*, **35**, 2090–2103.

Qiu, B., and S. Chen, 2010: Eddy-mean flow interaction in the decadally modulating Kuroshio Extension system. *Deep-Sea Res.*, **II.57**, 1098–1110.

Raftery, A. E., T. Gneiting, T., F. Balabdaoui, and M. Polakowski, 2005: Using Bayesian model averaging to calibrate forecast ensembles. *Mon. Wea. Rev.*, **133**, 1155–1174.

Rasmusson, E. M., and T. H. Carpenter, 1982: Variations in tropical sea surface tempera-
  ture and surface wind fields associated with the Southern Oscillation/El Niño. *Mon.
  Wea. Rev.*, **110**, 354–384.
Ren, H.-L., and F.-F. Jin, 2013: Recharge oscillator mechanisms in two types of ENSO.
  J. Climate, **26**, 6506–6523.
Rogers, J. C., 1981: The North Pacific Oscillation. *Int. J. Climatol.*, **1**, 39–57.
Ropelewski, C. F., and M. S. Halpert, 1987: Global and regional scale precipitation
  patterns associated with the El Niño/Southern Oscillation. *Mon. Wea. Rev.*, **115**,
  1606–1626.
Sadler, J., 1983: Tropical Pacific atmospheric anomalies during 1982-83. In *Proceedings of
  the 1982/83 El Niño/Southern Oscillation Workshop*, 1–10. Miami, NOAA Atlantic
  Oceanographic and Meteorological Laboratory.
Saha, S., and Coauthors, 2006: The NCEP climate forecast system. *J. Climate*, **19**,
  3483–3517.
Saha, S., and Coauthors, 2014: The NCEP climate forecast system version 2. *J. Climate*,
  27, 2185–2208.
Sarachik, E. S., and M. A. Cane, 2010: *The El Niño-Southern Oscillation Phenomenon.*
  Cambridge University Press.
Sardeshmukh, P. D., and B. J. Hoskins, 1985: The generation of global rotational flow by
  steady, idealized tropical divergence. *J. Atmos. Sci.*, **45**, 1228–1251.
Schlesinger, M. E., and N. Ramankutty, 1994: An oscillation in the global climate system
  of period 65–70 years. *Nature*, **367**, 723–726.
Schneider, N., and B. D. Cornuelle, 2005: The forcing of the Pacific decadal oscillation.
  J. Climate, **18**, 4355–4373.
Schneider, N., A. J. Miller, and D. W. Pierce, 2002: Anatomy of North Pacific decadal
  variability. *J. Climate*, **15**, 586–605.
Servain, J., I. Wainer, J. P. McCreary, and A. Dessier, 1999: Relationship between the
  equatorial and meridional modes of climatic variability in the tropical Atlantic.
  Geophys. Res. Lett., **26**, 485–488.
Stuecker, M. F., 2018: Revisiting the Pacific Meridional Mode. *Sci. Rep.*, **8**, 3216.
Suarez, M. J., and P. S. Schopf, 1988: A delayed action oscillator for ENSO. *J. Atmos. Sci.*,
  45, 3283–3287.
Taguchi, B., and Coauthors, 2007: Decadal variability of the Kuroshio Extension:
  Observations and an eddy-resolving model hindcast. *J. Climate*, **20**, 2357–2377.
Takahashi, K., A. Montecinos, K. Goubanova, and B. Dewitte, 2011: ENSO
  regimes: Reinterpreting the canonical and Modoki El Niño. *Geophys. Res. Lett.*,
  38, L10740.
Thomas, E. E., and D. J. Vimont, 2016: Modeling the mechanisms of linear and nonlinear
  ENSO responses to the Pacific Meridional Mode. *J. Climate*, **29**, 8745–8761.
Timmermann, A., and Coauthors, 2018: El Niño–Southern Oscillation complexity. *Nature*,
  559, 535–545.
Ting, M., Y. Kushnir, R. Seager, and C. Li, 2009: Forced and internal twentieth-century
  SST trend in the North Atlantic. *J. Climate*, **22**, 1469–1481.
Trenberth, K. E., 1997: The definition of El Niño. *Bull. Amer. Meteorol. Soc.*, **78**,
  2771–2777.
Trenberth, K. E., and J. W. Hurrell, 1994: Decadal atmosphere-ocean variations in the
  Pacific. *Clim. Dyn.*, **9**, 303–319.
Trenberth, K. E., and D. J. Shea, 2006: Atlantic hurricanes and natural variability in 2005.
  Geophys. Res. Lett., **33**, L12704.

Trenberth, K. E., and Coauthors, 1998: Progress during TOGA in understanding and modeling global teleconnections associated with tropical sea surface temperatures. *J. Geophys. Res.*, **103**, 14,291–14,324.

Van den Dool, H. M., 1994: Searching for analogues, how long must we wait? *Tellus*, **46A**, 314–324.

Van Oldenborgh, G. J., L. A. Raa, H. A. Dijkstra, and S. Y. Philip, 2009: Frequency of amplitude dependent effects of the Atlantic multidecadal overturning on the tropical Pacific Ocean. *Ocean Sci.*, **5**, 293–301.

Vimont, D. J., and J. P. Kossin, 2007: The Atlantic meridional mode and hurricane activity. *Geophys. Res. Lett.*, **34**, L07709.

Vimont, D. J., J. M. Wallace, and D. S. Battisti, 2003: The seasonal footprinting mechanism in the Pacific: Implications for ENSO. *J. Climate*, **16**, 2668–2675.

Wallace, J. M., and D. S. Gutzler, 1981: Teleconnections in the geopotential height field during the Northern Hemisphere winter. *Mon. Wea. Rev.*, **109**, 784–812.

Wang, B., and Coauthors, 2019: Historical change of El Niño properties sheds light on future changes of extreme El Niño. *Proc. Natl. Acad. Sci.*, **116**(45), 22512–22517.

Wang, C., 2001: A unified oscillator model for the El Niño-Southern Oscillation. *J. Climate*, **14**, 98–115.

Wang, X., H. Liu, and G. R. Foltz, 2017: Persistent influence of tropical North Atlantic wintertime sea surface temperature on the subsequent Atlantic hurricane season. *Geophys. Res. Lett.*, **44**, 7927–7935, doi:10.1002/2017GL074801.

Webster, P. J., and S. Yang, 1992: Monsoon and ENSO: Selectively interactive systems. *Quart. J. Roy. Meteorol. Soc.*, **118**, 877–925.

Weisberg, R. H., and C. Wang, 1997: A Western Pacific oscillator paradigm for the El Niño-Southern Oscillation. *Geophys. Res. Lett.*, **24**, 779–782.

Wilks, D. S., 2011: *Statistical Methods in the Atmospheric Sciences*. Academic Press.

Wyrtki, K., 1975: El Niño – The dynamical response of the ocean to atmospheric forcing. *J. Phys. Oceanogr.*, **5**, 572–584.

Wyrtki, K., 1985: Water displacements in the Pacific and the genesis of El Niño cycles. *J. Geophys. Res.*, **90**, 7129–7132.

Xie, S.-P., 1998: Ocean–atmosphere interaction in the making of the Walker circulation and equatorial cold tongue. *J. Climate*, **11**, 189–201.

Xie., S.-P. and S. G. H. Philander, 1994: A coupled ocean-atmosphere model of relevance to the ITCZ in the eastern Pacific. *Tellus*, **46A**, 340–350.

Xie, S.-P., and Y. Tanimoto, 1998: A pan-Atlantic decadal climate oscillation. *Geophys. Res. Lett.*, **25**, 2185–2188.

Xue, Y., and A. Leetmaa, 2000: Forecasts of tropical Pacific SST and sea level using a Markov model. *Geophys. Res. Lett.*, **27**, 2701–2704.

Yu, J.-Y., and S. T. Kim, 2010: Three evolution patterns of Central-Pacific El Niño. *Geophys. Res. Lett.*, **37**, L08706.

Yuan, T., and Coauthors, 2016: Positive low cloud and dust feedbacks amplify tropical North Atlantic multidecadal oscillation. *Geophys. Res. Lett.*, **43**, 1349–1356.

Zebiak, S. E., and M. A. Cane, 1987: A model El Niño-Southern Oscillation. *Mon. Wea. Rev.*, **115**, 2262–2278.

Zhang, H., A. Clement, and P. Di Nezio, 2014: The South Pacific Meridional Mode: A mechanism for ENSO-like variability. *J. Climate*, **27**, 769–783.

Zhang, H., P.-S. Chu, L. He, and D. Unger, 2019: Improving the CPC's ENSO forecasts using Bayesian model averaging. *Clim. Dyn.* **53**, 3373–3385.

Zhang, L., P. Chang, and L. Ji, 2009: Linking the Pacific meridional mode to ENSO: Coupled model analysis. *J. Climate*, **22**, 3488–3505.

Zhang, W., and Coauthors, 2016: The Pacific meridional mode and the occurrence of tropical cyclones in the western North Pacific. *J. Climate*, **29**, 381–398.

Zhang, Y., J. M. Wallace, and D. S. Battisti, 1997: ENSO-like interdecadal variability: 1900-93. *J. Climate*, **10**, 1004–1020.

Zhao, S., F.-F. Jin, X. Long, and M. A. Cane, 2021: On the breakdown of ENSO's relationship with thermocline depth in the central-equatorial Pacific. *Geophys. Res. Lett.*, **48**, e2020GL092335.

4

Climate Variability and Tropical Cyclones

4.1 Introduction

In this chapter, we focus on climate variability and tropical cyclone (TC) activity for five ocean basins, namely, the western North Pacific, eastern North Pacific, central North Pacific, South Pacific, and North Atlantic. For each basin, the discussion includes the background climatology, the modulation of TC activity by intraseasonal oscillations, the influence of ENSO and the PMM (or AMM) on interannual TC variations, and decadal TC variations. In addition, large-scale low-level flow patterns instrumental for TC geneses and equatorial waves that are regarded as TC precursors are described in Section 4.2.2. Observed changes in TC attributes such as frequency, intensity, translation speed, and poleward migration of the latitude of lifetime maximum intensity are also described at the end of this chapter.

4.2 The Western North Pacific and the South China Sea

4.2.1 Background Climatology

The western North Pacific (WNP, Fig. 4.1) is the only ocean basin where TC genesis is observed all 12 months of the year, although a majority of storms develop between June and November with the peak season being July to October. Approximately one third of global TCs are generated in the WNP in an average year – the most of any single basin. The average number of TCs (i.e., tropical storm and typhoons) during the typhoon season (June to November) is 25.4 in the WNP, based on 66 years of data from 1950 to 2015 (Patricola et al., 2018a). TC data were taken from the Joint Typhoon Warning Center in Honolulu, Hawaii. Of these 25.4 TCs, 16.8 reached typhoon intensity and 9.4 reached intense typhoon intensity (category 3 and above according to the Saffir–Simpson Hurricane Wind scale) (Patricola et al., 2018a). While the frequency of TCs decreases with

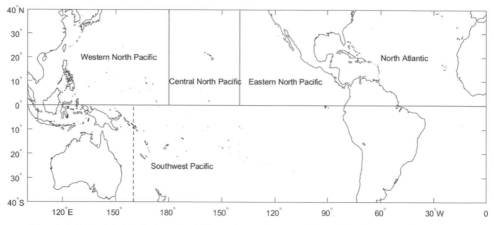

Fig. 4.1 Map highlighting the TC regions discussed in this chapter: the western North Pacific and South China Sea (SCS), eastern and central North Pacific, southwestern Pacific, and North Atlantic basin.

increasing intensity (e.g., the WNP has more tropical storms than category 5 typhoons), these stronger typhoons have longer life time (Chen et al., 2006).

4.2.2 *Large-Scale Flow Patterns, Equatorial Waves, and Tropical Cyclogenesis*

Large-scale conditions prior to tropical cyclogenesis in the WNP have been discussed previously (e.g., Frank, 1982; Harr and Elsberry, 1995a, 1995b; Briegel and Frank, 1997; Ritchie and Holland, 1999; Yoshida and Ishikawa, 2013). In particular, five distinctive flow patterns associated with cyclogenesis over the WNP have been proposed by Ritchie and Holland (1999): monsoon shear line; monsoon confluence zone; monsoon gyre; easterly waves; and Rossby wave energy dispersion. Using an eight years record with low-level data 72-h before TC genesis, Ritchie and Holland (1999) note that monsoon shear lines (42%) and monsoon confluence regions (29%) account for 71% of the 199 genesis cases examined, followed by easterly waves (18%), Rossby energy dispersion (8%), and monsoon gyres (3%). Moreover, upper-tropospheric troughs, when they are located approximately 2,000 km to the northwest of a low-level disturbance, may also trigger cyclogenesis by providing upper-level divergence and vorticity advection and consequently enhancing vertical motion (Briegel and Frank, 1997). Yoshida and Ishikawa (2013) updated the study by Ritchie and Holland (1999) using more recent data and found a somewhat different fraction for monsoon confluence regions, Rossby energy dispersion, and monsoon gyres.

We now demonstrate the basic flow characteristics and configurations for each pattern from Ritchie and Holland (1999). The monsoon shear line, the most

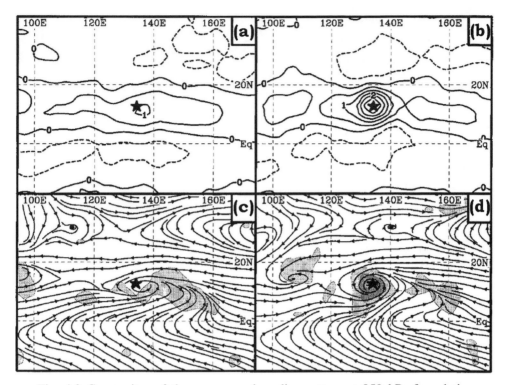

Fig. 4.2 Composites of the monsoon shear line pattern at 850 hPa for relative vorticity ($\times 10^{-5}$ s^{-1}) at (a) -72 h and (b) 0 h (genesis time), and streamlines and convergence ($\times 10^{-6}$ s^{-1}) at (c) -72 h and (d) 0 h. Shading in (c) and (d) indicates regions greater than 2×10^{-6} s^{-1}. The mean genesis location is denoted by a star. Source: Ritchie and Holland, 1999; ©American Meteorological Society. Used with permission.

dominant pattern, is characterized by low pressure to the west of the confluence area between the monsoon westerlies to the south and easterly trade winds poleward of the genesis location (Fig. 4.2). The strong meridional shear is conducive for generation of cyclonic vorticity and formation of tropical cyclogenesis. Upper-level (250-hPa) divergence is observed immediately to the south of the mean genesis location 72-h prior to the genesis, and strong divergence overlies the low-level disturbance by the time of cyclogenesis (not shown). For cyclogenesis in the monsoon confluence region, westerlies (easterlies) prevail to the west (east) of the genesis location, thus forming a distinctive low-level convergence zone (Fig. 4.3). Once this background condition is established, Rossby waves from the east accumulate energy and enhance cyclonic circulation in the confluence zone, as seen by easterly waves propagating into the confluence region prior to genesis. The low-level confluence region is accompanied by an upper-level trough, which enhances the low-level ridge to the northeast of the

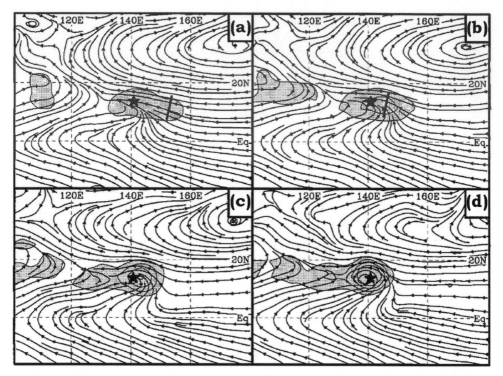

Fig. 4.3 Composites of the monsoon confluence region pattern as represented by streamlines and relative vorticity (4×10^{-6} s^{-1}) at 850 hPa at (a) –72, (b) –48, (c) –24, and (d) 0 h. Shading indicates values greater than 4×10^{-6} s^{-1}. In (a) and (b), an easterly wave approaching the mean genesis location (star) is denoted by a slanted solid line.

Source: Ritchie and Holland, 1999; ©American Meteorological Society. Used with permission.

genesis location, thereby maintaining persistent easterlies to the east of the region (not shown).

For easterly wave genesis, westward-propagating disturbances are evident before genesis. The low-level wind flow is characterized by westerlies to the west of genesis location at the genesis time (Fig. 4.4). Note that the main confluence region is located to the west of the genesis location. The convection associated with easterly waves tends to be short lived and less organized relative to the monsoon confluence region. For the flow pattern associated with Rossby energy dispersion, studies have shown that eastward energy dispersion from a mature TC can subsequently result in new cyclogenesis in the wake of a preceding TC (Frank, 1982; Harr and Elsberry, 1995a, 1995b; Holland, 1995; Fu et al., 2007).

Frank (1982) suggests that the pre-existing TC to the west of a new storm may increase low-level southwesterly inflow to the genesis location, enhancing convergence, deep uplift and convection in the trailing area before genesis

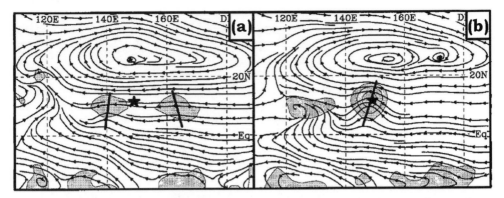

Fig. 4.4 Composites of the easterly wave pattern as represented by streamlines and relative vorticity (4×10^{-6} s^{-1}) at 700 hPa at (a) –72 and (b) 0 h. Shading indicates values greater than 4×10^{-6} s^{-1}. The trough of the easterly wave is denoted by a slanted solid line and the mean genesis location is indicated by a star.
Source: Ritchie and Holland, 1999; ©American Meteorological Society. Used with permission.

(Fig. 4.5). The wave energy dispersion pattern bears some similarity to the monsoon shear line and monsoon confluence region patterns (Figs. 4.2 and 4.3); however, the genesis trails southeastward of the preceding storm. The last characteristic circulation pattern is the monsoon gyre (Fig. 4.6). The monsoon gyre is a large low-level cyclonic vortex, accompanied by a large low-pressure center (Lander, 1994). Deep convective clouds rim the southern and eastern boundaries of the gyre. The monsoon gyre is the least frequent of the five distinctive flow patterns associated with cyclogenesis, with only 3% of TCs undergoing genesis via this mechanism (Ritchie and Holland, 1999). Lander (1994) also noted that a monsoon gyre is observed, on average, once every two years in the WNP. A monsoon gyre in August 1991 lasted for 20 days and was associated with the formation of six TCs (Lander, 1994).

There is a close connection between multiscale equatorial waves and TC genesis in various ocean basins (e.g., Frank and Roundy, 2006; Fu et al., 2007; Schreck and Molinari, 2009; Schreck et al., 2011; Schreck et al., 2012; Ventrice et al., 2012; Chen and Chou, 2014; Chen et al., 2018). These wave types include: MJO; equatorial Rossby (ER) waves; higher-frequency westward-moving waves such as tropical-depression (TD) type disturbances and mixed Rossby-gravity (MRG) waves; and eastward-propagating Kelvin (K) waves (Fig. 2.17). These waves act as precursors for TC genesis by enhancing upward motion and convection in regions of converging zonal winds, where wave energy tends to accumulate (Ritchie and Holland, 1999; Frank and Roundy, 2006). They may also increase low-level cyclonic vorticity, and/or alter the local vertical wind shear pattern. For the WNP, all wave types are active and TD-type disturbances are most

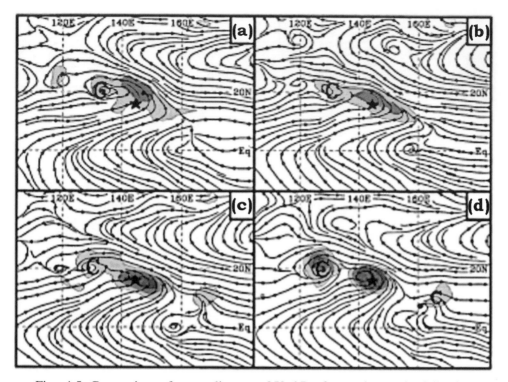

Fig. 4.5 Composites of streamlines at 850 hPa for cyclogenesis following Rossby energy dispersion from a preceding storm at (a) –72, (b) –48, (c) –24, and (d) 0 h. The preceding cyclone is denoted by "C." In (d) a secondary easterly wave development is also indicated by "C." The contour interval is 4×10^{-6} s^{-1}. Source: Ritchie and Holland, 1999; ©American Meteorological Society. Used with permission.

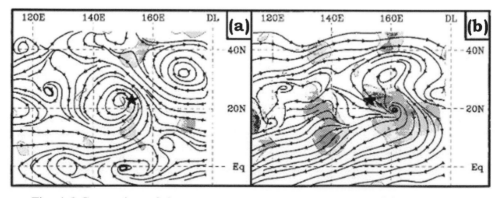

Fig. 4.6 Composites of the monsoon gyre pattern as represented by streamlines at (a) 850 hPa with convergence ($\times 10^{-6}$ s^{-1}) and (b) 250 hPa with divergence ($\times 10^{-6}$ s^{-1}). Shading indicates values greater than 2.5×10^{-6} s^{-1}. Source: Ritchie and Holland, 1999; ©American Meteorological Society. Used with permission.

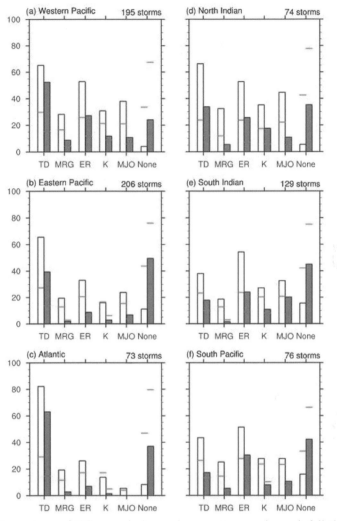

Fig. 4.7 Percentage of TC genesis by each wave types using rainfall thresholds of 2 mm d^{-1} (white bars) and 4 mm d^{-1} (gray bars) in the (a) western North Pacific, (b) eastern North Pacific, (c) Atlantic, (d) north Indian, (e) south Indian, and (f) south Pacific basins. Short horizontal lines denote the 99% significance level.

Source: Schreck et al., 2012; ©American Meteorological Society. Used with permission.

common, followed by westward-propagating ER waves and eastward-moving MJO when waves are identified using rainfall threshold of 2 mm d^{-1} (Schreck et al., 2012; Fig. 4.7). In the WNP, TCs are most likely to occur when equatorial waves interact with an environment resembling a monsoon trough.

Extending from a single wave-type approach, Chen and Chou (2014) considered TC genesis from a combination of equatorial waves over the WNP for

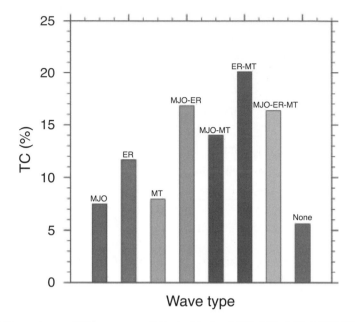

Fig. 4.8 Percentage of TC geneses attributed to the MJO, ER, MT, MJO-ER, MJO-MT, ER-MT, and MJO-ER-MT, and none types over the WNP.
Source: Chen and Chou, 2014; ©American Meteorological Society. Used with permission.

May–November from 1998 to 2010. For simplicity, the MRG wave and TD-type disturbances are jointly referred to as the MT wave. Single waves that were considered in their study included the MJO, ER, and MT waves. Multiple waves included the MJO-ER, MJO-MT, ER-MT, and MJO-ER-MT. The Tropical Rainfall Measuring Mission (TRMM) multi-satellite precipitation analysis (TRMM 3B42) was employed as a proxy for tropical convection. Chen and Chou (2014) noted that TC geneses attributed to multiple waves were approximately twice as likely to occur as those attributed to single waves (Fig. 4.8). In terms of single wave types, the ER wave occurred most frequently, while the contribution made by MJO and MT was comparable. This result was different from Frank and Roundy (2006) and Schreck et al. (2012) due to differences in the threshold criteria applied, the data used, and the way in which the single waves were amalgamated into multiple waves. Combining more than one wave type can alter the resulting wave behavior, propagation direction, and the phase relationship with TC genesis locations.

4.2.3 Modulation of WNP Tropical Cyclone Activity by Intraseasonal Oscillations

As described in Chapter 2, there are two types of tropical intraseasonal oscillations (ISO): the 30–60-day MJO and the 10–20-day quasi-biweekly oscillation

(QBWO). Gray (1979) noted that TC formations are not evenly distributed events, but tend to cluster in time with periods of two to three weeks of active formation followed by inactive periods of similar duration. Subsequent studies reveal that MJO modulated TC clustering in various ocean basins (e.g., Liebmann et al., 1994; Maloney and Hartmann, 2000a; Hall et al., 2001; Barrett and Leslie, 2009).

Based on an index that classified the MJO location into eight phases (Wheeler and Hendon, 2004), Klotzbach (2014) studied the impact of the MJO on worldwide TC activity. By definition, phases 2 and 3 were associated with enhanced convection over the Indian Ocean, phases 4 and 5 denote convective enhancement over the Maritime Continent, phases 6 and 7 were associated with enhanced convection over the Pacific Ocean, and phases 8 and 1 denote convective enhancement over the Western Hemisphere. Table 4.1 displays TC activity during June–November associated with each of the eight phases from 1979 to 2011 over the WNP. All values are normalized by the number of days that the MJO spent in each phase during June–November (with an amplitude greater than one). NS denotes named storms with sustained winds of 34 kt or greater. H denotes TCs with sustained winds of 64 kt or greater, and MH denotes major hurricanes with sustained winds of 96 kt or greater. Also shown are the percentage of accumulated cyclone energy (ACE) generated by TCs in each phase, the number of 24-h rapid intensification (RI) periods of at least 30 kt, and percent chance of a TC forming in each phase undergoing at least one RI event. ACE is the sum of the squares of the

Table 4.1. Named storms (NS), hurricanes (H), and major hurricanes (MH) in the western North Pacific during each phase of the MJO when the Wheeler and Hendon (2004) index is greater than one. Positive differences from each phase average that are significant at the 5% level are highlighted in boldface, and negative differences that are significant are italicized. Also shown is the percentage of basin-wide ACE in each phase, the number of 24-h periods of Rapid Intensification (RI) of at least 30 kt, and the percentage chance of an individual TC undergoing RI of at least 30 kt during its lifetime from 1979 to 2011. All values are normalized by the number of days that the MJO spends in each phase during June–November.

Phase	NS	H	MH	Basinwide ACE (%)	RI 24-h periods	RI chance (%)
1	9.5	7.3	3.8	12	19.2	49
2	*8.4*	*6.3*	*3.2*	*9*	*16.8*	**57**
3	*6.1*	*3.5*	*1.7*	*6*	*12.1*	43
4	9.0	*4.7*	2.9	*8*	15.3	35
5	**15.7**	9.3	4.6	14	23.5	36
6	**18.0**	**11.8**	4.8	**17**	27.6	34
7	**15.2**	**11.1**	**7.0**	**21**	**33.5**	50
8	9.5	6.6	3.7	12	22.6	**55**
Phase 1–8 avg.	11.6	7.7	3.9	12	21.2	43

Source: Klotzbach, 2014; ©American Meteorological Society. Used with permission.

6-hourly maximum sustained wind speed during the lifetime of a TC when maximum sustained winds ≥ 34 kt and is commonly integrated for all TCs in a season or year (Bell et al., 2000). TC activity is most enhanced in phases 6 and 7 when the convectively enhanced MJO is present over the Pacific Ocean. Conversely, TC activity is much reduced in phases 3–4 when VWS becomes strong, and low-level easterlies prevail. The number of 24-h RI periods in phase 7 is almost three times higher than that in phase 3. However, the chance of individual TCs undergoing RI during their lifetime was not significantly related to the number of 24-h periods of RI.

The MJO phase has been explained in terms of barotropic wave dynamics over the region where there is a strong low-level westerly anomaly. Maloney and Hartmann (2001) suggested that small-scale disturbances could grow by drawing barotropic eddy kinetic energy conversion from the time-mean flow during the MJO convective phase. These growing disturbances, together with other favorable conditions, could support tropical cyclogenesis. The barotropic conversion process is also identified as a major energy source for synoptic disturbances in the WNP (Lau and Lau, 1992). Likewise, this mechanism is also explained as Rossby wave accumulation through convergence of low-level zonal flows for the development of synoptic-scale disturbances into "tropical depression" type storms in the WNP in the boreal summer (Holland, 1995; Sobel and Bretherton, 1999).

During 1975–2010, Li and Zhou (2013a) found that 23% (20%) of TCs in the WNP are associated with the active MJO (QBWO). Therefore, the QBWO is also important for TC formation in the WNP. In the boreal summer, the MJO signal, which typically originates in the Indian Ocean, propagates northeastward to the WNP. Because the MJO is a planetary-scale mode, it can significantly influence basin-wide TC frequency and its northeastward genesis location in the WNP (Kim et al., 2008; Mao and Wu, 2010; Li and Zhou, 2013a). Li and Zhou (2013a) examined the impact of these two major elements of the ISO on TC activity in the WNP. A MJO cycle is partitioned into eight phases, with phases 1+2 and 7+8 being identified as convectively enhanced phases, and phases 3+4 and 5+6 as non-convective phases. Likewise, the QBWO cycle was also broken into eight phases.

Phases 7+8 of the MJO cycle feature an extensive band of negative OLR anomalies from the Indian Ocean to the tropical WNP and the majority of TCs coincide with the location of this band of enhanced convection in the WNP (Fig. 4.9d). A slight northeast movement of convection over the WNP is noted from phases 7+8 to 1+2 while convection is concurrently reduced over the Indian Ocean (Fig. 4.9a). The negative OLR anomalies over the WNP begin to decay in phases 3+4 and are replaced by positive anomalies in phases 5+6, corresponding to

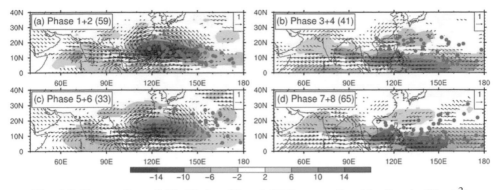

Fig. 4.9 Composites of 30–60-day filtered OLR anomalies (shading in W m^{-2}) and 850-hPa wind anomalies (vectors in m s^{-1}) during different MJO phases (a–d). Only anomalies exceeding the 5% significance level based on a Student's t test are shown. Solid circles denote the TC genesis positions, and the numbers in the parentheses denote the TC counts forming in each phase. A black and white version of this figure will appear in some formats. For the color version, refer to the plate section.

Source: Li and Zhou, 2013a; ©American Meteorological Society. Used with permission.

the suppressed convection stage (Fig. 4.9b and c). At the same time, the low-level anomalous circulation is dominated by a large anticyclone.

Tropical cyclones generally track around the periphery of a subtropical high (e.g., Harr and Elsberry, 1995a, 1995b), and their movement is primarily determined by the large-scale steering flow and a small propagation component (e.g., Wu and Wang, 2004). The steering flow of a TC was originally defined as an area-averaged vertically integrated wind around the storm (Chan and Gray, 1982). To simplify the analysis, the vertically integrated wind, usually from 850 hPa to 300 hPa, at discrete grid points of the domain, without following the storm, was used in many studies. The slowly varying MJO modulates the basic state of the large-scale environment, which then causes changes in TC activity. In the convectively enhanced MJO phase, TCs mainly track westward and north-westward in the WNP under the influence of the westward extension of the WNP subtropical high, leading to increased TC activity near the Philippines, Vietnam, and in southeast China (Fig. 4.10). However, there is little difference in the ratios of strong to weak TCs during different MJO phases (Liebmann et al., 1994; Li and Zhou, 2013b) although higher Accumulated Cyclone Energy (ACE) values are found in the convectively enhanced MJO phase relative to the non-convective phase. Intensity, duration, and frequency of storms in a season (year) all contribute to the computation of ACE (Camargo and Sobel, 2005). Longer-duration storms may accumulate larger ACE values than storms with a higher lifetime maximum

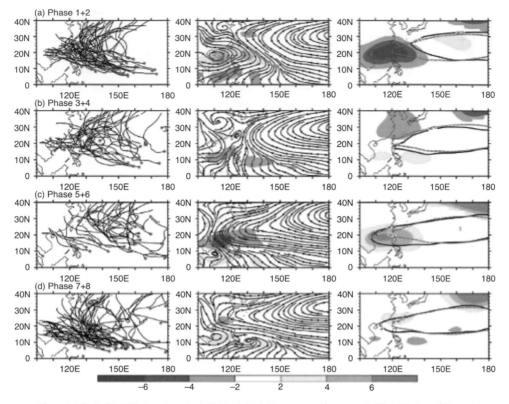

Fig. 4.10 (left) TC tracks; (middle) 850-hPa streamlines and 30–60 day filtered vorticity anomalies 10^{-6} s^{-1}; and (right) 500-hPa geopotential height anomalies (m) for different MJO phases (a–d). In the right-hand panels, dashed purple contours denote the 5,870 geopotential height level, and solid black contours denote the 5,870 geopotential height for different MJO phases. Only anomalies that exceed the 5% significance level based on a Student's t test are shown. A black and white version of this figure will appear in some formats. For the color version, refer to the plate section.

Source: Li and Zhou, 2013b; ©American Meteorological Society. Used with permission.

intensity but a shorter duration. In the convectively suppressed MJO phase, TC tracks are more likely to recurve over the open ocean so TC landfalls over East Asia or Southeast Asia are greatly reduced (Fig. 4.10b).

Because TC genesis events are marked by a distinct annual cycle, Huang et al. (2011) partitioned the analysis period into three seasons: May–June (early summer); July–September (peak season); and October–December (late season). The entire MJO cycle is also divided into eight phases, as in Wheeler and Hendon (2004). The modulation of TC genesis events by the MJO was strongest in the early season, followed by the peak season and then the late season. The strong modulation of TC activity by the MJO in the early season results from the

northward propagation of MJO from two routes: Sumatra to the South China
Sea and Papua New Guinea to the WNP. In the peak season, the modulation of
the MJO appears to be weaker due to the out of phase relationship between the
circulation in the monsoon trough region, the East Asian summer monsoon region,
and the WNP subtropical high-pressure region caused by a farther northward
propagation of the MJO into higher latitudes. As in the early season, midlevel
relative humidity contributes significantly to the MJO modulation of TC activity in
the peak season. In the late season, the MJO propagates eastward, and its
modulation is stronger west of 150°E because of the stronger MJO signal there.
The MJO's modulation of TC activity during the late season is weaker due to the
shift of the maximum convection towards the equator and away from the TC
genesis region.

The QBWO signal in the boreal summer appears to originate near the dateline in
the equatorial region and subsequently propagates northwestward through the
tropical WNP and the South China Sea (Fig. 2.14), resulting in the opposite-signed
TC modulation east and west of 150°E (Fig. 4.11d to c). In contrast to the MJO,
the impact of the QBWO is more localized because of its smaller spatial scale.
ACE is slightly higher during the QBWO convective phase, with the intensity
component of ACE appearing to be the most important driver of the increase in

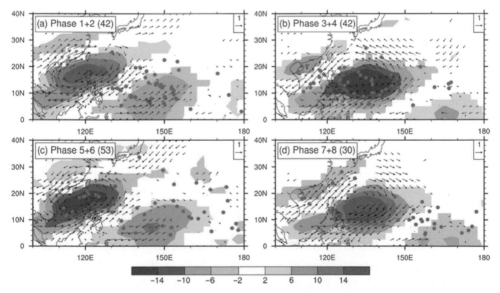

Fig. 4.11 As in Fig. 4.9, but for composites of 10–20-day filtered OLR anomalies
(shading in W m^{-2}) and 850-hPa wind anomalies (vector in m s^{-1}). A black and
white version of this figure will appear in some formats. For the color version,
refer to the plate section.
Source: Li and Zhou, 2013a; ©American Meteorological Society. Used with permission.

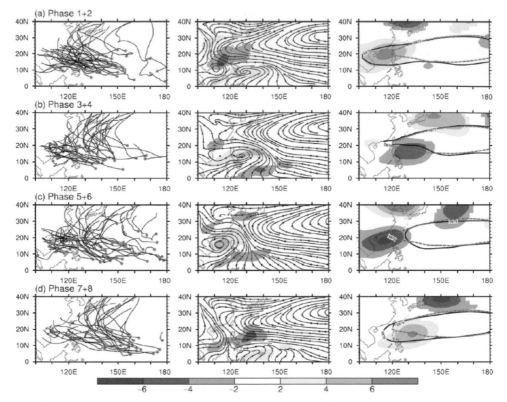

Fig. 4.12 As in Fig. 4.10, but for (left) TC tracks, (middle) 850-hPa streamlines and 10–20-day filtered vorticity anomalies in color, and (right) 500-hPa geopotential height anomalies (m) for different QBWO phases. A black and white version of this figure will appear in some formats. For the color version, refer to the plate section.

Source: Li and Zhou, 2013b; ©American Meteorological Society. Used with permission.

overall ACE (Li and Zhou, 2013a). By contrast, the frequency of TC genesis plays a major role in the MJO's modulation of WNP ACE. The QBWO also modulates TC landfalls in the Philippines and Japan. The WNP subtropical high strengthens and extends westward towards the South China Sea in the convective QBWO phase, favoring more TCs in Philippines and in the South China Sea (Fig. 4.12a and d). Conversely, in the convectively suppressed QBWO phase, the subtropical high retreats eastward, allowing TCs to move northward to affect Japan (Fig. 4.12b and c). While most studies have focused on the effects of the MJO and QBWO on TC activity separately, You et al. (2019) examined the joint effect of these two ISO phenomena on TC activity over the WNP. They noted that TC genesis frequency was mainly modulated by the MJO signal while TC genesis location was more influenced by the QBWO. Through the strength of the monsoon trough and displacement of the western Pacific subtropical high, the ISO modulates thermodynamic conditions, leading to changes in frequency and location of TC genesis.

4.2.4 Influence of ENSO, Indian Ocean Sea Surface Temperatures, and Atlantic Ocean Sea Surface Temperatures on WNP Tropical Cyclone Activity

Numerous studies have examined the relationship between TC activity in the WNP and ENSO (Lander, 1994; Chen et al., 1998; Chan, 2000; Chia and Ropelewski, 2002; Wang and Chan, 2002; Chu, 2004; Wu et al., 2004, 2018; Camargo and Sobel, 2005; Camargo et al., 2007a; Chen and Tam, 2010; Kim et al., 2011a; Zhan et al., 2011; Li and Zhou, 2012; Zhang et al., 2012; Jin et al., 2013; Patricola et al., 2018a; Song et al., 2020; Zhao et al., 2020). These studies have investigated the relationship between ENSO and WNP TC genesis location, frequency of occurrence, tracks, intensity, life span, and landfall location. Early studies noted that TC genesis positions tend to shift southeastward (northwestward) and that TCs tend to have longer (shorter) life spans during El Niño (La Niña) typhoon seasons. TCs are also more likely to recurve farther eastward during El Niño seasons. In contrast, TCs forming during La Niña seasons tend to track more westward, thus increasing the probability of landfall over China. These changes are attributed to variations in the large-scale circulation relevant to TC activity such as changes of the position and intensity of the subtropical high, low-level equatorial westerlies and the attendant monsoon trough.

In addition, a clustering method built on a mixture regression model or the fuzzy c-Means has been applied to historical TC tracks over the WNP to objectively classify TC tracks into seven or eight types (Camargo et al., 2007c; Chu et al., 2010; Kim et al., 2011b). Camargo et al. (2007c) found that two of the seven clusters that they identified were related to El Niño events. These two clusters had genesis positions that were shifted southeastward during El Niño seasons. These TCs also became more intense. The most dominant clusters using the Camargo et al. (2007c) were recurving systems, which tend to occur more often during La Niña events. Using empirical orthogonal function (EOF) analysis, Zhao et al. (2020) studied summertime WNP TC track density anomalies from 1970 to 2012. The leading EOF mode was essentially characterized by a basin-wide pattern with a concentration of TC tracks over the Philippines Sea. This mode was related to the ENSO-like pattern. The second EOF mode exhibited a tri-pole pattern and appeared to be modulated by global SST patterns.

During El Niño empirical orthogonal function, in accordance with the eastward displacement of equatorial westerlies, the WNP monsoon trough extends eastward

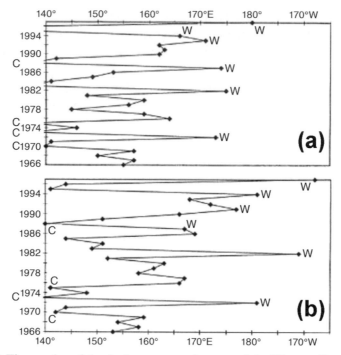

Fig. 4.13 Time series of the farthest eastward extent of the Western North Pacific monsoon trough for (a) summer (June–August) and (b) autumn (September–November) means. W's (C's) denote El Niño (La Niña) composites.
Source: Clark and Chu, 2002, *Journal of the Meteorological Society of Japan.*

and can sometimes reach past the dateline into the central Pacific in response to anomalous convective heating near the dateline (Fig. 4.13; Wu et al., 2012).

This response is associated with positive sea surface temperature anomalies (SSTA) in the central/eastern equatorial Pacific and the associated altered Pacific Walker circulation during El Niño. The monsoon trough is characterized by a directional shear zone with monsoon westerlies on the equatorward side and trade-wind easterlies on the poleward side. The enhanced heating in the central and eastern equatorial Pacific induces strong equatorial low-level westerly anomalies and the formation of a pair of anomalous off-equatorial cyclonic circulations on its northwestern and southwestern flanks in a Gill/Matsuno-type Rossby wave response (Gill, 1980). Accordingly, TC genesis also shifts to the southeastern quadrant of the WNP where weaker vertical wind shear and stronger low-level cyclonic vorticity are found. As a result, those TCs can persist for longer periods in the WNP warm pool region, gaining more thermal energy from the underlying warm pool of the ocean, thereby increasing both their lifespan and intensity (Wang and Chan, 2002; Camargo and Sobel, 2005; Zhan et al., 2011). Camargo and Sobel (2005) used ACE as a measure of TC intensity and studied variations of ACE in the WNP with respect to ENSO phases. They found higher ACE per year during El Niño than in La Niña, and the lifespan contribution appeared to be more important relative to either the intensity or the frequency of storms in the WNP (Camargo and Sobel, 2005).

Li and Zhou (2012) clustered TC intensities into three groups and showed that more super typhoons (category 4 and 5, $\geq$130 kt) develop during El Niño events and fewer during La Niña, while the converse is true for the weakest TC group (e.g., tropical storms and tropical depressions). This result is consistent with Wang and Chan (2002), Camargo and Sobel (2005), and Zhan et al. (2011). For the weaker typhoon group (category 1–3), its frequency decreases when El Niño transitions to La Niña (Li and Zhou, 2012). Chen et al. (2018) investigated the combined impacts of ENSO and the MJO on productivity of TC genesis (P_{TCG}) over the WNP. The P_{TCG} is defined as the percentage value of precursory tropical disturbances evolving into tropical storms or typhoons. In Chen et al. (2018), ENSO states are classified into warm, cold, and normal years based on the Niño-3.4 SSTA. They noted that ENSO likely exerts more influence on P_{TCG} patterns than the MJO in the combined phases, and the important environmental variables modulated by ENSO and MJO were low-level relative vorticity and mid-level relative humidity.

Because El Niño is known to have at least two types, subsequent studies focused on TC characteristics associated with EP and CP El Niño events, as well as La Niña events. In addition to EP and CP events, several studies have suggested a third event of El Niño events, called a mixed El Niño (ME) (Kug et al., 2009; Yu and Kim, 2013; Timmermann et al., 2018). The ME event is characterized by a maximum ocean warming along the equator between 120°W and 150°W, roughly coinciding with the Niño-3.4 region. During an EP El Niño typhoon season (JASO), TC genesis is enhanced in the southeastern quadrant of the WNP (east of 140°E and south of 15°N) but suppressed in the northwestern quadrant of the WNP (Fig. 4.14a).

The eastward displacement of equatorial westerlies and the monsoon trough, as well as weak vertical wind shear (VWS) and a convective heating anomaly near the dateline, contribute to increased TC activity to the east of the climatological mean TC genesis location during EP events (Fig. 4.14a). The TC genesis anomaly pattern seen in Fig. 4.14a is also well reflected in the corresponding TC track density anomaly map (Fig. 4.15a). TC tracks in EP events display a dipole pattern with higher TC occurrences in the southeastern part and lower occurrences in the northwestern part of the WNP. Of particular note is that the entire East Asian coast experiences fewer TC landfalls in EP summers. TC landfall is influenced by both TC genesis position and steering flow. By breaking the entire typhoon season (JJASO) into summer (JJA) and autumn (September/October), Zhang et al. (2012)

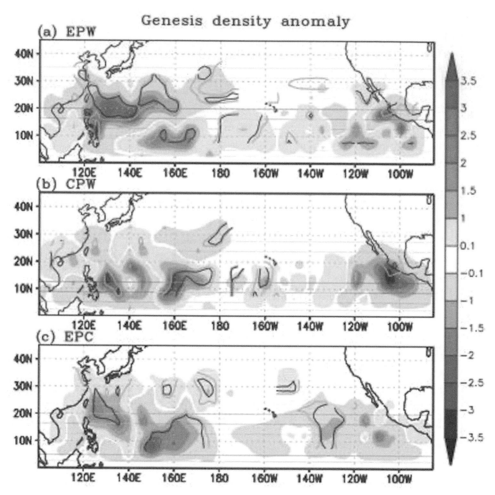

Fig. 4.14 Composite of TC genesis density anomalies (×10) in JASO over the Pacific for (a) EP El Niño, (b) CP El Niño, and (c) La Niña events. Light (dark) contours denote statistical significance at the 10% (5%) level. A black and white version of this figure will appear in some formats. For the color version, refer to the plate section.

Source: Kim et al., 2011a; ©American Meteorological Society. Used with permission.

also found fewer typhoons making landfall in the Philippines, Indochina, China, Japan and Korea during both summer and autumn during EP events relative to CP events. For the entire peak TC season (JJASO), more landfalls occurred over Japan and Korea during a CP El Niño year. Based on a two-sample permutation procedure (Chu and Wang, 1997), Chen (2011) noted significantly higher TC frequency over the South China Sea in a CP summer. This increased TC activity over the South China Sea was associated with an elongated cyclonic anomaly over the western WNP in response to a broad convective anomaly.

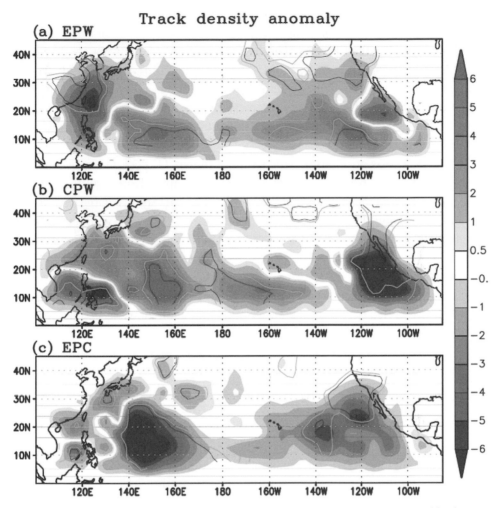

Fig. 4.15 Composite of TC track density anomalies in JASO over the Pacific for (a) EP El Niño, (b) CP El Niño, and (c) La Niña events. Light (dark) contours denote statistical significance at the 10% (5%) level. A black and white version of this figure will appear in some formats. For the color version, refer to the plate section.

Source: Kim et al., 2011a; ©American Meteorological Society. Used with permission.

A strong monsoon trough favors the growth of westward-moving synoptic-scale wave disturbances through barotropic energy conversion, by which tropical waves gain energy from the mean flow. This interaction has been proposed as a possible forcing mechanism for TC formation (Wu et al., 2012) that may also yield more intense typhoons (Zhan et al., 2011). The growth of perturbations, together with favorable large-scale environmental conditions associated with the eastward extension of the monsoon trough, is conducive to higher frequency of TC genesis in the southeastern part of the WNP during strong El Niño years (Wu et al., 2018). On the other hand, an anomalous anticyclone is observed in the northwestern quadrant of the WNP in EP years. This anomalous anticyclone is not conducive for TC genesis (Figs. 4.14a and 4.15). In CP years (Fig. 4.14b), the overall pattern of genesis position is somewhat similar to that during EP years (Fig. 4.14a). However, the major difference between these two extreme El Niño types is a band of enhanced genesis protruding from the southeastern WNP northwestward to southern Japan (Fig. 4.14b).

Similar to genesis positions, a wide swath of higher TC occurrences following a northwestward track from the southeastern WNP to Japan, Korea, southeast China coast, and Taiwan occurs during CP years (Fig. 4.15b). A large anomalous cyclonic circulation dominates over the WNP during CP years, with an extensive trough extending northwestward from 10°N, 150°E to 30°N, 120°E, favoring TC formation over a substantial portion of the WNP (Song et al., 2020). During CP TC season (JJASON), more TCs recurved near the East Asian coast and threatened Japan, Korea, and northern China. Fewer TCs tracked through the Philippines and the South China Sea where anomalous cooling and anomalous descending motion may have played a role in suppressing TC activity (Kim et al., 2011a; Song et al., 2020). In accordance with the westward shift in CP-induced heating, low-level westerly anomalies moved northward and dominated the tropical WNP (Figs. 4.16b and 4.17b). As a result, an anomalous cyclone is observed south of Japan and east of Taiwan (Chen and Tam, 2010; Kim et al., 2011a; Jin et al., 2013; Wu et al., 2018), favoring TCs tracking into the East China Sea and consequently threatening the East Asian coasts (Jin et al., 2013). Using regional climate model experiments, Jin et al. (2013) demonstrated that during CP summers northern off-equatorial warming, rather than equatorial warming, effectively induced anomalous steering flow that enhanced TC activity over East Asia. For the ME phase, the spatial patterns of genesis anomalies and track anomalies are similar to that of EP, although the magnitude of these anomalies is weaker and less significant during ME (Song et al., 2020).

TC landfalls also display seasonal variation during EP and CP events. For example, Zhang et al. (2012) noticed a difference in landfall characteristics from summer (JJA) to autumn (September/October) in CP years. For both seasons, a northward shift in TC genesis locations is noted in CP years compared to EP years. Due to a westward shift in the subtropical high, enhanced easterly steering flow tends to increase landfalls over East Asia during CP summers, particularly in Japan and Korea. In contrast, in the autumn of CP years, TC landfall is generally suppressed over East Asia, particularly southeast Asia and the Philippines, because of the prevalence of westerly steering flow anomalies, which counteract the prevailing westward or northwestward motion taken by most TCs.

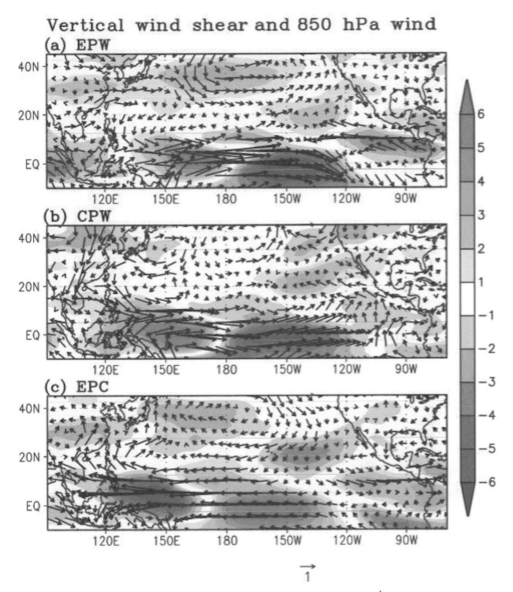

Fig. 4.16 Composite of vertical wind shear anomalies (m s^{-1} in shading) and 850-hPa wind vector (m s^{-1}) in JASO over the Pacific for (a) EP El Niño, (b) CP El Niño, and (c) La Niña events. The vertical wind shear is defined as the difference in the zonal wind component between 200 and 850 hPa. A black and white version of this figure will appear in some formats. For the color version, refer to the plate section.

Source: Kim et al., 2011a; ©American Meteorological Society. Used with permission.

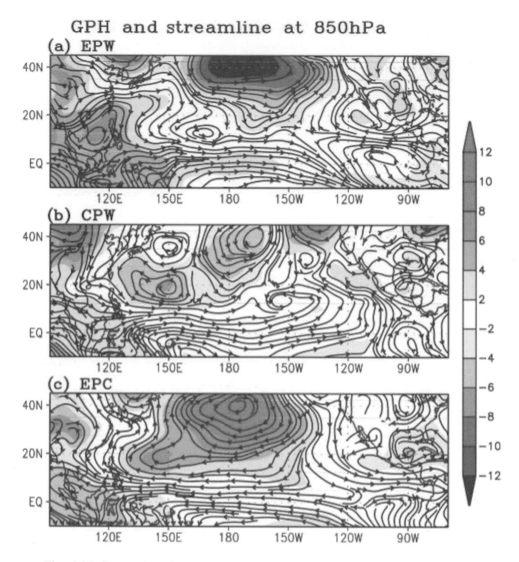

Fig. 4.17 Composite of geopotential height anomalies (m in shading) and wind anomalies at 850 hPa in JASO over the Pacific for (a) EP El Niño, (b) CP El Niño, and (c) La Niña events. A black and white version of this figure will appear in some formats. For the color version, refer to the plate section.
Source: Kim et al., 2011a; ©American Meteorological Society. Used with permission.

For La Niña years (Fig. 4.14c), the pattern is opposite of El Niño years with more TC genesis occurring to the west of 140°E and a local maximum in the Philippine Sea. Tropical storm and typhoon occurrences in the eastern portion of the WNP are suppressed by anomalously strong shear near the dateline, while reduced VWS in the western WNP favors enhanced TC activity in a relatively narrow band to the west of 140°E (Figs. 4.14c and 4.16c). Both a westward shift in the mean genesis location and easterly steering anomalies (Zhang et al., 2012) increase the threat to Japan, southeastern China, Taiwan, and the northern Philippines from typhoon during La Niña seasons, compared to EP seasons (Fig. 4.15a and c).

In recent years, some studies have investigated not only the geographic location of the maximum El Niño warming (e.g., central or eastern Pacific), but also the magnitude of the warming (Patricola et al., 2018a; Wu et al., 2018). In general, larger overall warming is associated with an EP event relative to a CP event. When the peak in the anomalous SST warming is over the central equatorial Pacific, simulated TC activity such as ACE and the total number of typhoons and number of intense typhoons is enhanced, relative to SST warming over the eastern equatorial Pacific (Patricola et al., 2018a). By doubling anomalous CP warming to make it similar in magnitude to that of EP warming, the change in simulated TC activity in the WNP is even more dramatic, e.g., the number of typhoons and intense typhoons increased by 59% and 120%, respectively, relative to the climatology simulation. Given the warmer background SST and greater atmospheric instability in the central Pacific, a small warming there can effectively induce significant changes in WNP TC activity (Patricola et al., 2018a).

Wu et al. (2018) also found that WNP TC activity is very sensitive to the intensity of the CP SST warming. For strong EP or CP warming, the monsoon trough is imbedded in a broad confluence zone and extends farther eastward. Weaker VWS and stronger low-level cyclonic vorticity also enhance the likelihood for TC genesis in the southeastern quadrant of the WNP. During moderate EP years, the monsoon trough is weaker, while during weak EP years the trough retreats westward and becomes much weaker. Wu et al. (2018) suggested that strong El Niño events could affect TC activity over the WNP, while moderate or weak EP events do not.

Previous studies have examined the relationship between TC activity over the WNP and Pacific Ocean warming. WNP TC activity can also be influenced by Indian Ocean SST. For example, Indian Ocean warming during the decaying summer of an El Niño can enhance the formation of the Philippine anticyclone (Xie et al., 2009). Moreover, the frequency of all TCs and weak TCs appears to be highly related to the East Indian Ocean (EIO) SSTA, while the east–west shift of the mean TC genesis location and frequency of intense typhoons are mainly determined by ENSO (Zhan et al., 2011). That is, the EIO SSTA does not appear to exert significant effects on where WNP TCs form. When the EIO is anomalously warm (cold), a reduced (enhanced) land–sea thermal contrast is established, which results in a weaker (stronger) WNP and South China Sea (SCS) summer monsoon as well as a weaker (deeper) monsoon trough. In the meantime, persistent EIO warming after an El Niño onset can elevate local tropospheric temperature via moist adiabatic processes associated with deep convection and subsequently excite a warm tropospheric Kelvin waves along the equator to the east, leading to surface divergence and an anomalous anticyclone off of the equator in the tropical WNP (Xie et al., 2009). These anomalies are unfavorable for TC genesis and contribute to a significant decrease in TC genesis over the WNP in the summer following a strong El Niño (Du et al., 2011).

The SST gradient between the southwestern Pacific and western Pacific warm pool in the boreal spring may also modulate WNP TC frequency by creating low-level easterly anomalies over the equatorial central–western Pacific (Zhan et al., 2013). The easterly anomalies induce local upwelling and a rising thermocline to the east (see Chapter 3), thereby cooling SSTs in the central Pacific. Given the anomalous equatorial cooling and invoking Gill's model, low-level easterlies to the west of the cooling source will be further enhanced, inducing an anticyclonic shear and circulation pattern over the WNP, suppressing WNP TC genesis in summer.

Can the tropical Atlantic SSTA influence TC activity in the WNP? Through observations and model simulations, an anomalously warm tropical Atlantic Ocean has been shown to result in low-level easterly anomalies over the western Indian Ocean. This weakens the Indian summer monsoon and warms local ocean temperature due to both decreased evaporation and ocean downwelling (Wang et al., 2009). Given the direct impact of Indian Ocean SSTA on the WNP atmospheric circulation due to their close proximity (e.g., Xie et al., 2009), Atlantic SSTA may also indirectly affect WNP TC activity through remote teleconnections (Huo et al., 2015). Based on the simultaneous correlation between the frequency of the WNP TCs in JASO and the SST through the three ocean basins, the largest negative significant correlation is found in the tropical North Atlantic and the tropical Indian Ocean, while a smaller negative correlation is found in the WNP (Yu et al., 2016). These findings imply that tropical North Atlantic and Indian Ocean SSTA are important for TC activity in the WNP, while local SST is less crucial for TC variability in the WNP. Chan and Liu (2004) also found that increases in local SST did not significantly impact WNP TC activity.

An anomalously cold tropical Atlantic Ocean, through a Gill-type response, induces low-level westerly anomalies to the east over the tropical Indian Ocean. These low-level westerly anomalies enhance the Indian summer monsoon and cool in-situ ocean temperature due to increased evaporation and ocean mixing. The cool SSTA in the Indian Ocean further enhance the monsoonal circulation and induce anomalous westerlies in the tropical WNP. This cold SSTA also results in low-level cyclonic vorticity anomalies and lower pressure, favoring more TC genesis in the WNP (Yu et al., 2016). Therefore, the northern Indian Ocean may act as a relay region through which tropical Atlantic SSTA can remotely influence WNP TC activity. Zhang et al. (2018) also highlighted a link between Atlantic SST and WNP TCs via a modulation of the Pacific–Atlantic Walker Circulation.

4.2.5 Modulation of WNP Tropical Cyclone Activity by the Pacific Meridional Mode

While ENSO clearly modulates TC variability in the WNP, the Pacific Meridional Mode (PMM) has also been postulated to exert an influence on WNP TC activity (Zhang et al., 2016; Gao et al., 2018; Hong et al., 2018). The positive PMM pattern is characterized by anomalous warming in the subtropical eastern North Pacific and anomalous cooling in the southeastern part of the tropical eastern Pacific (Chapter 3). The positive PMM phase has been shown to be conducive for higher TC frequency in the WNP, especially in the eastern part of the basin through reductions in zonal VWS (Zhang et al., 2016). An anomalous low-level cyclonic circulation is forced by anomalous heating in the subtropical central Pacific during the positive PMM phase via a Gill-type response under asymmetric heating (Gill, 1980). The cyclonic flow in the northwestern portion of the WNP implies a weaker subtropical high and an elongated and perhaps deeper monsoon trough, both of which favor WNP TC genesis.

Hong et al. (2018) compared the impact of PMM-like and ENSO-like SSTA on the mean WNP TC genesis location for two very strong El Niño events: 2015–2016 and 1997–1998. They found that the mean genesis location (MGL) in the summer (JJA) 2015 shifted eastward by 10° longitude relative to that in 1997, whereas the southward shift (~ several degrees latitude) during these two events is approximately equal. The 2015–2016 El Niño was comparable to 1997–1998 in terms of the Niño-3.4 index, but the amplitude of the PMM-like SSTA in summer 2015 was twice as large as that in summer 1997. Their numerical experiments indicate that the positive PMM-like SSTA forced a zonal overturning circulation anomaly with ascending motion in the subtropical central Pacific and descending motion in the subtropical western Pacific. This pattern likely shifted the MGL in 2015 eastward. In comparison, the positive ENSO-like SSTA likely led to a southward shift of the MGL in both strong El Niño events. The PMM-like and ENSO-like SSTA jointly contributed to the unprecedented southeastward shift of the MGL in 2015; however, the pronounced eastward displacement of the MGL was likely caused by the PMM-like SST forcing.

Using data from the period 1990–2016, Gao et al. (2018) noted a southeast shift of intense typhoons in positive PMM phases (six cases) relative to the negative phases (two cases) when ENSO was in neutral conditions. The off-equatorial heating anomalies in the eastern North Pacific during the positive PMM phase induce low-level westerlies and upper-level easterlies over the MDR of the WNP through a Gill–Matsuno type Rossby wave response. Consequently, VWS is reduced and low-level relative vorticity is enhanced, two key dynamical conditions favorable for TC genesis. When storms form farther southeastward in the WNP during the positive PMM phase, they may persist longer and gain more energy from the underlying warm ocean before making landfall on their northwestward track or recurving northward around the western flank of the subtropical high. Thus, when the PMM is in the positive phase, storms tend to become more intense and storm duration tends to increase. It should be noted that the small number of negative PMM phase events could be an issue in determining the WNP TC response in these events.

4.2.6 Stratospheric Quasi-biennial Oscillation and Tropical Cyclone Activity

Previous studies have suggested that the stratospheric quasi-biennial oscillation (QBO) modulates TC activity in the Atlantic and WNP (Gray, 1984; Chan, 1995; Collimore et al., 2003; Ho et al., 2009). The QBO is a quasi-periodic oscillation of the zonal wind in the tropical stratosphere with a period of 28–29 months. The wind regime alternates between easterlies and westerlies and propagates downward from the lower stratosphere to the tropical tropopause. Changes in vertical wind shear, static stability, and tropopause dynamics associated with the QBO phases are thought to be the driver of TC variations. When the QBO is in its westerly phase, more TCs occurred over the East China Sea, likely caused by changes in the extratropical circulation (Ho et al., 2009). However, after removing the ENSO effect, Camargo and Sobel (2010) found that the influence of the QBO on TC activity over the WNP was not statistically significant.

4.2.7 Decadal to Interdecadal Tropical Cyclone Variations

Tropical cyclone activity in the WNP also exhibits decadal or interdecadal variability. This includes changes in TC genesis frequency for various portions of the basin, as well as changes in both track and intensity and TC landfall rates (Liu and Chan, 2008, 2020; Tu et al., 2009; Hsu et al., 2014; Park et al., 2014; He et al., 2015; Zhan et al., 2017; Zhao et al., 2018; Shan and Yu, 2021). Rodionov (2004) used a Student's t test and a fixed cutoff length to implement a regime shift detection algorithm. This same technique has been used by many others to identify decadal variations of TC genesis frequency over the WNP, although the original application was intended for a Gaussian-like data set, not for Poisson-like TC records. Applying this method to peak typhoon season (July to October) TC records from 1979 to 2012, He et al. (2015) found two regime shifts occurring in 1989 and 1998, and consequently the full time period in their study was broken into three periods. The first (1979–1988) and the third (1998–2012) periods were marked by lower TC frequency while the second epoch (1989–1997) was characterized by higher frequency. To facilitate the analysis, He et al. (2015) contrasted the active second epoch, 1989–1997 (called P1), with the inactive third epoch, 1998–2012 (named P2). Their analysis revealed a decrease in TC genesis frequency over the southern WNP (5°N–20°N, 105°E–170°E) from P1 to P2 and a concomitant increase over the northern WNP (20°N–25°N, 115°E–155°E) (Fig. 4.18). This suggested a poleward migration of TC genesis positions since 1998, which is consistent with the findings of Tu et al. (2009) who noted a northward shift of typhoon tracks over the WNP–East Asian region since 2000 and an increase of typhoon frequency over the Taiwan–East China Sea region. Decadal changes in TC genesis frequency were linked with corresponding variations in tropical Indo-Pacific SST (He et al., 2015). Both the tropical central Pacific cooling and the tropical Indian Ocean warming induce an anticyclonic anomaly in the WNP , which is unfavorable for southern WNP genesis in P2. The local low-level cyclonic anomaly, mainly responsible for the increase in northern WNP TC genesis, was likely forced by local SST warming or a regional anti-Hadley circulation anomaly. Zhao et al. (2018) inferred that natural variability such as the Interdecadal Pacific Oscillation contributed to the lower TC frequency in the WNP since 1998. Takahashi et al. (2017) postulated that the observed decline in TC frequency over the southeastern WNP during 1992–2011 might be due to changes in sulfate aerosol emissions.

He et al. (2015) also studied decadal changes in TC tracks over the WNP and classified them into three prevailing patterns: westward-moving, northwestward-moving, and northeastward recurring tracks. Relative to P1 (1989–1997), the northwestward-moving pattern became the most dominant mode in P2 while the other two TC track types became less frequent during P2 (1998–2012; Fig. 4.19). They further divided the WNP into four subregions: the Philippines Sea; South China Sea; Japan/east Japan; and southeast China–Okinawa. The last subregion is centered over Taiwan and the East China Sea so it is referred to as the Taiwan–ECS area. The first three subregions exhibited a reduction in TC occurrence frequency from P1 to P2 while the last subregion had an increase in TC occurrence frequency. For the South China Sea, the reduction is a result of a decrease in local TC genesis and a declining number of storms that came from the east (Fig. 4.20a). For the Philippines Sea, the reduction is mainly caused by a decrease in local TC formation (Fig. 4.20b). The change in Japan/east Japan is mainly due to a decrease in TCs from the south (Fig. 4.20c). For the Taiwan–ECS region, the increase in occurrence frequency results from both an increase in TCs tracking northwestward and a longer local TC duration (Fig. 4.20d).

Besides the peak season, late-season (October–December) typhoon activity over the WNP also exhibits decadal variability. Some of these late-season typhoons

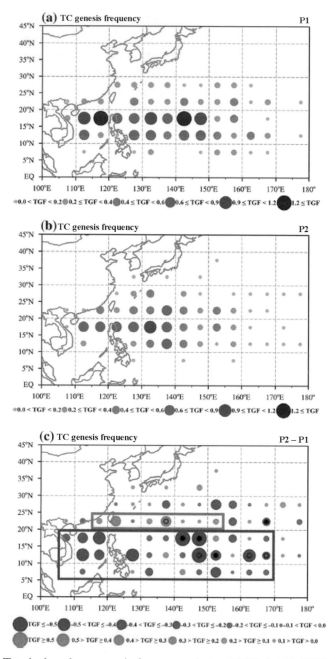

Fig. 4.18 Tropical cyclone genesis frequency during JASO for (a) 1989–1997 (P1), (b) 1998–2012 (P2), and (c) their difference (P2 – P1). The rectangles in (c) denote regions with pronounced changes in genesis frequency. The large (small) rectangle at the bottom (top) indicates a decrease (increase). A black and white version of this figure will appear in some formats. For the color version, refer to the plate section. Source: He et al., 2015; ©www.creativecommons.org/licenses/by/4.0/. Used with permission from Springer Nature.

have been catastrophic. For example, Super typhoon Megi in late October 2010 caused enormous damage ($735.9 million) to the Philippines, Taiwan, and China. Two years later, Super typhoon Bopha (category 5) in late November 2012 killed over 1,100 people and caused $1.04 billion in damage in the Philippines. This is followed by another powerful Super typhoon Haiyan, which devastated the central Philippines in early November 2013 (Lin et al., 2014). Haiyan was the deadliest Philippine typhoon in modern history, killing more than 5,000 people. Although the mean number of typhoons that occurred in the late season is fewer than that during the peak season, given the frequent intense typhoons during the latter period and the major damage inflicted, it is also of importance to understand the long-term variations in late-season typhoon activity.

The Bayesian paradigm under a one change-point hypothesis was used by Hsu et al. (2014) to infer abrupt shifts in late-season typhoon activity over the WNP during the period 1979–2011. They detected a change point in 1995 (Fig. 4.21). Their methodology involved the following approaches: (1) seasonal TC counts were modeled by a Poisson distribution where the Poisson parameter was codified by its conjugate gamma distribution; and (2) a hierarchical approach involving three layers – data, parameters, and hypothesis – was formulated to determine the posterior probability of the shift in time. For technical details, see Chu and Zhao (2004), Zhao and Chu (2006), and Chu and Zhao (2011). The Bayesian analysis provides the probability statement of change-points and is an advantage over a deterministic approach (e.g., Student t test), because the uncertainty inherent in statistical inferences is quantitatively expressed in the probability statement, and probability is the universal language of uncertainty.

TC counts, life spans, and ACE in the late-season during the 1995–2011 epoch significantly decreased relative to that which occurred during the 1979–1994 epoch. This decrease is also seen when individual months within the late-season are considered. Figure 4.22 shows epochal changes in the spatial distribution of typhoon genesis and the 850-hPa wind field (Hsu et al., 2014). The low-level cyclonic circulation between 130°E and 160°E was more enhanced in the early epoch than in the late epoch (not shown). The weakened cyclonic circulation (or an anticyclonic anomaly) appears to lead to a significant decrease in typhoon genesis over the southeastern WNP and the South China Sea from 1995 to 2011 (Fig. 4.22c). This decadal decrease appears to contribute to the decreasing trend of the TC landfalls in South China since the late 1990s (Shan and Yu, 2021).

To examine how different environmental factors contributed to epochal changes quantitatively, a modified version of the tropical cyclone genesis potential index (GPI) of Emanuel and Nolan (2004) was used. The negative vorticity anomaly was the leading contributor to the GPI decrease over the southeastern WNP during 1995–2011. A suite of sensitivity experiments based on an AGCM was conducted

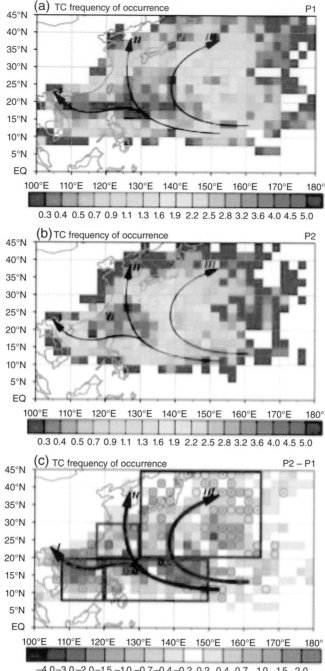

Fig. 4.19 Tropical cyclone occurrence frequency during JASO for (a) 1989–1997 (P1), (b) 1998–2012 (P2), and (c) their difference (P2 – P1). The thick black arrows in (a) and (b) denote prevailing TC tracks. In (c), blue (red) arrows denote less frequent (more frequent) tracks from P1 to P2. The rectangles in (c) indicate

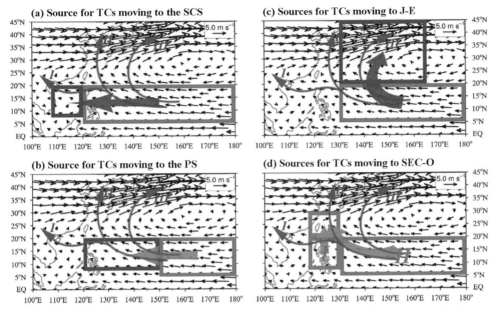

Fig. 4.20 Major genesis sources (green box) for TCs moving into (a) the South China Sea, (b) the Philippine Sea, (c) Japan and east of Japan, and (d) Taiwan and the East China Sea. Vectors denote the climatological mean of the steering flow. The purple rectangles in (a), (b), and (c) denote regions with significant decrease in TC occurrence frequency, while the orange box in (d) indicates a significant increase in TC occurrence frequency. The brown solid curves in each panel denote the three prevailing background TC tracks. Thick arrows denote the TC movement direction from each source region, with blue (red) denoting a decreasing (increasing) tendency. A black and white version of this figure will appear in some formats. For the color version, refer to the plate section.
Source: He et al., 2015; ©www.creativecommons.org/licenses/by/4.0/. Used with permission from Springer Nature.

to examine the impact of different ocean basins to the epochal changes in environmental conditions. The ensemble simulations suggest that recent SST changes to a more La Niña-like state induced unfavorable conditions for typhoon genesis over the southeastern WNP (e.g., decreasing anomaly) since 1995. The cooling in the eastern Pacific also enhanced the anticyclonic anomaly over the southeastern WNP. The weak Indian Ocean warming, on the other hand, contributed insignificantly to the circulation anomaly that was related to the decadal decrease in typhoon genesis over the southeastern WNP.

Caption for fig. 4.19 (*cont.*). regions with pronounced changes in occurrence frequency. A black and white version of this figure will appear in some formats. For the color version, refer to the plate section.
Source: He et al., 2015; ©www.creativecommons.org/licenses/by/4.0/. Used with permission from Springer Nature.

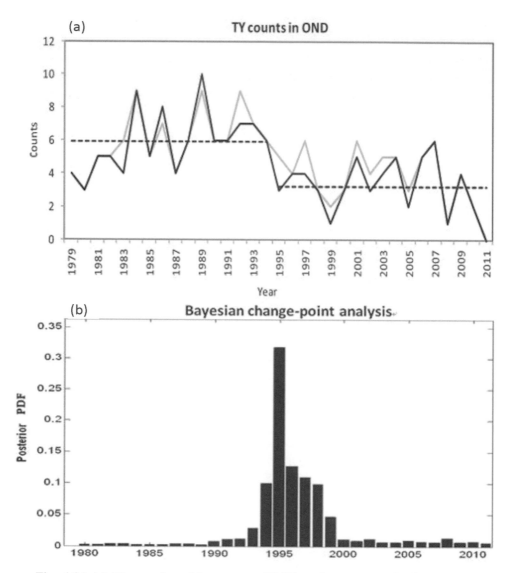

Fig. 4.21 (a) Time series of late-season (OND) typhoon counts in the western North Pacific from 1979 to 2011. Thick and gray lines denote the data from the RSMC Tokyo and the JTWC datasets, respectively. Dashed lines denote the mean for the 1979–1994 epoch and the 1995–2011 epoch, respectively. (b) The posterior probability mass function of the change-point as a function of time (years) is based on RSMC Tokyo data.

Source: Hsu et al., 2014; ©American Meteorological Society. Used with permission.

In the foregoing, TC genesis frequency in the WNP has exhibited decadal changes with higher numbers from 1979 to the mid or late 1990s and lower number since that time. In contrast to TC numbers, intense typhoons ($V_{max} \geq 96$ kt), defined as category 3 and above on the Saffir–Simpson Hurricane Wind scale, have seen a significant increase over the WNP since 1970s (Webster et al., 2005; Elsner et al., 2008). Recently, defining intense typhoons as TC reaching category 4–5 intensity (1-minute averaged maximum sustained wind $\geq$114 kt), Liu and Chan (2020) found a higher proportion of intense typhoons during 2002–2016, relative to the period 1975–1986 over the WNP. The proportion of intense typhoons was defined as the ratio of the annual number of intense typhoons to the annual number of TCs (tropical storms and typhoons). They attributed this decadal change to a longer intensification duration and a higher intensification rate. Consequently, storms have a better chance of developing into intense typhoons since the early 2000s. The decadal increase of intense typhoon occurrence frequency in the WNP since the late 1990s is associated with a La Niña-like sea surface temperature pattern (Shan and Yu, 2020). On the sub-basin scale, there is an increase in intense typhoons in the Philippines Sea and a concomitant decrease in the southeastern portion of the WNP when the period 1999–2018 is compared to an earlier period (1979–1998). Reduction in low-level relative vorticity and enhanced VWS are two major environmental factors suppressing typhoon activity in the southeastern WNP during the more recent period (Shan and Yu, 2020). The increased SST over the western part of the WNP promotes increased TC formation and also the likelihood of intensification into an intense typhoon due to changes in potential intensity, which is the theoretical maximum intensity a TC can attain for a given thermodynamic state (Emanuel, 1988).

4.3 The Eastern North Pacific

4.3.1 Background Climatology

Based on HURDAT2 from the US National Hurricane Center (Landsea and Franklin, 2013), the average annual number of tropical cyclones (TCs) (tropical storms and hurricanes) in the eastern North Pacific (ENP) from 1971 to 2018 is 15. The average annual number of TCs is the second highest for any TC basin around the globe, trailing only the WNP. On average, of these 15 TCs, 9 become hurricanes and 4 become major hurricanes (category 3, 4, and 5 on the Saffir–Simpson Hurricane Wind scale). The ENP basin is defined to extend from 140°W to the Pacific coast of Central America (Fig. 4.1). One peculiar feature of the ENP TC is that the formation location is confined to a relatively smaller domain, especially meridionally, relative to other basins (Fig. 4.23). Specifically, there is a high concentration of TC formation points between 120°W and the Pacific coast of Central America along the axis of the monsoon trough (Fig. 4.24). Unlike the western North Pacific, TCs in the ENP do not occur during the cool season and the broad hurricane season runs from 15 May to 30 November.

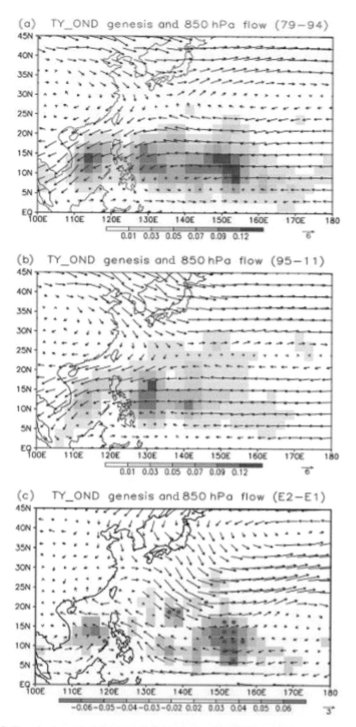

Fig. 4.22 Tropical cyclone genesis frequency (shading, number per season) and 850-hPa wind fields (vectors in m s^{-1}) during OND for (a) 1979–1994,

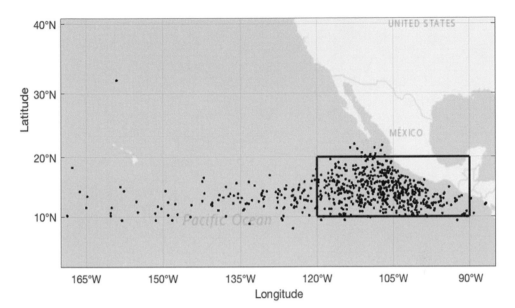

Fig. 4.23 Initial detection locations of named tropical cyclones over the eastern and central North Pacific during the hurricane season (June–November) from 1966 to 2019. Most storms formed in the region between the Pacific coast and 120°W as denoted by the black rectangle.

In the austral winter, strong southeast trades from the South Pacific cross the equator and turn into low-level southwesterlies in the ENP. This low-latitude southwest monsoonal flow meets the easterly trade winds in the subtropics and forms the monsoon trough where SSTs are warm ($\geq 28°C$) and vertical wind shear (VWS) is weak. The main development region (MDR) is defined as the belt between 10°N and 20°N (Fig. 4.23), whereas the Intertropical Convergence Zone (ITCZ) is located on the southernmost of this belt. The long-term seasonal average of VWS over the ENP MDR is about 6 m s^{-1}, which is smaller than the critical vertical shear line (10 m s^{-1}) for intensification (Wu and Chu, 2007). The relative humidity in the mid-troposphere (500 hPa) also exhibits a local maximum (>50%) over the MDR. Long-term seasonal mean low-level relative vorticity at 925 hPa is about 4×10^{-6} s^{-1} over the MDR. Climatological mean SSTs during the peak season (July–September) are >28°C over most of the MDR, with the standard deviation of the seasonal mean SST being extremely small (<0.5°C). Given warm and stable SSTs, weak VWS, moist mid-troposphere, and the presence of strong

Caption for fig. 4.22 (*cont.*). (b) 1995–2011, and (c) their difference (1995–2011 minus 1979–1994). Red dots in (c) denote differences that are significant at the 5% level. A black and white version of this figure will appear in some formats. For the color version, refer to the plate section.
Source: Hsu et al., 2014; ©American Meteorological Society. Used with permission.

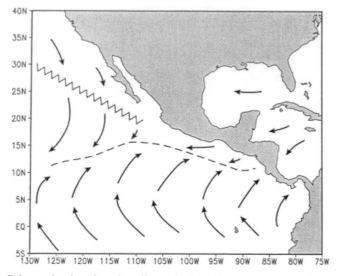

Fig. 4.24 Schematic showing the climatological August mean surface circulation in the eastern North Pacific. The monsoon trough axis is denoted by a broken line, and the subtropical ridge axis is denoted by a zigzag line. Low-level wind directions are indicated by arrows.
Source: Chu, 2004; ©Columbia University Press. Used with permission.

cyclonic relative vorticity, seasonal mean environmental conditions are conducive for TC formation over the MDR. These results agree with McBride (1995) and Molinari et al. (2000), who suggested that the highest frequency of cyclogenesis per unit area in the world is found over a compact region in the ENP.

When TCs are active in the ENP, they tend to be simultaneously inactive over the North Atlantic, and vice versa (e.g., Elsner and Kara, 1999). This dipole pattern in TC activity between the ENP and North Atlantic is also observed on sub-seasonal timescales as seen in the TC–MJO relationship in both basins (Barrett and Leslie, 2009). Once these TCs form, they tend to track westward or northwestward over the cooler water of the ENP and gradually lose their strength. Some TCs occasionally recurve northeastward and strike Mexico, with lingering effects such as heavy rain and flooding occurring in the southwest United States (e.g., Collins et al., 2016).

4.3.2 *Modulation of ENP Tropical Cyclone Activity by Intraseasonal Oscillations*

The most dominant equatorial waves in the ENP are TD-type disturbances and ER waves (Fig. 4.7). Maloney and Hartmann (2000a) composited a lifecycle of MJO phases based on the 20–80 day bandpass-filtered 850-hPa zonal wind averaged from 5°N to 5°S during May–November, 1979–1995. They found that

MJO-associated low-level westerly wind anomalies that are initiated from convective regions of the WNP propagate eastward as Kelvin waves into the ENP. Given the background easterly trade wind flows in the tropics, the encroachment of equatorial westerly wind anomalies promotes increased meridional shear of the zonal winds. As a result, low-level cyclonic shear forms on the northern side of the maximum winds, with maximum cyclonic vorticity occurring during the period with the strongest westerly wind anomalies (Fig. 4.25a). Through interactions with the boundary layer, low-level convergence and upward motions

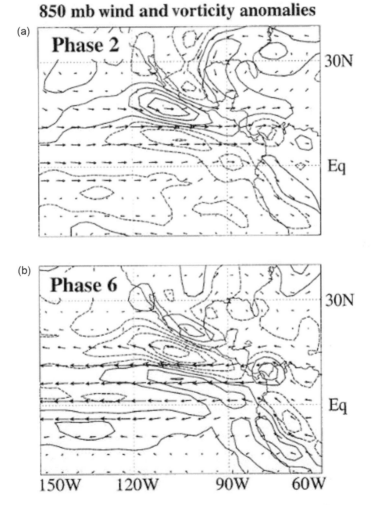

Fig. 4.25 Bandpass wind anomalies and relative vorticity anomalies at 850 hPa during May–November 1979–1995 for phases 2 and 6 (a and b). Maximum vectors are 3 m s^{-1}. Contour intervals are 1.2×10^{-6} s^{-1}. Negative contours are dashed. Source: Maloney and Hartmann, 2000a; ©American Meteorological Society, Used with permission.

occur in the ENP MDR with the arrival of low-level westerly anomalies. These changes in local environmental conditions are favorable for TC genesis and intensification over the ENP. On the other hand, during periods of strong equatorial easterly anomalies, anticyclonic vorticity anomalies prevail and ENP TCs are suppressed (Fig. 4.25b).

Maloney and Hartmann (2000a) found that more than twice as many named TCs occurred when equatorial low-level wind anomalies were westerlies compared with easterlies. Moreover, ENP hurricanes are four times more likely when the MJO is in westerly phases than in easterly phases. Maloney and Hartmann (2000a) suggested that during the MJO convective phase VWS is more important for tropical cyclogenesis than in the non-convective phase. However, based on a case study, Aiyyer and Molinari (2008) found that during the convective phase of MJO, low-level convergence and vorticity are stronger but VWS is also higher. This strong VWS was caused by anomalous low-level westerly winds and upper-level easterly winds. Thus, they found competing influences among convergence, vorticity, and VWS on cyclogenesis. In other words, variations of VWS in the ENP are not necessarily in favor of cyclogenesis during the MJO convective phase. For the ENP, Klotzbach (2014) found that TC activity was enhanced in phases 1, 7, and 8 when VWS was weaker (Table 4.2). This is also confirmed recently by Okun (2021). The ACE and number of 24-h RI periods are similarly increased during these phases relative to their average values. Reduced TC activity is noted in phases 3–5. In general, enhanced (suppressed) TC activity is associated with the passage of a convectively active (convectively inactive) phase of the MJO. Changes in the percentage chance of TCs undergoing RI during their lifetime during the eight phases of MJO were rather small, with the exception of phase 7.

Table 4.2. As in Table 4.1 but for the eastern North Pacific.

Phase	NS	H	MH	Basinwide ACE (%)	RI 24-h periods	RI chance (%)
1	**12.4**	**7.3**	2.9	15	17.5	37
2	9.0	4.8	2.9	11	14.0	35
3	7.7	4.0	*0.8*	7	*7.3*	37
4	7.3	3.7	2.0	8	13.0	41
5	*4.9*	*2.2*	*1.3*	5	*5.1*	27
6	9.3	4.8	2.4	12	11.4	34
7	**13.0**	**7.3**	**4.5**	**20**	**22.4**	**50**
8	**14.0**	**8.6**	**4.5**	**21**	**23.3**	41
Phase 1–8 avg.	9.6	5.2	2.6	12	13.9	38

Source: Klotzbach, 2014; ©American Meteorological Society. Used with permission.

4.3.3 Influence of ENSO and Pacific Meridional Mode on ENP Tropical Cyclone Activity

On interannual time scales, Whitney and Hobgood (1997) did not find an ENSO impact on overall ENP TC frequency. They found that the average annual number of TCs (named storms) is 15.1 during El Niño years and 15.0 for non-El Niño years, based on records from 1963 to 1993. Collins and Mason (2000) and Ralph and Gough (2009), however, pointed out the need to study the ENP by sub-regions because environmental parameters affecting TC activity are different east and west of 116°W. Jien et al. (2015) used a net tropical cyclone (NTC) activity index and power dissipation index (PDI) to study the impact of three different ENSO regimes (i.e., El Niño, La Niña, and neutral) on TC activity by sub-regions of the ENP. NTC represents an overall measure of the net tropical cyclone by considering named storms, hurricanes, and major hurricanes together. The PDI is the sum of the maximum one-minute wind speed cubed, at 6-h intervals, for periods when TCs are at least tropical storm strength. Their dividing line between the eastern and western portion of the ENP was at 112°W, and their MDR was bounded by 10°N to 20°N. Only the western main development region (WMDR) of the ENP experienced a significant difference ($p < 0.05$) in the NTC and PDI indices during El Niño versus La Niña years (Jien et al., 2015).

Changes in PDI can be further represented by contributions from storm intensity, duration, and frequency separately (Camargo and Sobel, 2005). Only the storm intensity of the PDI in the MDR was significantly correlated with ENSO signal (Jien et al., 2015). When the ENP was divided into two sub-regions, all three storm factors are highly sensitive to ENSO signal in the western sub-region. That is, El Niño years saw more frequent, longer storm duration, and more intense TCs in the WMDR relative to La Niña years, and this enhancement is probably attributed to by thermodynamic effects (Jien et al., 2015; Collins, 2010).

Although the WMDR TC response in the ENP is sensitive to ENSO, previous studies did not find a linkage between TC activity in the eastern main development region (hereafter called EMDR). However, Fu et al. (2017) showed reduced (enhanced) TC activity near the Pacific coast of Central America during El Niño (La Niña). They also found that fewer TCs formed near the coast of Central America but more were generated just to the south and west of this region within the EMDR during El Niño years. The main reason for this finding was that large area-integrated storm measures such as TC counts or PDI were used in earlier studies, while the study by Fu et al. (2017) investigated TC activity using a relatively high resolution 2° × 2° box. The area-integrated analyses used in earlier studies tended to cancel out EMDR changes in TC metrics.

Fig. 4.26 Topographical map depicting the mountain gaps in Central America. The arrows denote the direction of the flow through the mountain gaps.
Source: Chelton et al., 2000; Holbach and Bourassa, 2014; ©American Meteorological Society. Used with permission.

Tropical storms and hurricanes in the ENP are also triggered by tropical easterly waves from the North Atlantic (e.g., Rappaport et al., 1998) or gap winds from the Sierra Madre mountains (e.g., Romero-Centeno et al., 2003; Alexander et al., 2012; Holbach and Bourassa, 2014; Fu et al., 2017). Gap winds occur when northerly winds associated with midlatitude storms over the Gulf of Mexico are funneled intermittently through the mountains over the Gulf of Tehuantepec or easterly trade winds from the Caribbean Sea are channeled through the gap over the Gulf of Papagayo (Fig. 4.26). Holbach and Bourassa (2014) and Fu et al. (2017) suggested that strong low-level jets through topographically narrow gaps near the Isthmus of Tehuantepec and Gulf of Papagayo are particularly important in changing low-level relative vorticity. Climatologically, two low-level relative vorticity dipoles are found along the Pacific coast of Central America (Fu et al., 2017). That is, one pair of cyclonic-anticyclonic vorticity maxima occurs over the Gulf of Tehuantepec and another pair occurs over the Gulf of Papagayo.

There are three mechanisms by which the gap winds may generate surface vorticity in the ENP (Holbach and Bourassa, 2014). The first mechanism is by horizontal wind shear so that regions of cyclonic (anticyclonic) vorticity are found on the eastern (western) flank of the gap wind jet (Fig. 4.27a). The second mechanism is through the additional effect of curvature (Fig. 4.27b), and the third

(a)   Horizontal shear   (b) Horizontal shear + curvature   (c) Horizontal shear + monsoon trough

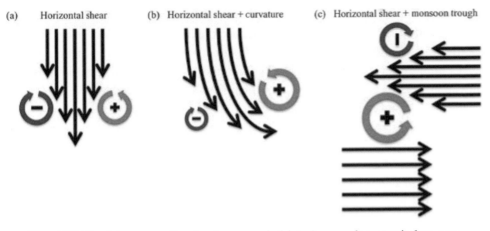

Fig. 4.27 Vorticity generation by the gap wind jet. Arrows denote wind vectors. An open circle with a plus (minus) sign denotes regions of positive (negative) vorticity for the NH.
Source: Holbach and Bourassa, 2014; ©American Meteorological Society. Used with permission.

is due to the presence of monsoon westerlies to the south of the jet (Fig. 4.27c). Also noteworthy is that more TCs tend to form when the monsoon trough is near the gap wind jet than when the ITCZ is located near 8°N . Because ENSO modulates the strength of these gap winds (Romero-Centeno et al., 2003), the enhanced gap winds result in stronger anticyclonic curvature along the jet path and thus contribute to anticyclonic vorticity anomalies along the Pacific coast of Central America north of 10°N during El Niño (Fu et al., 2017). In the meantime, mid-tropospheric specific humidity is also reduced in the same area. These two adverse environmental conditions suppress TC activity near the coast during El Niño, and the former appears to be the primary cause for EMDR TC variability. For La Niña years, the gap winds are reduced, leading to positive relative vorticity anomalies in the EMDR. Mid-tropospheric moisture increases concurrently. These two conditions favor TC genesis near the coast in the EMDR during La Niña.

Are there any changes in ENP TC activity in response to two different types of ENSO? During JASO of an EP event (Fig. 4.14a), with the exception of coastal regions, genesis frequency in the ENP between 90°W and 130°W increased, in response to favorable environmental conditions as the low-latitude ENP becomes anomalously warm (Collins and Mason, 2000; Camargo et al., 2007a; Kim et al., 2011a; Jien et al., 2015). The reduction in VWS over the MDR also favors enhancement of TC activity during EP events (Fig. 4.16a). During a CP event (Fig. 4.14b), TC genesis to the east of 130°W is reduced.

Changes in TC track density are also distinctly different between EP and CP events. During EP events, TC track density increased over the eastern and central Pacific (Fig. 4.15a). This increase appears in a very large domain extending from the Central American coast to the dateline and from southern California to the Hawaiian Islands. The increasing hurricane risks in the central North Pacific during an EP summer is consistent with Chu and Wang (1997), Clark and Chu (2002), and Klotzbach and Blake (2013). Decreased track density and genesis off of the Central American coast coincide with an anomalous anticyclone over the same area. In a CP hurricane season, a large area of suppressed TC track from the eastern to central Pacific just to the north of Hawaii is found (Fig. 4.15b), in conjunction with a band of increased vertical wind shear (Fig. 4.16b) and descending motion (Kim et al., 2011a). There is also a band of increasing track density to the south of Hawaii during the CP hurricane season, related to the westward shift of warm SST anomalies over the central Pacific. This increase in track density is also accompanied by a reduction in vertical wind shear and a westward shift of an anomalous cyclone near 150°W (Figs. 4.17b, 4.16b, and 4.15b). For La Niña events, termed Eastern Pacific cooling (EPC) by Kim et al. (2011a), TC genesis and track over the ENP are greatly reduced (Figs. 4.14c and 4.15c) due to anomalously cool water associated with a La Niña event and an increase in VWS over the ENP MDR (Fig. 4.16c).

The effect of ENSO warming on ENP TC activity is felt not only during the El Niño developing year but also during the following year. Invoking the discharge–recharge mechanism of El Niño (Section 3.1.5), Jin et al. (2014) noted that the discharge of the boreal winter El Niño anomalous ocean heat content takes about 6–9 months to reach the ENP MDR as a result of the time involved in ocean heat transport. This time lag corresponds to the the boreal summer and autumn of the second year following an El Niño event and are seasons when TCs in the ENP are climatologically active. Consequently, this additional heat supply from the equatorial subsurface ocean fuels TCs intensifying into major hurricanes in the year following El Niño.

By partitioning ENSO events further into EP and CP types (Section 3.1.6), Boucharel et al. (2016) highlighted the importance of ocean subsurface heat on intensifying hurricane activity in the ENP following EP events. Corresponding changes in vertical wind shear and relative humidity were actually detrimental to hurricane activity following an EP event. By comparison, following CP El Niño events, vertical wind shear and relative humidity became more favorable for hurricane activity. This finding highlights the vital role of the oceans on ENP hurricane activity for post EP events and the atmosphere on ENP hurricane activity for post CP events. Moreover, TCs tend to track farther west and north toward the Hawaiian Islands in the CNP following EP El Niño, while they are more likely to remain in the ENP following CP El Niños (Boucharel et al., 2016).

The 2015 ENP hurricane season was very active, with nine major hurricanes. Hurricane Patricia was the strongest hurricane in the ENP on record with maximum sustained winds estimated to reach 185 kts around October 23, 2015 (Avila, 2016). Patricia was also the most powerful hurricane in the Western Hemisphere on record, causing significant damage to Mexico ($460 million US). Because the hyperactive ENP 2015 hurricane season coincided with the very strong 2015 El Niño event, it was initially thought that this extreme TC season was triggered by tropical Pacific warming associated with El Niño development. However, Murakami et al. (2017) and Hong et al. (2018) suggested that subtropical warming associated with internal variability of the positive PMM in 2015 leads to more favorable environmental conditions for TC genesis over the ENP MDR, as opposed to directly from tropical Pacific warming.

Compared to the very strong 1997 El Niño, the mean genesis location of ENP TCs shifted westward by 20° longitude during the summer of 2015 , although the Niño-3.4 SSTA in 2015 were comparable to that of 1997 (Hong et al., 2018). During the summer of 2015 , in addition to equatorial warming, a strong PMM-like SST pattern emerged in the subtropical ENP. Results from numerical experiments suggested that the positive PMM-like SSTA pattern induced an east–west zonal circulation anomaly in the subtropical North Pacific with anomalous ascending (descending) motion in the subtropical central (western) Pacific (Hong et al., 2018). This anomalous vertical motion and combined with anomalously warm ocean temperatures drove a westward shift in mean TC genesis location. In addition to the subtropical warming, long-term anthropogenic warming was shown to boost the odds of an extremely active 2015 TC season in the ENP and CNP (Murakami et al., 2017).

4.3.4 Decadal Tropical Cyclone Variations

Hurricane activity over the ENP has undergone decadal variations and it is desirable to know when the change is likely to have occurred and possibly develop a means for predicting its future variation once a shift can be identified from a sound statistical analysis. Furthermore, many diagnostic studies are deeply rooted on the basis of comparing active and inactive (or positive and negative) phases of climate states when the knowledge of a regime shift in the system is given. To this end, Zhao and Chu (2006) used a Bayesian change-point analysis to identify shifts in annual major hurricane counts over the ENP from 1972 to 2003. Major hurricanes refer to category 3 or higher on the Saffir–Simpson Hurricane Wind which corresponds to maximum sustained wind speeds exceeding 50 m s^{-1}.

A triple hypothesis space concerning annual hurricane rate is considered: "a no change in the rate"; "a single change in the rate"; and "a double change in the rate." Zhao and Chu (2006) found that major hurricane activity over the ENP likely underwent decadal variations with two abrupt shifts occurring around 1982 and 1999. They classified their study period from 1972 to 2003 into three epochs: an inactive 1972–1981 epoch, an active 1982–1998 epoch, and an inactive 1999–2003 epoch (Fig. 4.28). The Bayesian model also provides a means for predicting decadal (or longer-term) hurricane variations. A lower number of major hurricanes was predicted for the next decade (2004–2013) given the last inactive epoch of hurricane activity. This decadal prediction, using data up to 2003, verified well with 2004–2013 averaging 3.0 major hurricanes per year, which is less than

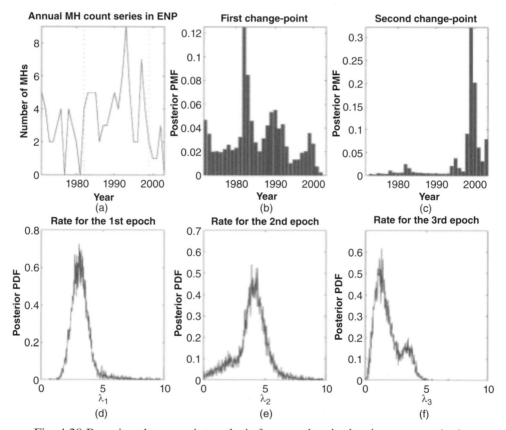

Fig. 4.28 Bayesian change-point analysis for annual major hurricane counts in the ENP. (a) Time series of annual major hurricane frequency over the ENP for 1972–2003; (b) The posterior probability mass function for the first change-point; (c) The posterior probability mass function for the second change-point; (d), (e), and (f) The posterior probability density function for the rate of each epoch.
Source: Zhao and Chu, 2006; ©American Meteorological Society. Used with permission.

the 1972–2003 average of 4.0 major hurricanes per year and much less than the active 1982–1998 epoch, which averaged 5.1 major hurricanes per year.

4.4 The Central North Pacific

4.4.1 Background Climatology

The central North Pacific (CNP) covers an area between the dateline and 140°W and north of the equator (Fig. 4.1). This domain coincides with the area of responsibility for the Central Pacific Hurricane Center, an entity of the US National Weather Service Forecast Office in Honolulu, Hawaii. Tropical storm and hurricane counts for the CNP include storms that formed within the domain of the CNP as well as storms that formed in the eastern North Pacific and subsequently propagated into the CNP. In this regard, two types of TCs are counted here (e.g., TCs that entered the CNP from the ENP and those that formed in situ in the CNP).

The time series of the annual number of TC in the CNP from 1966 to 2017 is shown in Fig. 4.29. The mean annual number of TCs is 2.9 and the standard deviation is 2.8, which is almost identical to the mean, signifying large variability in TC frequency from year to year. The TC records prior to 1966 are considered less reliable because satellite observations were not in sufficient quantities over this ocean region. One notable feature is a tendency for a relative maximum of CNP TC occurrences during EP El Niño years (e.g., 1972, 1982, 1997, and 2015) and a minimum of CNP TC occurrences during La Niña years (e.g., 1973, 1999). In Fig. 4.29, there is also an indication of decadal variations with fewer TCs during 1966–1981, more TCs during 1982–1994, and fewer TCs during 1995–2010. CNP TCs have increased again in the past 10 years. Moreover, a quasi-biennial oscillation is also evident in the TC time series, particularly in the late 1970s, 1990s and 2000s. Although this is a simple time series, it reflects a multitude of various climate forcing on TC activity.

For the CNP, the 2015 season was also the most active, with 14 TCs occurring, with 9 of them becoming hurricanes. Another astounding number is the frequency of major hurricanes in 2015. Climatologically, the mean annual number of major hurricanes in the CNP is 0.7 but 2015 saw five major hurricanes. As a result, the ACE in 2015 was seven times more than the average value with 124.6×10^4 kt^2 occurring in 2015 (the mean value is 17.8×10^4 kt^2). Particularly noteworthy in 2015 is that on August 30, three category four hurricanes (Ignacio, Jimena, and Kilo) were spinning in the CNP and the western portion of the ENP simultaneously. This marked the first time on record that three major hurricanes were spiraling concurrently in the central/eastern Pacific. This was also the first

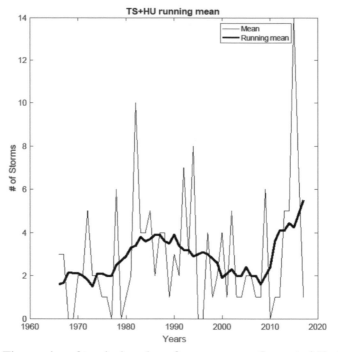

Fig. 4.29 Time series of tropical cyclone frequency over the central North Pacific for 1966–2017. The "movmean" in MatLab is used to plot the running mean series. A 10-yr running mean is computed over a sliding window that is centered about the current and previous elements. The window size is automatically truncated at the endpoints when there are not enough elements to fill the window. When the window is truncated, the average is taken over only the elements that fill the window.

time since 1966 that the Hawaiian Islands were sandwiched concurrently by two major hurricanes (Ignacio and Kilo) in the CNP.

In August 2018, Hurricane Lane, a category 5 storm, was located just a short distance from Oahu, Hawaii. Because of its intensity, torrential rainfall, and close proximity to the islands, hurricane warnings were issued by the NWS Forecast Office in Honolulu. The consistent warnings over several days caused a large degree of anxiety as Hawaii residents prepared food and water supplies for a long period of time due to the isolated geographic position of the Islands in the central Pacific (Nugent et al., 2020). Adding to the anxiety was the possibility of power and phone outages, the threat to property damage and lives, and the lack of enough shelters. Following Lane, tropical storm Olivia made a double landfall on Maui in September 2018, with heavy rainfall and flash flooding over Hawaii, Maui, Molokai, and Oahu. Olivia was the third direct landfall in the Hawaiian Islands in last five years.

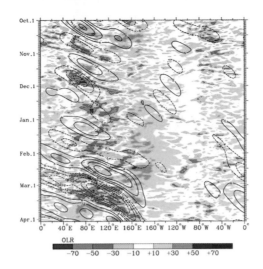

Fig. 2.1 Time–longitude diagram of OLR anomalies (shading) and MJO-filtered OLR (contour) averaged between 10°S and 10°N from October 1, 2011 to April 2, 2012. Negative anomalies are in blue at 20 W m^{-2} interval. The contour interval is 6 W m^{-2}. A black and white version of this figure will appear in some formats. Source: Kiladis et al., 2014; ©American Meteorological Society. Used with permission.

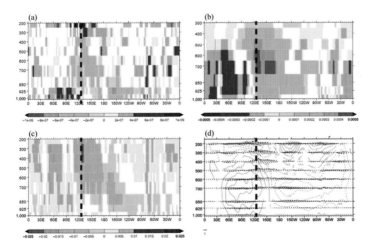

Fig. 2.3 Longitude–vertical cross sections of composite fields in reference to MJO center near 125°E (thick dashed line). (a) divergence (s^{-1}), (b) specific humidity (kg kg^{-1}), (c) vertical velocity (Pa s^{-1}), and (d) zonal wind and vertical velocity. In (a) convergence (divergence) is green or blue (yellow or red). In (b), positive (negative) humidity is green or blue (yellow or red), In (c), rising (sinking) motion in green or blue (yellow or red). Note that the vertical velocity (Pa s^{-1}) is multiplied by –100 to yield scaling compatible with the v (meridional) wind (m s^{-1}). A black and white version of this figure will appear in some formats. Source: Sperber, 2003; ©American Meteorological Society. Used with permission.

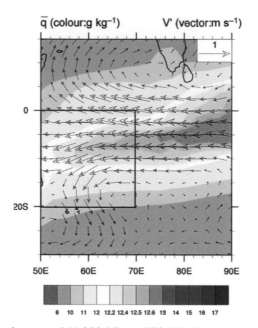

Fig. 2.5 Vertically integrated (1,000-hPa to 700-hPa) intraseasonal wind and low-frequency background state specific humidity averaged during the initiation period (day −25 to day −15). A black and white version of this figure will appear in some formats.
Source: Zhao et al., 2013; ©American Meteorological Society. Used with permission.

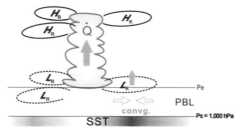

Fig. 2.6 Schematic diagram showing a Kelvin–Rossby wave couplet in response to the MJO convection with heating (Q). Also shown are the boundary-layer convergence (convg.) induced by atmospheric wave dynamics and SST. Dashed ellipses with L_R and L_K denote low-pressure anomalies (L) associated with Rossby and Kelvin waves response, respectively. Solid ellipses with H_R and H_K denote the high-pressure anomalies (H) associated with Rossby and Kelvin waves response in the upper troposphere. Red and blue shading denote positive and negative SST anomalies, respectively. Solid green arrows denote anomalous ascending motion. Ps and Pe denote pressures at the bottom and top of the atmospheric boundary layer, respectively. A black and white version of this figure will appear in some formats.
Source: Hsu and Li, 2012; ©American Meteorological Society. Used with permission.

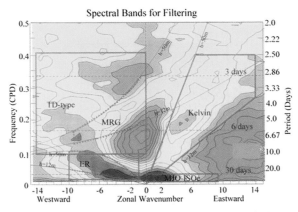

Fig. 2.17 Normalized wavenumber-frequency spectrum of OLR adapted from Wheeler and Kiladis, 1999. Thick black lines encompass the regime of filter bands. Thin dashed lines denote shallow-water dispersion curves in a dry, motionless atmosphere for the equivalent depths indicated by labels. A black and white version of this figure will appear in some formats.

Source: Frank and Roundy, 2006; ©American Meteorological Society. Used with permission.

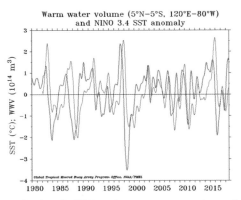

Fig. 3.8 Time series of Niño-3 SST and warm water volume from 1980 to 2017. A black and white version of this figure will appear in some formats.

Source: NOAA/PMEL.

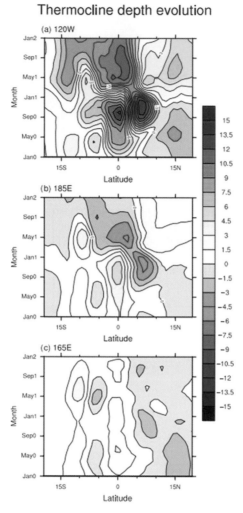

Fig. 3.10 Zonally averaged thermocline depth during El Niño events from year 0, year 1 (event peak), to year 2 for the Niño-3 region (5°N–5°S, 150°W–90°W) as indicated by 120W at top (a), the Niño-4 region (5°N–5°S, 160°E–150°W) as indicated by 185E in (b), and a region displaced 20°to the west of the Niño-4 region in the western Pacific in (c). The depth of the 15°C (m) is used as a proxy for thermocline depth. A black and white version of this figure will appear in some formats.
Source: Capotondi et al., 2015; ©American Meteorological Society. Used with permission.

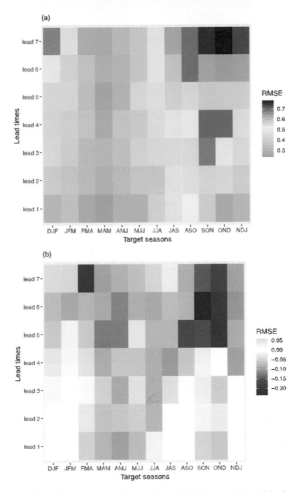

Fig. 3.13 (a) RMSE of statistical-dynamical BMA models, (b) the difference of RMSE between the statistical-dynamical models and statistical BMA models for various leads (months) and target seasons. A black and white version of this figure will appear in some formats.
Source: Zhang et al., 2019; ©www.creativecommons.org/licenses/by/4.0/.

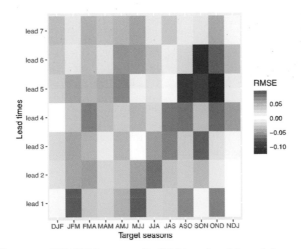

Fig. 3.14 Difference of RMSE between the BMA and multi model ensemble average (MMEA). A black and white version of this figure will appear in some formats.
Source: Zhang et al., 2019; ©www.creativecommons.org/licenses/by/4.0/. Used with permission from Springer Nature.

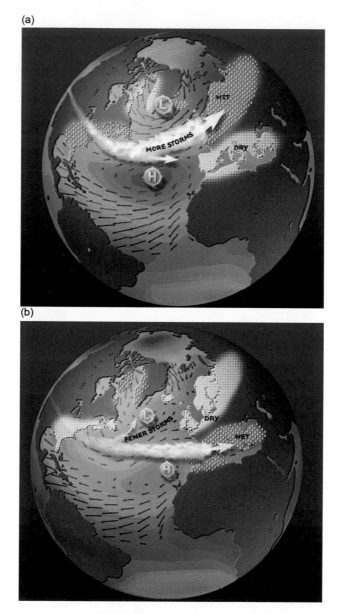

Fig. 3.17 Changes in surface pressure and weather patterns associated with the (a) positive phase of the North Atlantic Oscillation (NAO) and (b) negative phase of the NAO. A black and white version of this figure will appear in some formats. Source: www.ldeo.columbia.edu/res/pi/NAO/.

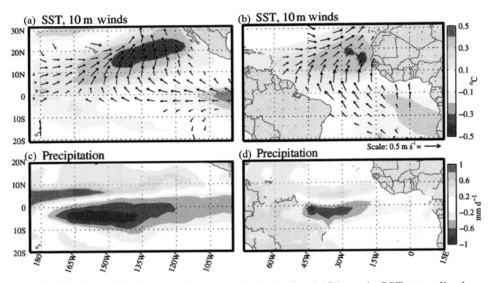

Fig. 3.18 (a) and (b) Regression maps of the leading MCA mode SST normalized expansion coefficient on SST and 10-m wind vectors in the Pacific (left) and Atlantic (right). Same in (c) and (d) but for precipitation (mm d^{-1}). A black and white version of this figure will appear in some formats.

Source: Chiang and Vimont, 2004; ©American Meteorological Society. Used with permission.

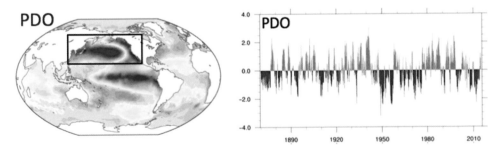

Fig. 3.20 Spatial pattern of the Pacific Decadal Oscillation (PDO) (left) derived from SSTs and time series of the PDO (right) from 1870 to 2014. A black and white version of this figure will appear in some formats.

Source: www.climatedataguide.ucar.edu/.

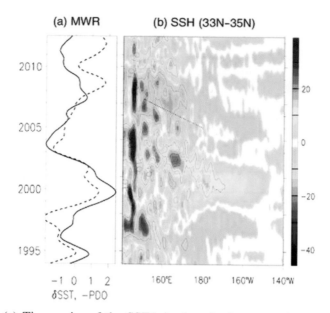

(a) MWR   (b) SSH (33N–35N)

Fig. 3.23 (a) Time series of the SSTA in the mixed-water region (MWR, solid) and the PDO index (with sign reversed, dashed). The MWR spans from the coast of Japan to 150°E, and 36°N–42°N. Both indices are normalized by their respective standard deviations. (b) Satellite-derived sea surface height anomalies (cm), averaged over 33°N–35°N. The dotted line marks a slow westward phase speed of the ocean process. A black and white version of this figure will appear in some formats.
Source: Newman et al., 2016; ©American Meteorological Society. Used with permission.

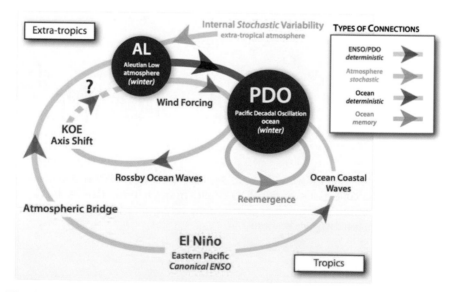

Fig. 3.24 Schematic showing processes involved in the PDO. A black and white version of this figure will appear in some formats.
Source: Newman et al., 2016; © American Meteorological Society. Used with permission.

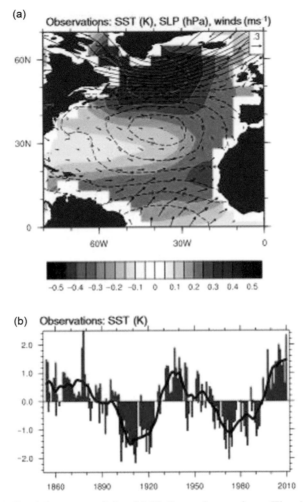

Fig. 3.25 (a) Spatial pattern of the AMO from observations. The AMO index is defined as the average SST for 0°–60°N and 80°W –0°, after detrending to isolate the natural variability. In (a), the AMO index is regressed to SST (shaded), SLP (contours), and surface winds (vectors). (b) Time series of annual mean anomalies of the standardized AMO index (colored bars) and the ten-year running average of the index (black line). A black and white version of this figure will appear in some formats.

Source: Clement et al., 2015; ©Science. Used with permission.

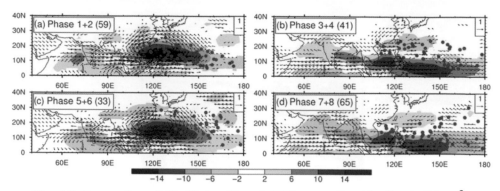

Fig. 4.9 Composites of 30–60-day filtered OLR anomalies (shading in W m^{-2}) and 850-hPa wind anomalies (vectors in m s^{-1}) during different MJO phases (a–d). Only anomalies exceeding the 5% significance level based on a Student's t test are shown. Solid circles denote the TC genesis positions, and the numbers in the parentheses denote the TC counts forming in each phase. A black and white version of this figure will appear in some formats.

Source: Li and Zhou, 2013a; ©American Meteorological Society. Used with permission.

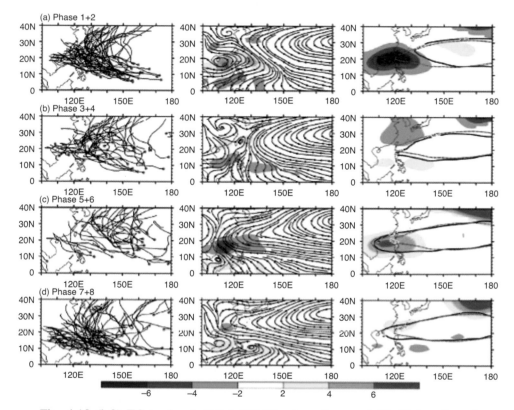

Fig. 4.10 (left) TC tracks; (middle) 850-hPa streamlines and 30–60 day filtered vorticity anomalies 10^{-6} s^{-1}; and (right) 500-hPa geopotential height anomalies (m) for different MJO phases (a–d). In the right-hand panels, dashed purple contours denote the 5,870 geopotential height level, and solid black contours denote the 5,870 geopotential height for different MJO phases (a–d). Only anomalies that exceed the 5% significance level based on a Student's t test are shown. A black and white version of this figure will appear in some formats.

Source: Li and Zhou, 2013b; ©American Meteorological Society. Used with permission.

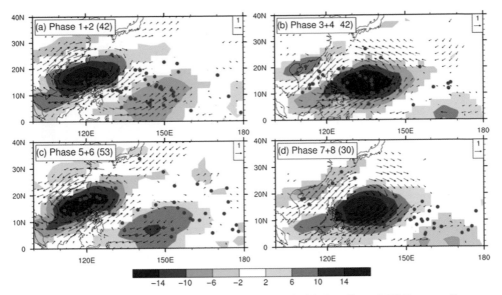

-14   -10   -6   -2   2   6   10   14

Fig. 4.11 As in Fig. 4.9, but for composites of 10–20-day filtered OLR anomalies (shading in W m⁻²) and 850-hPa wind anomalies (vector in m s⁻¹). A black and white version of this figure will appear in some formats.

Source: Li and Zhou, 2013a; ©American Meteorological Society. Used with permission.

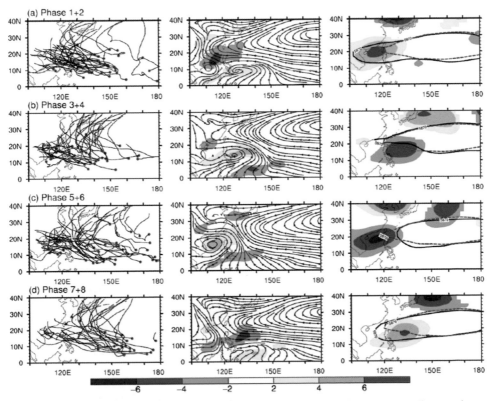

-6   -4   -2   2   4   6

Fig. 4.12 As in Fig. 4.10, but for (left) TC tracks, (middle) 850-hPa streamlines and 10–20-day filtered vorticity anomalies in color, and (right) 500-hPa geopotential height anomalies (m) for different QBWO phases. A black and white version of this figure will appear in some formats.

Source: Li and Zhou, 2013b; ©American Meteorological Society. Used with permission.

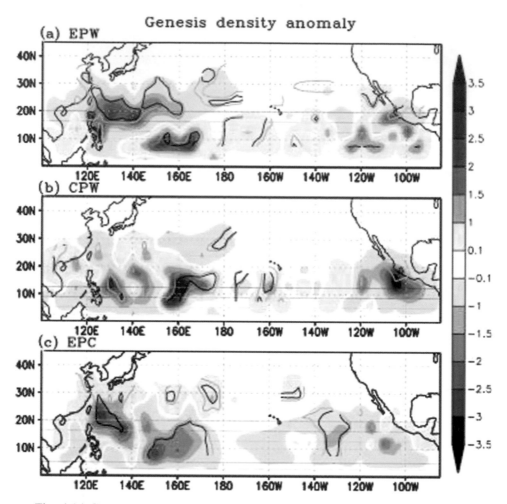

Fig. 4.14 Composite of TC genesis density anomalies (×10) in JASO over the Pacific for (a) EP El Niño, (b) CP El Niño, and (c) La Niña events. Light (dark) contours denote statistical significance at the 10% (5%) level. A black and white version of this figure will appear in some formats.

Source: Kim et al., 2011a; ©American Meteorological Society. Used with permission.

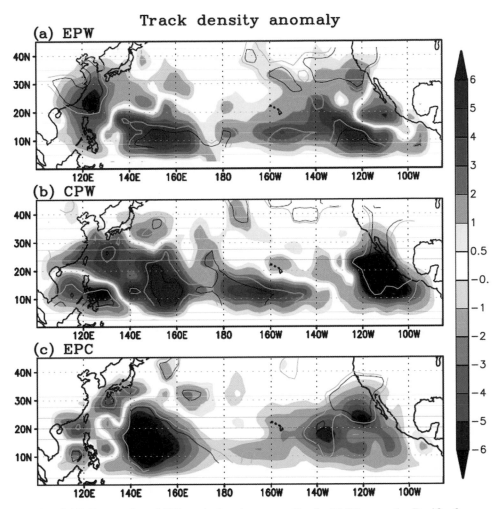

Fig. 4.15 Composite of TC track density anomalies in JASO over the Pacific for (a) EP El Niño, (b) CP El Niño, and (c) La Niña events. Light (dark) contours denote statistical significance at the 10% (5%) level. A black and white version of this figure will appear in some formats.

Source: Kim et al., 2011a; ©American Meteorological Society. Used with permission.

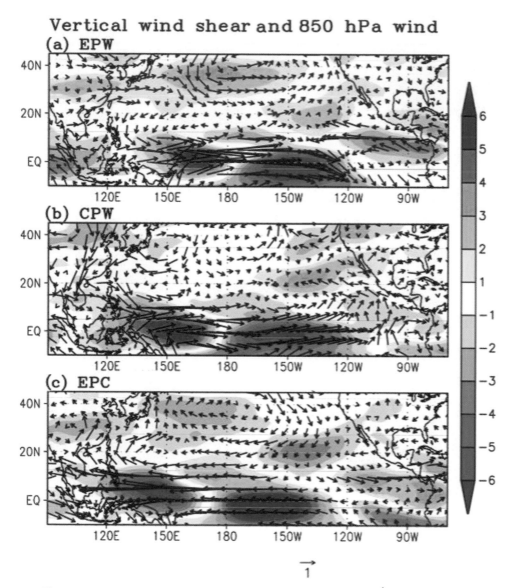

Fig. 4.16 Composite of vertical wind shear anomalies (m s^{-1} in shading) and 850-hPa wind vector (m s^{-1}) in JASO over the Pacific for (a) EP El Niño, (b) CP El Niño, and (c) La Niña events. The vertical wind shear is defined as the difference in the zonal wind component between 200 and 850 hPa. A black and white version of this figure will appear in some formats.

Source: Kim et al., 2011a; ©American Meteorological Society. Used with permission.

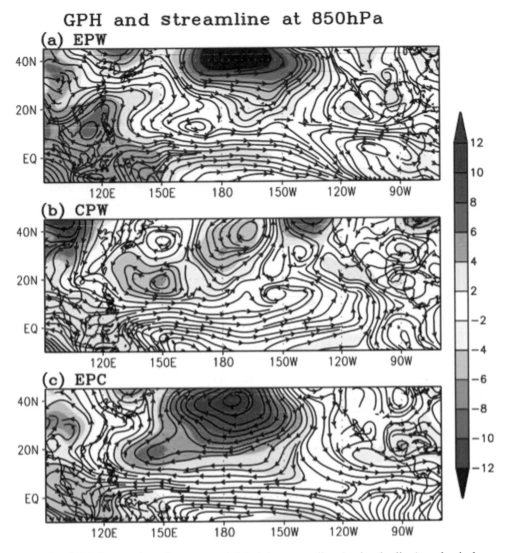

Fig. 4.17 Composite of geopotential height anomalies (m in shading) and wind anomalies at 850 hPa in JASO over the Pacific for (a) EP El Niño, (b) CP El Niño, and (c) La Niña events. A black and white version of this figure will appear in some formats.

Source: Kim et al., 2011a; ©American Meteorological Society. Used with permission.

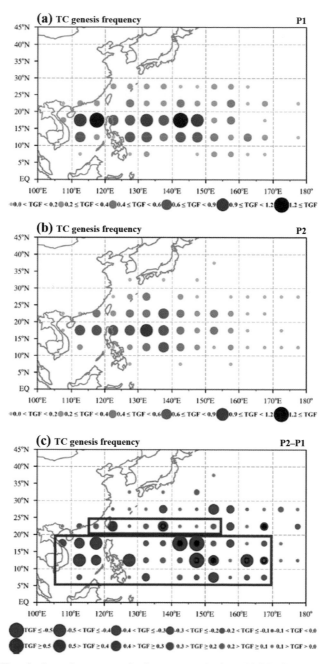

Fig. 4.18 Tropical cyclone genesis frequency during JASO for (a) 1989–1997 (P1), (b) 1998–2012 (P2), and (c) their difference (P2 – P1). The rectangles in (c) denote regions with pronounced changes in genesis frequency. The large (small) rectangle at the bottom (top) indicates a decrease (increase). A black and white version of this figure will appear in some formats.

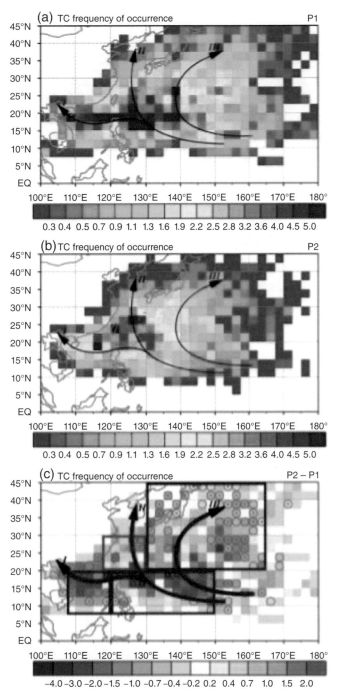

Fig. 4.19 Tropical cyclone occurrence frequency during JASO for (a) 1989–1997 (P1), (b) 1998–2012 (P2), and (c) their difference (P2 − P1). The thick black arrows in (a) and (b) denote prevailing TC tracks. In (c), blue (red) arrows denote less frequent (more frequent) tracks from P1 to P2. The rectangles in (c) indicate regions with pronounced changes in occurrence frequency. A black and white version of this figure will appear in some formats.

Source: He et al., 2015; ©www.creativecommons.org/licenses/by/4.0/. Used with permission from Springer Nature.

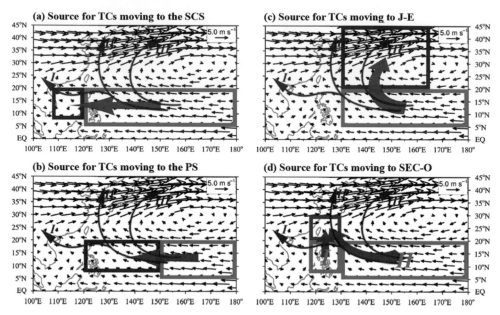

Fig. 4.20 Major genesis sources (green box) for TCs moving into (a) the South China Sea, (b) the Philippine Sea, (c) Japan and east of Japan, and (d) Taiwan and the East China Sea. Vectors denote the climatological mean of the steering flow. The purple rectangles in (a), (b), and (c) denote regions with significant decrease in TC occurrence frequency, while the orange box in (d) indicates a significant increase in TC occurrence frequency. The brown solid curves in each panel denote the three prevailing background TC tracks. Thick arrows denote the TC movement direction from each source region, with blue (red) denoting a decreasing (increasing) tendency. A black and white version of this figure will appear in some formats.

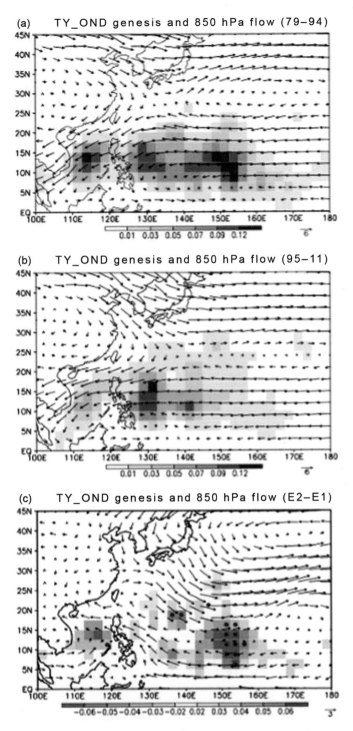

Fig. 4.22 Tropical cyclone genesis frequency (shading, number per season) and 850-hPa wind fields (vectors in m s^{-1}) during OND for (a) 1979–1994, (b) 1995–2011, and (c) their difference (1995–2011 minus 1979–1994). Red dots in (c) denote differences that are significant at the 5% level. A black and white version of this figure will appear in some formats.

Source: Hsu et al., 2014; ©American Meteorological Society. Used with permission.

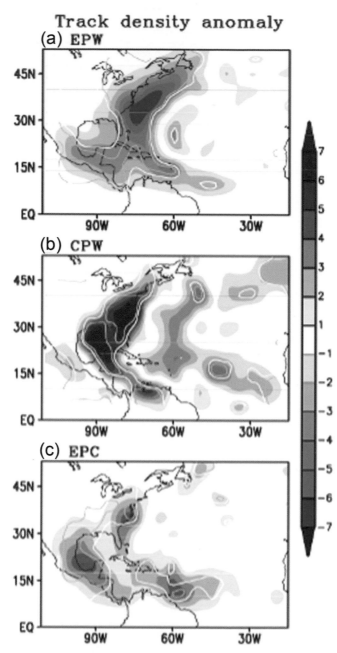

Fig. 4.44 Composites of TC track density anomalies during August to October for (A) canonical El Niño (EPW), (B) central Pacific El Niño (CPW), and (C) La Niña (EPC). Light (dark) contours show statistical significance at the 90% (95%) level. Units: number per $5° \times 5°$ grid box per year multiplied by 0.1. A black and white version of this figure will appear in some formats.

Source: Kim et al., 2009; ©Science. Used with permission.

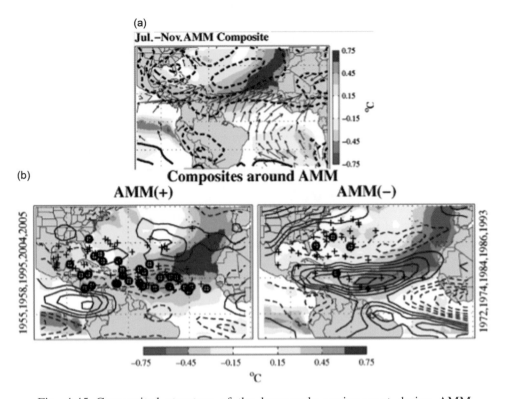

Fig. 4.45 Composited structure of the large-scale environment during AMM positive and negative events: (a) SST (shaded), 925-hPa height (contour, 2 m), and 925-hPa wind (vectors) anomalies around the 5-yr with the largest positive AMM index values minus the anomalies around the 5-yr with the largest negative AMM index values. For 925-hPa heights, solid contours denote positive anomalies, dashed contours denote negative anomalies. (b) Tropical cyclogenesis points for (left) the five strongest and (right) five weakest AMM years. Superimposed is the composites of SST (shaded) and vertical wind shear (contours) anomalies. Crosses show the genesis points for all storms that reached tropical storm strength. Storms that reached major hurricane strength (≥ 49 m s^{-1}) also have a circle around their genesis point. Solid (dashed) shear contours denote positive (negative) values. The contour interval is 0.25 m s^{-1}, and the zero contour has been omitted. Shear was calculated every 6 h as the amplitude of the vector difference between the layer mean winds in the 300–150-hPA and 925–700-hPa layers, and means were formed around the hurricane season using monthly means. Number on the sides are the years during the AMM years. A black and white version of this figure will appear in some formats.

Source: Kossin and Vimont, 2007; ©American Meteorological Society. Used with permission.

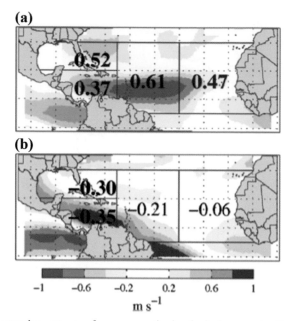

Fig. 4.46 Regression maps of mean vertical wind shear onto the standardized (a) AMM index and (b) Nino-3.4 index. Units: m s^{-1} per standard deviation of the respective time series. The listed number is the correlation between the number of storm days within each region and the respective index. Statistically significant correlations are listed in boldface. A black and white version of this figure will appear in some formats.

Source: Kossin and Vimont, 2007; ©American Meteorological Society. Used with permission.

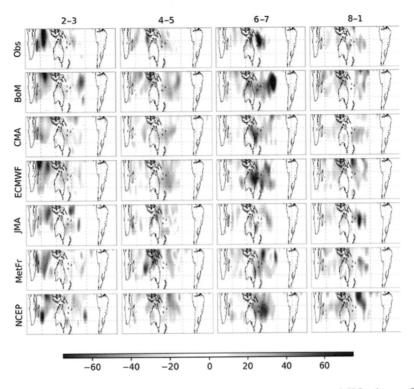

Fig. 5.2 Ensemble mean TC genesis anomalies (%) at every two MJO phases for 2–3, 4–5, 6–7, and 8–1 in the Southern Hemisphere from the observations (top row) and the week 2 reforecasts from six S2S models. A black and white version of this figure will appear in some formats.

Source: Lee et al., 2018; ©American Meteorological Society. Used with permission.

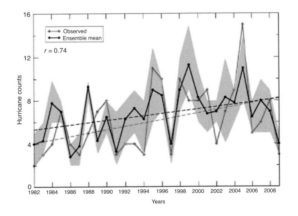

Fig. 5.6 Time series of observed (red solid line) and the ensemble mean (black solid line) of hurricane counts over the North Atlantic from 1982–2009. The shading region denotes the ensemble spread using the bias correction. The dashed lines are linear trends. A black and white version of this figure will appear in some formats.

Source: LaRow, 2013; ©American Meteorological Society. Used with permission.

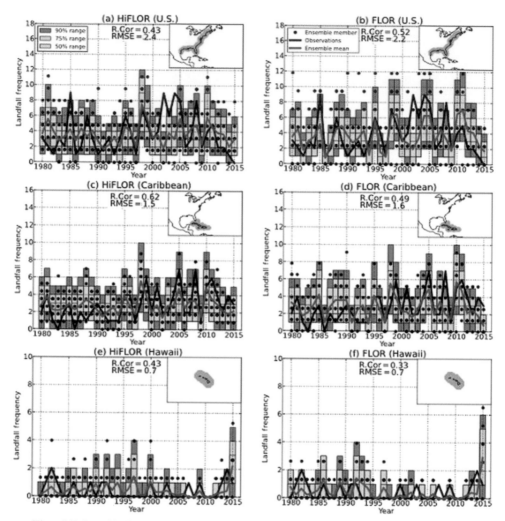

Fig. 5.7 Landfalling TC frequency during July–November 1980–2015 for the retrospective forecasts initialized in July using HiFLOR (left) and FLOR (right) for (a), (b) United States, (c), (d) Caribbean islands, and (e), (f) Hawaiian Islands. The black lines denote the observed statistics, the green lines are the mean forecast, and shading indicates the confidence intervals computed by convolving interensemble spread based on the Poisson distribution. The black dots denote the forecast value from each ensemble member. R.Cor and RMSE indicate the rank correlation coefficient and root-mean-square error between the black and green lines. A black and white version of this figure will appear in some formats.

Source: Murakami et al., 2016a; © American Meteorological Society. Used with permission.

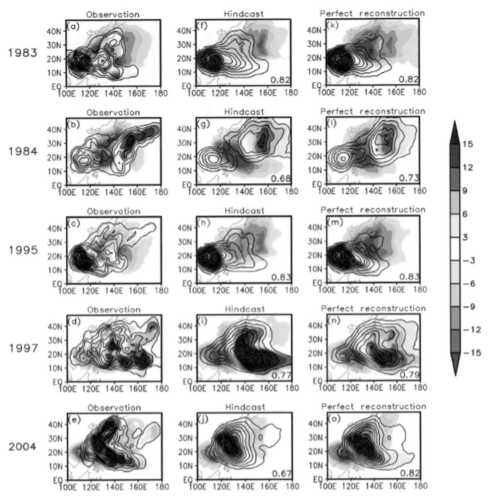

Fig. 5.10 (a–o) Final forecasting map of the TC track density for five years with the most skillful hindcasts: (left column) observation, (middle column) maps constructed from the hindcasts of the hybrid model, and (right column) maps constructed using the observed TC frequency in the seven patterns. The seasonal means of the probability of the TC tracks are in contours and their anomalies in shading. The contour interval is 5% and the zero line is omitted. The pattern correlation coefficient with the observation is shown in the bottom-right corner. A black and white version of this figure will appear in some formats.

Source: Kim et al., 2012; ©American Meteorological Society. Used with permission.

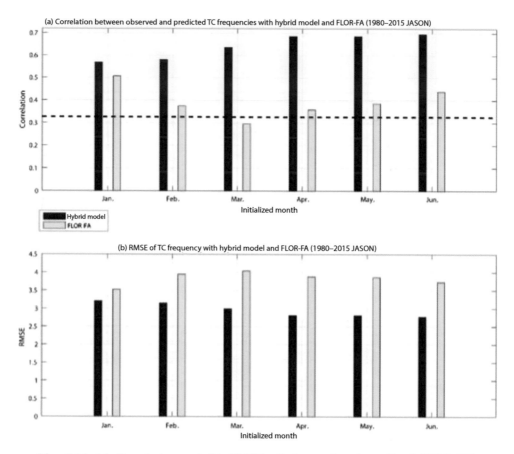

Fig. 5.11 (a) Correlation and (b) RMSE of observed and predicted WNP TC frequencies using the hybrid model (blue) and FLOR-FA (yellow) for initialization months from January to June after LOOCV. The black dashed line in (a) denotes the value for which the correlation coefficients are significantly different from zero at the 5% level. A black and white version of this figure will appear in some formats.

Source: Zhang et al., 2017; ©American Meteorological Society. Used with permission.

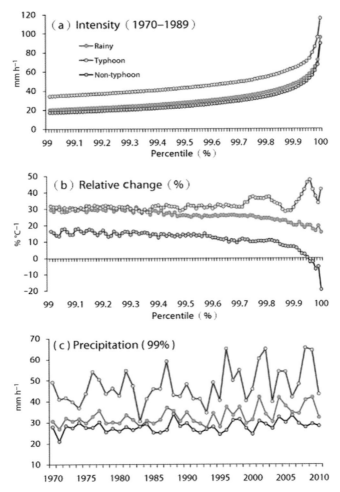

Fig. 6.4 (a) Climatological rainfall intensity for extremes (the 99th percentile) during 1970–1989 in Taiwan with a bin width of 0.01%. The unit for rainfall intensity is mm h⁻¹. (b) The relative changes in the intensity of rainfall extremes between 1990–2009 and 1970–1989, normalized by the global-mean surface temperature difference. (c) Time series of rainfall intensity averaged over the entire 99th percentile in 1970–2010. Green, red, and blue curves denote total, typhoon, and non-typhoon rainfall, respectively. A black and white version of this figure will appear in some formats.

Source: Tu and Chou, 2013; © CC By license.

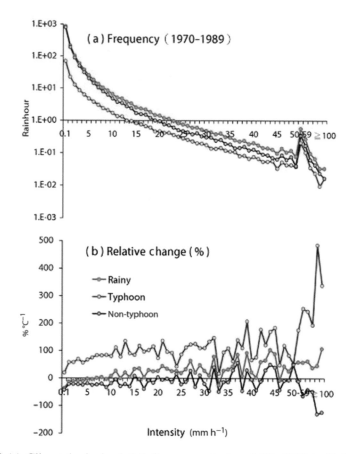

Fig. 6.5 (a) Climatological rainfall frequency during 1970–1989 in Taiwan. The bin width of rainfall intensity along the horizontal axis is 1 and 10 mm h^{-1} for intensity smaller and greater than 50 mm h^{-1}, respectively. The unit for frequency along the vertical axis is per station per year. (b) The relative changes in rainfall frequency between 1990–2009 and 1970–1989, normalized by the global-mean surface temperature difference between these two epochs. For both (a) and (b), green, red, and blue curves denote total, typhoon, and non-typhoon rainfall, respectively. A black and white version of this figure will appear in some formats. Source: Tu and Chou, 2013; © CC BY license.

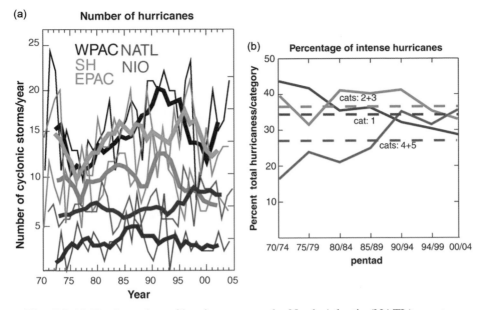

Fig. 7.3 (a) Total number of hurricanes over the North Atlantic (NATL), western North Pacific (WPAC), eastern North Pacific (EPAC), North Indian Ocean (NIO), and Southern Hemisphere (South Indian Ocean plus South Pacific) for the period 1970–2004. (b) The percentage of the total number of hurricanes in each category class at the global scale in 5-yr periods. The dashed lines show the 1970–2004 average numbers in each category. A black and white version of this figure will appear in some formats.

Source: Adapted from Webster et al., 2005; ©Science. Used with permission.

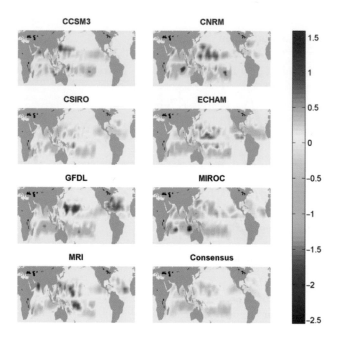

Fig. 7.4 Projected future change in storm genesis density by a downscaling model by Emanuel et al. (2008) using seven different large-scale parameters from different climate models. The ensemble mean is shown in the bottom right panel. Units: TC number per 5° latitude-longitude square per year. A black and white version of this figure will appear in some formats.
Source: Adapted from Emanuel et al., 2008; ©American Meteorological Society. Used with permission.

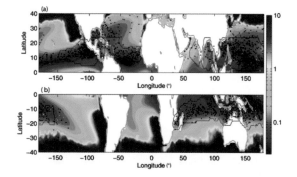

Fig. 7.7 (a) July–October mean ventilation index for the Northern Hemisphere and (b) December–March ventilation index for the Southern Hemisphere averaged over 1990–2009 using interim European Center for Medium-Range Weather Forecasts (ECMWF) Re-Analysis (ERA-Interim) data (Dee et al., 2011). Black dots are tropical cyclogenesis points over the same period. The black outline demarcates the main genesis regions, which are considered to be equatorward of 25°. A black and white version of this figure will appear in some formats.
Source: Adapted from Tang and Emanuel, 2012; ©American Meteorological Society. Used with permission.

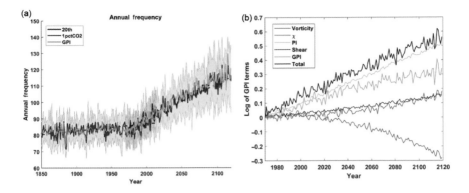

Fig. 7.8 (a) Annual global number of simulated tropical cyclones. Solid curves represent multi-model means and shading indicates one standard deviation up and down. Dashed lines show linear regression trends. Blue indicates the historical period 1850–2014 while red shows the 1% yr^{-1} CO$_2$ increase experiment beginning in 1970. Green curves show the multi-model mean, globally summed genesis potential index (GPI). (b) Each term contribution to the total GPI in terms of the right side of Eq. 7.9. The contributions are from vorticity, χ, potential intensity (PI), and vertical wind shear. The black curve shows their sum. A black and white version of this figure will appear in some formats.

Source: Adapted from Emanuel, 2021; ©American Meteorological Society. Used with permission.

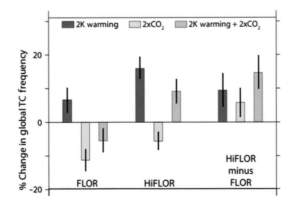

Fig. 7.9 Response of global tropical cyclone numbers in the idealized experiments: leftmost bars are for the FLOR model; the second set of bars for the HiFLOR model; and rightmost bars show the difference between HiFLOR and FLOR. The blue bars show the response to a combined uniform 2K warming and a CO$_2$ doubling, the gray bars show the response to CO$_2$ doubling with fixed SST, and the red bars show the response to uniform 2K warming. Black lines show the 95% confidence interval on the change. A black and white version of this figure will appear in some formats.

Source: Adapted from Vecchi et al., 2019; ©creativecommons.org/licenses/by/4.0/.

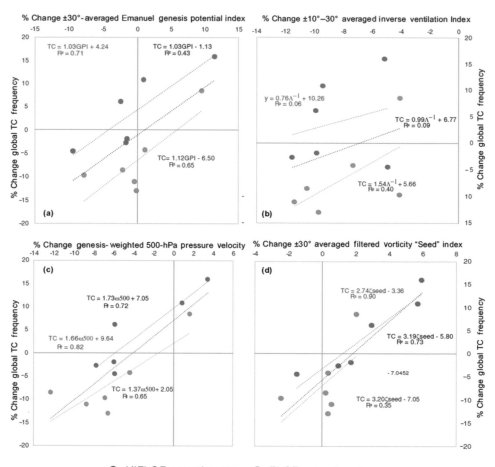

% Change ±30°-averaged Emanuel genesis potential index

% Change ±10°–30° averaged inverse ventilation Index

% Change genesis-weighted 500-hPa pressure velocity

% Change ±30° averaged filtered vorticity "Seed" index

● HiFLOR experiments   ● FLOR experiments

Fit across HiFLOR exps.   Fit across FLOR exps.   Fit across all exps.

Fig. 7.10 Fractional response in global number of tropical cyclones versus fractional response in spatially averaged large-scale parameters. Orange symbols show the response of HiFLOR and gray symbols denote those for FLOR. The dotted lines indicate the linear least-squares regression fit with the covariance indicated by R^2. Orange lines show regression for HiFLOR points, gray for FLOR points, and blue for all data combined. Each symbol is the response of one idealized experiment relative to the present-day control experiment (e.g., doubling CO_2 experiments, +2K uniform SST experiments, and the combined experiments). Fractional response of simulated global tropical cyclone numbers is compared with (a) tropical-mean, Emanuel (2013) GPI (Eq. 7.8) response , (b) ± 10–30° averaged inverse Tang and Emanuel ventilation index (Eq. 7.6 with minus sign), (c) 500 hPa pressure velocity, and (d) tropical cyclones "seed" index (Li et al. 2010). A black and white version of this figure will appear in some formats.

Source: Adapted from Vecchi et al., 2019; ©creativecommons.org/licenses/by/4.0/.

4.4.2 Modulation of CNP Tropical Cyclone Activity by Intraseasonal Oscillations

By extending the CNP eastward to 120°W (i.e., 120°W-dateline), Klotzbach and Blake (2013) studied the impact of ENSO and the MJO on CNP TC activity. They noted that TCs were more likely to occur in the CNP during an El Niño hurricane season (JJASO) rather than in a La Niña hurricane season. This difference was mainly attributed to anomalous changes in several environmental conditions such as warmer SSTs, lower SLP, increased mid-tropospheric moisture content, and anomalous ascending motion in El Niño years.

Klotzbach and Blake (2013) also divided the MJO cycle into eight phases using the Wheeler and Hendon (2004) index. The MJO also altered the large-scale circulation by favoring more TC activity during convectively enhanced phases of the MJO (phases 7 and 8) when anomalous ascending motion occurred over the CNP and ENP. Statistically significant reductions in zonal vertical wind shear and increases in 700-hPa relative humidity were seen during these two phases, relative to phases 4 and 5 when MJO convection is located over the Maritime Continent. A combined MJO–ENSO index showed even stronger relationships with TC activity over the CNP during phase 7 than either ENSO or the MJO individually.

4.4.3 ENSO and CNP Tropical Cyclone Activity

Based on data from 1966 to 1997, Clark and Chu (2002) showed that the average number of TCs in the CNP during El Niño hurricane season (June–November) was 5.5, compared to 1.8 during La Niña seasons. Even when the focus was restricted to a smaller region near Hawaii, the difference in the annual mean number of TCs between El Niño and non-El Niño years was still statistically significant at the 5% level based on a two-sample permutation test (Chu and Wang, 1997).

To illustrate the difference in large-scale environmental conditions conducive to TC development between extreme climatic events, Fig. 4.30 shows the low-level vorticity field in autumn (September–November) for El Niño (a) and La Niña (b) composites. The band of cyclonic (positive) relative vorticity in the El Niño composite is two to three times greater in the region from 160°E to 150°W to the south of Hawaii relative to the La Niña composite. This increase in cyclonic vorticity is mainly attributed to the eastward extension of the monsoon trough during El Niño years (Fig. 4.13). In addition, during El Niño, strong equatorial westerlies often extend from the western Pacific to the central Pacific. The presence of equatorial westerlies and easterly trade winds in the subtropics perturbs the climatological vorticity field, creating strong cyclonic shear and cyclonic relative vorticity in the MDR. This increase in low-level cyclonic vorticity, when coupled

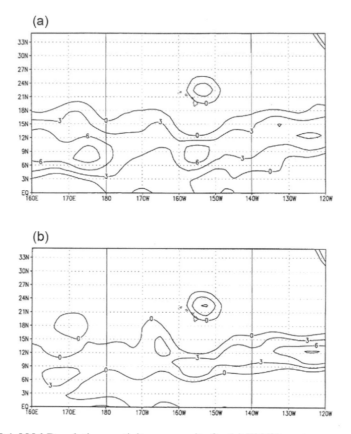

Fig. 4.30 1,000 hPa relative vorticity composite for (a) El Niño autumns and (b) La Niña autumn. Units are 10^{-6} s^{-1} and the contour interval is 3. Only positive values are contoured. The domain of the central North Pacific is delineated by thin solid lines.
Source: Clark and Chu, 2002, *Journal of the Meteorological Society of Japan.*

with other favorable environmental conditions such as a decrease in vertical wind shear (Fig. 4.31) and a possible increase in low-level moisture due to boundary layer moisture convergence in the TC spin-up process, are the likely drivers for more TC formation in the CNP. This result is consistent with Camargo et al. (2007a) who divided the genesis potential index (GPI) into four components (vorticity, VWS, potential intensity, and midlevel relative humidity) and found that low-level vorticity anomalies contributed most significantly to the total GPI change during El Niño years in the CNP, followed by vertical wind shear anomalies.

The westward shift in TC genesis location in the ENP during El Niño years (Irwin and Davis, 1999; Fu et al., 2017) also tends to propagate TCs farther west into the central Pacific. In addition, the decrease in vertical shear over the CNP MDR in El Niño years (Fig. 4.31) reduces the climatologically unfavorable

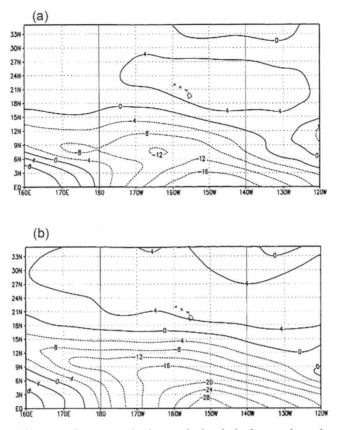

Fig. 4.31 Difference in tropospheric vertical wind shear when the La Niña composite is subtracted from the El Niño composite for (a) summer and (b) autumn. Units are m s^{-1} and the contour interval is 4. Positive and negative contours are given by solid and dashed lines, respectively.
Source: Clark and Chu, 2002, *Journal of the Meteorological Society of Japan.*

conditions for TCs and makes it more likely for TCs to propagate into the CNP. Longer TC lifetime and stronger intensity are found for EP and CP years relative to neutral and La Nina years (Okun, 2021).

4.4.4 Decadal Tropical Cyclone Variations

Elsner et al. (2000) developed a model for detecting change-points in Atlantic hurricane activity. Chu (2002) used a similar log-linear regression method to model shifts in annual TC frequency over the CNP for the period 1966–2000 and found two significant change-points at the $\alpha = 0.05$ level. The first change-point occurred in 1982 and the second shift occurred in 1995. As a result, the entire 35-yr record was partitioned into three epochs, 1966–1981 (inactive), 1982–1994

Table 4.3. Mean and variance of the observed, annual number of TCs in the CNP for the two epochs: 1966–1981 and 1982–1994. The 95% confidence intervals for the mean and variance are obtained by a bootstrap resampling method.

Epoch	Mean number of the annual number of cyclones from observations	The 95% confidence interval for the mean of the annual number of cyclones from bootstrap	Variance of the annual number of cyclones from observations	The 95% confidence interval from the variance of the annual number of cyclones from bootstrap
1966–1981	1.88	(1.06, 2.75)	3.05	(0.92, 5.18)
1982–1994	4.13	(3.15, 5.69)	6.06	(1.09, 10.81)

Source: Chu, 2002; © American Meteorological Society. Used with permission.

(active), and 1995–2000 (inactive). The mean number of annual TC counts for the early epoch (1966–1981) was 1.9 and its true value lies between 1.06 and 2.75 with 95% confidence as determined by the bootstrap technique (Table 4.3). The mean value in the second epoch (1982–1994) was 4.1. The last epoch (1995–2000) was of relatively short duration and was thus omitted. Independently, Chu and Zhao (2004) developed a Bayesian paradigm to show a high posterior probability of an abrupt shift in TC activity in 1982.

Because of the limited sample size in each epoch, a bootstrap resampling technique was applied to draw inferences about the TC statistics for each epoch. A bootstrap method operates on a construction of a large number of artificial data batches of the same size as the original dataset using sampling with replacement from the original data (Zwiers, 1990; Chu and Wang, 1997; Bove et al., 1998). Relative to the first epoch, larger variability is seen in the second epoch (Table 4.3). The variance of the annual mean number of TCs doubles from 3.1 during the first epoch to 6.1 during the second epoch. In the following, we will contrast TC-related changes in large-scale environments over the North Pacific during the first two epochs.

Because July–September is the most active portion of the CNP TC season, large-scale circulation differences for this season between the aforementioned two epochs are analyzed. To show the differences between the two epochs, the mean circulation from the first epoch (1966–1981) is subtracted from the second epoch (1982–1994). To help demonstrate the point that large-scale circulation changes are not only unique during the peak season, difference maps in the preseason (May–June) are also presented. As in the ENP, the latitudinal band 10°N–20°N over the CNP is called the main development region (MDR).

Relative to the first epoch, a large area of statistically significant SST warming covers the tropical Pacific and extends northeastward to the subtropical eastern North Pacific in the second epoch (Fig. 4.32a). In addition, the area enclosed by the 26.5°C isotherm in the eastern/central MDR expands northward in the recent epoch. Sea surface temperatures greater than 26.5°C are generally considered to eclipse the threshold necessary for TC formation. Warmer SSTs in the second epoch allow TCs to maintain their strength and also induce low-level wind convergence. Warmer SSTs are also seen in the preseason in the second epoch, particularly in the ENP and eastern tropical CNP (Fig. 4.32b).

Many studies have shown that mid-tropospheric moisture is directly related to tropical cyclogenesis and that a drier atmosphere tends to suppress deep convection and hence TCs (e.g., Camargo et al., 2007a). Figure 4.33a displays

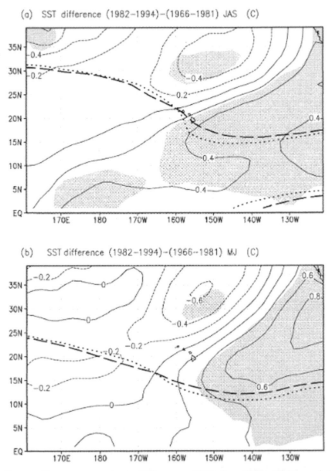

Fig. 4.32 Sea surface temperature difference (°C) for 1982–1994 and 1966–1981 over the North Pacific for (a) JAS and (b) MJ. Negative values are dashed. Contour interval is 0.2°C. The seasonal mean 26.5°C isotherm during 1982–1994 and 1966–1981 is denoted by a heavy broken and a dotted line, respectively. Shading indicates regions where the difference in the mean between the two epochs is statistically significant at the 5% level.
Source: Chu, 2002; ©American Meteorological Society. Used with permission.

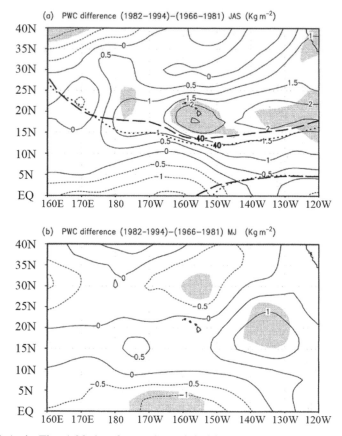

Fig. 4.33 As in Fig. 4.32, but for total precipitable water (TPW). Contour interval is 0.5 kg m^{-2}. The seasonal mean value of 40 kg m^{-2} during the 1982–1994 and 1966–1981 epochs is denoted by a heavy broken and dotted line, respectively. In May and June (b), the seasonal mean value of TPW is less than 40 kg m^{-2} and is not shown.
Source: Chu, 2002; ©American Meteorological Society. Used with permission.

tropospheric precipitable water (TPW) differences between the two epochs. An increase in TPW is found in a large band of the tropics in the second epoch, and a statistically significant increase with a maximum value as large as 2 kg m^{-2} is noted surrounding the Hawaiian Islands. There is also an indication of a maximum value near 20°N that protrudes from the ENP into the CNP. This increase in TPW is likely attributable to increased moisture fluxes from the ocean surface as the SST has become warmer in the second epoch (Fig. 4.33a). Less subsidence drying, possibly resulting from lower SLP and a weaker trade wind inversion (Knaff, 1997) could also account for this increase in TPW. The crucial moist TPW value, represented by the 40 kg m^{-2} isoline, also moved farther poleward from the first to the second epoch. The increase in TPW in the MDR was already established in the preseason and persisted and intensified in the peak season.

In the CNP MDR, persistent easterly trade winds at low levels combine with upper-level westerlies in the boreal summer and autumn to produce strong climatological VWS. Strong VWS disrupts organized deep convection (due to the ventilation effect), which inhibits intensification of incipient tropical disturbances. Climatologically, strong VWS and the absence of a monsoon trough in the MDR keep the CNP from being a TC-prone area. The difference in VWS between the two epochs is shown in Fig. 4.34a. The CNP MDR had a decrease in VWS in the second epoch. Of particular note is a steep meridional gradient in the VWS decrease with a local minimum as large as 4 m s^{-1} just to the south of Hawaii. This decrease is substantial when viewed from the context of the difference in multiyear

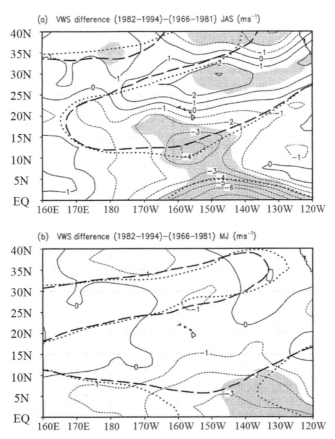

Fig. 4.34 As in Fig. 4.32 but for vertical wind shear. Contour interval is 1 m s^{-1}. The seasonal mean 10 m s^{-1} vertical shear line during the 1982–1994 and 1966–1981 epochs is denoted by a heavy broken and dotted line, respectively. Source: Chu, 2002; ©American Meteorological Society. Used with permission.

seasonal mean. The area with the sharpest decrease in the recent epoch is also statistically significant. The crucial vertical shear line, as represented by the 10 m s^{-1} isotach, has also shifted poleward in the second epoch, implying a larger area of weaker VWS in the MDR in the second epoch. Weaker VWS in the tropical and a portion of the subtropical North Pacific also characterized the preseason in the second epoch (Fig. 4.34b). Thus, weaker VWS in the MDR persisted and intensified from the preseason to the peak season.

Climatologically, a band of low-level cyclonic vorticity stretches from 160°E to 120°W over the southern half of the MDR during the peak season (not shown). Anticyclonic relative vorticity dominates an area poleward of 15°N, north of the band of cyclonic vorticity. In Fig. 4.35a, a large portion of the MDR is dominated

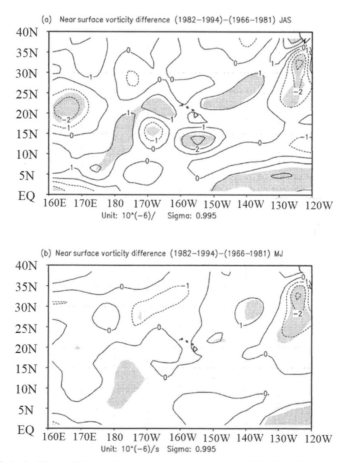

Fig. 4.35 As in Fig. 4.32, but for relative vorticity at 1,000 hPa. Contour interval is 1×10^{-6} s^{-1}.

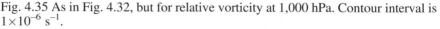

Source: Chu, 2002; ©American Meteorological Society. Used with permission.

by positive difference in relative vorticity in the second epoch, indicating either increased cyclonic vorticity south of 15°N or reduced anticyclonic vorticity north of 15°N. This positive difference is also significant in two smaller areas: one to the south of Hawaii and the other near the dateline. As discussed previously, an enhancement in low-level cyclonic vorticity is related to changes in near-surface wind fields. Presumably, this dynamical spin-up, when coupled with other favorable environmental conditions such as an anomalously SST, lower SLP, a deep moist layer, and weaker VWS are all instrumental for increasing TC genesis in the CNP MDR in the second epoch. The positive difference in relative vorticity is also seen in the preseason in the second epoch (Fig. 4.35b), although this difference is rather small. The increase in low-level cyclonic vorticity is consistent with warmer SST, lower SLP (not shown), the deeper moist layer, and a reduction in VWS in the MDR in the second epoch.

To investigate the impact of tropical versus subtropical warming on TC genesis over the CNP and ENP, Murakami et al. (2017) composited years with subtropical SSTA greater than 0.5 standard deviations but tropical SSTA that were less than +0.5 standard deviations as one group (SUB). Another group, TRP, consisted of tropical SSTA with more than +0.5 standard deviations but subtropical SSTA that were less than +0.5 standard deviations. For SUB, eight years were chosen and for TRP, nine years were used. They noted that all environmental conditions in the SUB composite, including SST, ocean thermal energy potential, mid-level relative humidity, VWS, and 850-hPa relative vorticity, were more favorable for cyclogenesis relative to the TRP composite over the CNP MDR. This result suggests the dominant role of subtropical warming associated with the positive PMM phase in increasing TC activity over the CNP.

4.5 The South Pacific

4.5.1 Background Climatology

The tropical South Pacific (SP) Ocean encompasses a vast region, extending from the Gulf of Carpentaria off of the northern Australia coast eastward through the dateline to the west coast of South America (~70°W) and from the equator to ~25°S (Fig. 4.1). Included in this area are eastern Australia, Papua New Guinea, the Solomon Islands, Vanuatu, Fiji, Tonga, Samoa, American Samoa, the Cook Islands, French Polynesia, and other small island nations and territories (Fig. 4.36). This region is vulnerable to TCs from year to year, with many small islands and atolls that are also subject to extreme sea levels (Church et al., 2006, 2010; Walsh et al., 2012; Chowdhury et al., 2014; Chowdhury and Chu, 2015). Based on data from 1970 to 2010, the region where TC frequency is highest in the South Pacific

Table 4.4. Annual tropical cyclone frequency in South Pacific islands by country based on data for 1970–2010.

Geographic grouping	TC frequency
Vanuatu	3.9
New Caledonia	3.4
Fiji	3.3
Tonga	2.7
Wallis and Futuna	2.4
Niue	2.2
Solomon Islands	2.2
Samoa	2.0
Tuvalu	1.8
Southern Cook Islands	1.7
Gulf of Carpentaria	1.5
Papua New Guinea	1.5
Tokelau	1.2
New Zealand	1.1
Austral Islands	1.0
Northern Cook Islands	1.0
Society Islands	0.9
French Polynesia	0.7
Tuamotu	0.4
Pitcairn Island	0.3
Western Kiribati	0.2
Marquesas	0.1
Eastern Kiribati	0.0

Source: Diamond et al., 2013; ©American Meteorological Society. Used with permission.

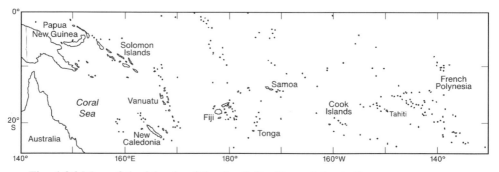

Fig. 4.36 Map of the islands of the South Pacific and Australia.

is to the west of dateline, between Vanuatu, New Caledonia, and Fiji (Diamond et al., 2013). This is reflected in Table 4.4, which lists annual TC frequencies by country. As the season progresses from the early season (November–January) to the late season (February–April), the maximum in TC activity shifts westward from Vanuatu to the Coral Sea (Fig. 4.36).

4.5.2 Modulation of South Pacific Tropical Cyclone Activity by Intraseasonal Oscillations

ER waves are the dominant equatorial waves in the SP, followed by TD-type disturbances (Fig. 4.7). The modulation of tropical cyclogenesis by MJO, MRG and Kelvin waves is roughly the same, less than half of that compared to ER waves if a threshold of 2 mm d^{-1} is applied. The MJO–TC modulation in the Australian region has been studied by Hall et al. (2001), in the Fiji region by Chand and Walsh (2010), and for the entire Southern Hemisphere by Camargo et al. (2009) and Klotzbach (2014). Hall et al. (2001) noted that significantly more TCs formed during the convectively active phase of the MJO, and this modulation was most pronounced to the northwest of Australia. Low-level relative vorticity anomalies appeared to contribute the most to the MJO–TC relationship that they observed. They also found increased modulation of TC activity by the MJO during El Niño event. Chand and Walsh (2010) found that dynamical parameters (low-level relative vorticity, VWS, and upper-level divergence) were primarily responsible for the enhanced TC genesis rate during the convective phase of the MJO in the Fiji region (5°S–25°S, 170°E–170°W). By examining the MJO modulation of tropical cyclogenesis using the Genesis Potential Index (GPI), however, Camargo et al. (2009) found that a thermodynamic parameter (mid-level relative humidity) was the single most important factor in the GPI, followed by vorticity. The contribution made by two other terms (potential intensity and VWS) was relatively small. This conclusion appears at odds with Hall et al. (2001) and Chand and Walsh (2010). The results from Camargo et al. (2009) were based upon two independent reanalysis products (NCEP–NCAR and ERA–40). Using a mixture Gaussian model, Ramsay et al. (2012) clustered TC tracks in the Southern Hemisphere into seven groups. Clusters 1–3 were associated with storms in the South Indian Ocean, clusters 4–6 with storms in the Australian region, and cluster 7 was storms in the South Pacific Ocean (165°E–150°W). Cluster 7 experienced significantly more TCs when the MJO was active over the Western Pacific and the Western Hemisphere, and was the cluster most strongly modulated by the MJO among all seven groups.

For the South Pacific (Table 4.5), enhanced TC activity was favored in phases 7 and 8, particularly for hurricane-strength TCs (Klotzbach, 2014). The ACE generated during these two phases was three times higher than that during phases 4 and 5 of the MJO. Nearly three times as many 24-h RI periods were observed in phase 8 compared to phases 3 and 4. Another notable difference during these two contrasting phases of the MJO is the eastward shift of TC formation location as the MJO progresses eastward. For example, a total of 22 hurricanes formed east of 175°W in phases 7 and 8, in contrast to only three hurricanes in phases 4 and 5.

Table 4.5. As in Table 4.1 but for the South Pacific.

Phase	NS	H	MH	Basinwide ACE (%)	RI 24-h periods	RI chance (%)
1	5.0	2.8	1.6	11	6.9	44
2	3.9	1.6	1.0	6	5.2	27
3	3.5	2.0	1.1	12	3.5	31
4	4.4	2.2	1.0	6	4.4	22
5	4.1	1.9	0.5	7	5.9	40
6	**8.5**	4.5	1.8	16	11.3	35
7	**10.3**	**5.2**	**3.1**	**22**	8.9	28
8	7.3	**4.8**	2.5	20	**13.4**	45
Phase 1–8 avg.	6.0	2.9	1.6	12	7.4	34

Source: Klotzbach, 2014; © American Meteorological Society. Used with permission.

4.5.3 ENSO and South Pacific Ocean Tropical Cyclone Activity

A number of studies have pointed out the relationship between TC activity in the South Pacific and ENSO. By correlating preseason (July–September) sea level pressure in Darwin, Australia and tropical cyclone days during the cyclone season (October–April) around the Australian region (105°E–165°E), Nicholls (1979) noted a strong negative correlation ($r = -0.68$), which is significant at the 5% level. Because Darwin is the western center of the Southern Oscillation, this strong and negative correlation implies a reduction in cyclone days preceded by anomalously high pressure at Darwin, which is typical of an El Niño event. Higher sea level pressures and anomalous sinking motion associated with the weakened Pacific Walker circulation are unfavorable for TC formation over northern Australia during El Niño years. Because of this strong lead–lag correlation, Nicholls (1985) subsequently suggested that seasonal TC activity in the Australian region was predictable using Darwin pressures in advance of the TC season.

The geographical distribution of TC activity over the southwest Pacific is also influenced by ENSO. For example, during El Niño years, reduced TC frequency in the Coral Sea and increased activity near the dateline has been noted (Hastings, 1990; Evans and Allan, 1992; Kuleshov et al., 2008). In this case, a decrease in TC activity in one region is compensated by an increase in another region. Tropical cyclone tracks in the southwest Pacific became more zonal during El Niño years but also occur closer to the coast of Queensland, Australia. These tracks also persisted farther southward, enhancing risks for coastal crossings during La Niña years (Evans and Allan, 1992).

Over the South Pacific, TC incidence tends to shift northeastward during El Niño years and southwestward during La Niña years (Revell and Goulter, 1986; Basher and Zheng, 1995; Chand and Walsh, 2009). This typical TC tendency is exemplified by the very strong 1982–1983 El Niño. For instance, the climatological median location for South Pacific TC genesis is 14°S, 170°E near Vanuatu. This point shifted northeastward to 11°S, 162°W during the 1982–1983 TC season (Chu, 2004). This is a remarkable eastward displacement of 3,000 km and an equatorward movement of 330 km from its climatological median position. Camargo et al. (2007a) also noted a pronounced eastward shift of genesis potential anomalies in the South Pacific during El Niño years compared to La Niña years. Another noteworthy feature is the eastward shift of the genesis location with time, consistent with the displacement of the monsoon trough, equatorial westerlies, and the South Pacific Convergence Zone (SPCZ). When TC tracks in the Southern Hemisphere are clustered into distinct groups, TC genesis events in some groups also show a pronounced shift associated with warm and cold events. For instance, TC genesis in cluster 7 (165°E–150°W in the South Pacific) exhibits equatorward and eastward (poleward and westward) shifts during El Niño (La Niña) years (Ramsay et al., 2012).

The monsoon trough occurs near large landmasses such as Asia and Australia due to land–sea monsoonal effects. The climatological position of the monsoon trough in February extends from near 12°S, 170°E westward through the Coral Sea. From December 1982 through May 1983, 11 TCs were named in the South Pacific, together with three unnamed tropical depressions which occurred in the far eastern South Pacific (Fig. 4.37). During these six months, six hurricane-strength TCs struck French Polynesia, a region that is not considered to be TC-prone (Sadler, 1983). Also noteworthy in El Niño years is greater activity during the late season, consistent with the findings of Diamond et al. (2013).

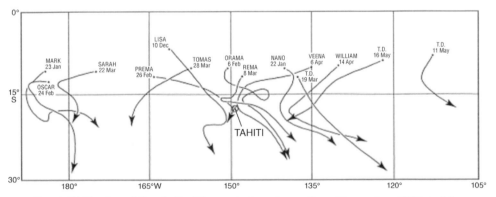

Fig. 4.37 Tracks of South Pacific tropical cyclones from December 1982 to May 1983. An asterisk indicates the origin points of tropical depressions (adapted from Sadler, 1983).

Source: Chu, 2004; © Columbia University Press. Used with permission.

The SPCZ is characterized by a band of cloudiness, low-level convergence, and precipitation with southeast trades from migratory anticyclones to the south and easterly to northeasterly winds from the eastern South Pacific anticyclone (Vincent, 1994; Folland et al., 2002). Generally, the SPCZ extends diagonally from the Solomon Islands (~10°S, 165°E) in the western Pacific warm pool southeastward towards the Cook Islands (Fig. 4.36). However, the orientation and location of the SPCZ depend on the season and the phase of ENSO. The SPCZ is often distinct from the monsoon trough off of the northeastern Australia but can sometimes merge together into one horizontal band of convergence.

In another very strong El Niño event (1997–1998), the median location of genesis for all named storms in the South Pacific was 12.2°S, 170.2°W, which is a 2,200 km shift eastward from its climatological median position (Chu, 2004). The 1997–1998 season was the most active TC season on record in the South Pacific, with 16 named storms. These TCs tracked as far eastward as 130°W (Fig. 4.38). The SPCZ moved farther eastward to the northern Cook Islands.

Consequently, the Cook Islands and French Polynesia, which are not usually affected by TCs, were frequented by seven TCs. To the west of 170°E, there was a westward track between 10°S and 15°S, a feature that is not uncommon during El Niño years (Evans and Allan, 1992; Dowdy et al., 2012).

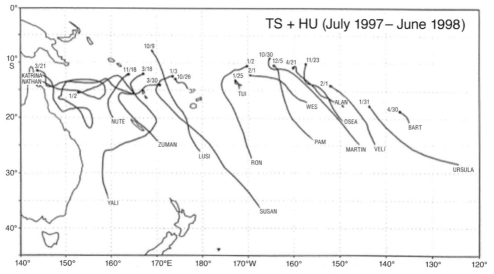

Fig. 4.38 Tracks of South Pacific tropical cyclones from July 1997 to June 1998. Origin points are denoted by dots, and the month and date for each tropical cyclone formation are indicated by numbers. Termination points of track are marked by the name of the storms. Tropical depressions are omitted.
Source: Chu, 2004; ©Columbia University Press. Used with permission.

Given the huge impact of two previously very strong El Niño events on TC activity in the South Pacific, one would guess that the very strong 2015–2016 El Niño in the South Pacific would also be extremely active? During that TC season, the South Pacific suffered the most damage on record, with a total of 50 deaths and $1.4 billion USD in damage. Although there are a total of eight named storms during 2015–2016, this number of named storms is not high compared to other two comparable warm events (1982–1983 and 1997–1998). However, the most intense TC in the Southern Hemisphere on the record, Winston, occurred in February 2016 with maximum 10-minute sustained wind speeds estimated at 175 mph (280 km h^{-1}). Winston was a category 5 severe tropical cyclone that made landfall near Suva at peak strength and wreaked havoc on Fiji. It also affected Tonga and Vanuatu. During the 2015–2016 season, most TCs in the South Pacific formed to the east of 160°E, consistent with Kuleshov et al. (2008), Chand and Walsh (2009), and Dowdy et al. (2012) who noted a maximum in genesis location between 160°E and the dateline during El Niño years.

Chand and Walsh (2009), hereafter referred to as CW09, focused on the Fiji region (Fiji, Tonga, and Samoa) and analyzed 122 TCs from 1970–1971 to 2005–2006. They constructed a smooth probability density estimate of TC genesis positions for three phases of ENSO (El Niño, neutral, and La Niña) using a kernel density estimation method. They also used a mixture regression model to cluster TC tracks in three types. This mixture model was used by Camargo et al. (2007b) and Chu et al. (2010) to group historical TC trajectories into seven or eight track types in the WNP.

During El Niño events, TC genesis is enhanced near the dateline, extending from north of Fiji eastward to Samoa (Fig. 4.39b). The maximum density that they found was centered at 10°S, 180°. With regard to TC paths, there are three preferred path types during El Niño events (CW09). Two of the three types were regarded as straight moving and the third one was recurving (Fig. 4.40a). TCs that formed poleward of 10°S and west of the dateline during El Niño tended to track southeastward to the north of Fiji and Tonga by predominantly northwesterly steering flow (700–500 hPa mean flow). The TCs that formed to the east of the dateline were steered by westerly currents and tended to track eastward and north of Samoa (Fig. 4.40a). Recurving TCs that are generated in deep tropical northeasterly flow tended to head southwestward to Vanuatu and then turned southeastward to between New Caledonia and Fiji (Fig. 4.40a).

During the neutral phase of ENSO, the majority of TCs formed between 10°S–15°S to the north of Fiji with an extension eastward to the south of Samoa (Fig. 4.39d). Tropical cyclone tracks during neutral years were similar to those in the El Niño phase, except for a slight poleward displacement (Fig. 4.40c). For the La Niña phase, fewer TCs were observed relative to El Niño and neutral phases. The genesis positions were also displaced southwestward with two centers of TC

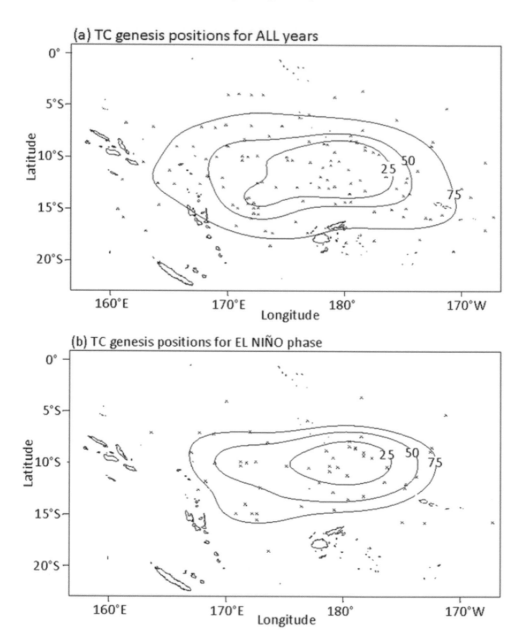

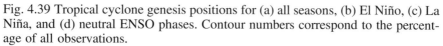

Fig. 4.39 Tropical cyclone genesis positions for (a) all seasons, (b) El Niño, (c) La Niña, and (d) neutral ENSO phases. Contour numbers correspond to the percentage of all observations.

Source: Chand and Walsh, 2009; ©American Meteorological Society. Used with permission.

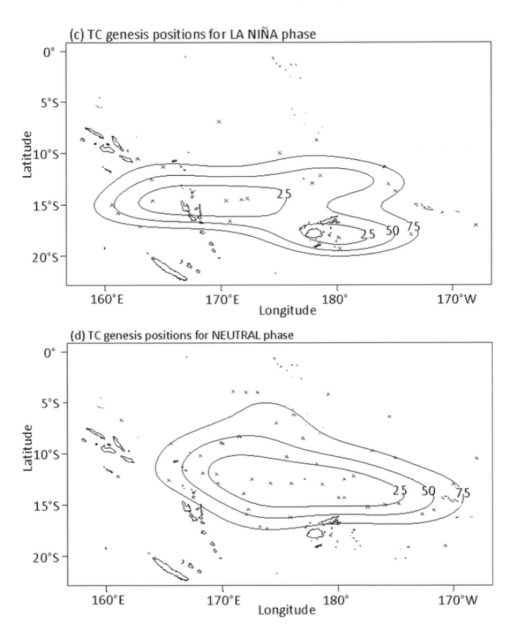

Fig. 4.39 (*cont.*)

activity. One of these TC centers was located around 15°S, 170°E near Vanuatu, and the other was near Fiji (Fig. 4.40c). During La Niña events, only one typical path was identified. Most storms formed between the Solomon Islands and Vanuatu in the southwest Pacific and tracked southeastward over Fiji and Tonga, posing little or no threat to Samoa.

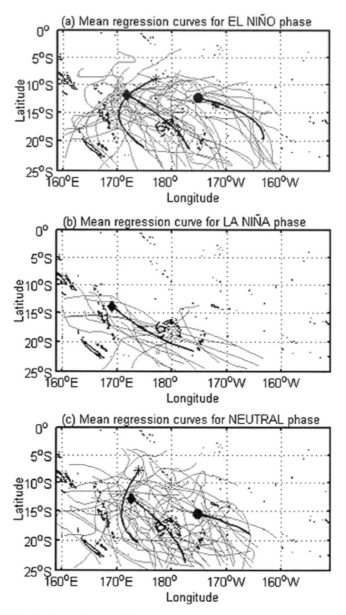

Fig. 4.40 Tropical cyclone tracks (gray) and the mean regression curve (bold) associated with each cluster for (a) El Niño, (b) La Niña, and (c) neutral ENSO phases. Initial positions of the mean regression curves are marked as cluster 1 (diamond), cluster 2 (circle), and cluster 3 (asterisk).

Source: Chand and Walsh, 2009; ©American Meteorological Society. Used with permission.

In CW09, three phases of ENSO (El Niño, La Niña, and neutral) were identified. Subsequently, Chand et al. (2013) classified the state of ENSO into four regimes: canonical El Niño, canonical La Niña, positive–neutral, and negative–neutral, and studied the impact of these four phases on southwest Pacific TCs (5°S–25°S, 170°E–170°W). They found an average of 4.3 TCs yr^{-1} for positive–neutral and 4.0 TCs yr^{-1} for canonical El Niño. In comparison, they found 2.2 yr^{-1} for La Niña and 2.4 yr^{-1} for neutral–negative. Therefore, increased numbers of TCs are found in the southwest Pacific for positive–neutral and El Niño phases relative to negative–neutral and La Niña regimes. The majority of TCs form between 170°E and 175°W and 10°S and 15°S during positive–neutral phases and the El Niño phase but the genesis position during El Niño phases also expands eastward to ~155°W (Fig. 4.41a and c). Chand et al. (2013) attributed this change to the extension of favorable low-level cyclonic vorticity and low VWS eastward across the dateline during an El Niño regime. For La Niña and negative–neutral phases, the genesis maxima shift westward compared to the El Niño case (Fig. 4.41b and d), and the large-scale environmental conditions are less favorable for TC development, particularly east of the dateline.

Chand and Walsh (2011) also investigated the influence of ENSO on TC intensity in the Fiji region. They measured intensity by accumulated cyclone energy (ACE). They found that ACE is strongly related to ENSO equatorward of

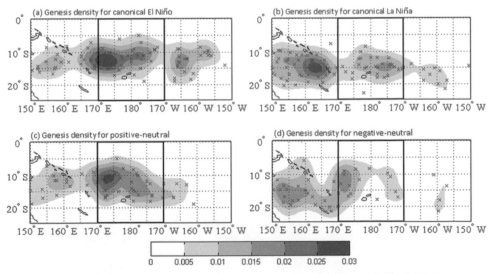

Fig. 4.41 Mean number of genesis events per year and per 2.5°×2.5° boxes (shading) using kernel smoothing and actual genesis positions (crosses) of TCs for different ENSO phases over the South Pacific basin. The central south Pacific between 170°E and 170°W is indicated with thicker black lines.

Source: Chand et al., 2013; ©American Meteorological Society. Used with permission.

15°S; that is, there were much larger ACE values in this region during El Niño events than during La Niña events. Relative to a La Niña event, TCs that tracked poleward of 15°S during El Niño phases weakened rapidly due to very strong VWS south of the Fiji Islands in conjunction with less favorable thermodynamic conditions. As a result, poleward of 15°S, TCs dissipate more rapidly during an El Niño than during a La Niña phase. Conversely, TCs that are poleward of 15°S during a La Niña phase maintain their intensity for a longer period of time.

4.6 The North Atlantic

4.6.1 Background Climatology

The North Atlantic Ocean generates more named storms in an average season than all TC basins except for the western North Pacific and the eastern North Pacific (Schreck et al., 2014) (Fig. 4.1). Figure 4.42 shows the average cumulative number of all named TCs (maximum wind speed ≥ 34 kt), all hurricanes (≥ 64 kt), and hurricanes that were category 3 or stronger on the Saffir–Simpson Hurricane Wind Scale (≥ 96 kt, called major hurricanes) in the North Atlantic. The annual mean number of named TCs over the North Atlantic is 12.1 over the period 1981–2010. Of these 12.1 named TCs, 6.4 become hurricanes and 2.7 become major hurricanes.

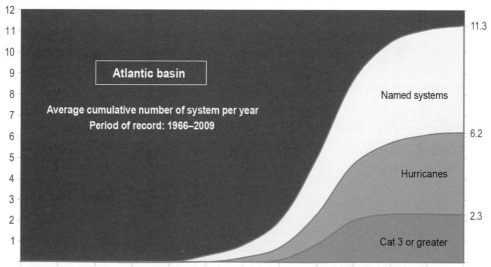

Fig. 4.42 Cumulative mean number of named storms, hurricanes, and category 3 or stronger hurricanes (major hurricanes) in the North Atlantic over the period 1966–2009.

Over 95% of the named storms that form are generated from June 1 and November 30. Thus, June–November is defined as the hurricane season in the North Atlantic by the National Hurricane Center. The National Hurricane Center is the division of the National Oceanic and Atmospheric Administration tasked for monitoring and forecasting Atlantic hurricane activity. Although the average number of major hurricanes is only ~20% of the number of named TCs, in the North Atlantic, ~85% of total storm-related normalized damage has been caused by major hurricanes over the United States for the period 1900–2017 (Weinkle et al., 2018).

4.6.2 Modulation of North Atlantic Tropical Cyclone Activity by Intraseasonal Oscillations

As in the western North Pacific (Section 4.2), TC activity in the North Atlantic is also significantly influenced by the MJO. Although modulation of convective activity by the MJO is relatively weak in the North Atlantic, the MJO has been shown to significantly modify of the North Atlantic low-level wind circulation, which in turn leads to modulation of TC activity, especially in the Gulf of Mexico and Caribbean Sea (Maloney and Hartmann, 2000b). Figure 4.43 displays composites of anomalous wind circulations at 850 hPa (Fig. 4.43a) and TC tracks (Fig. 4.43b) during positive phases (upper panels, A) and negative phases of the MJO (bottom panels, B). During the positive MJO phases, an anomalous low-level cyclonic circulation over the Gulf of Mexico is observed, resulting in an active period for TCs in the Gulf of Mexico and Caribbean via enhanced low-level relative vorticity. Negative phases of the MJO suppress TC formation in this same region given anticyclonic circulation anomalies in the western North Atlantic. Klotzbach (2010) further examined the relationship between MJO and TC activity in the North Atlantic. He showed that vertical wind shear and mid-level relative humidity were significantly different between active (phases 1–2) and inactive (phases 6–7) phases of the MJO by about 4 m s^{-1} and 2%, respectively. Specifically, more than three times the number of major hurricanes were observed across the North Atlantic during active MJO phases than inactive MJO phases in the observations.

4.6.3 Interannual Variability in North Atlantic Tropical Cyclones

There is also significant interannual variation of TC activity in the North Atlantic. ENSO is one of the strongest drivers of interannual variability in North Atlantic TC activity. The ENSO warm phase (El Niño) leads to lower hurricane numbers (Gray, 1984, 1993), fewer hurricanes striking the continental USA (Gray, 1984;

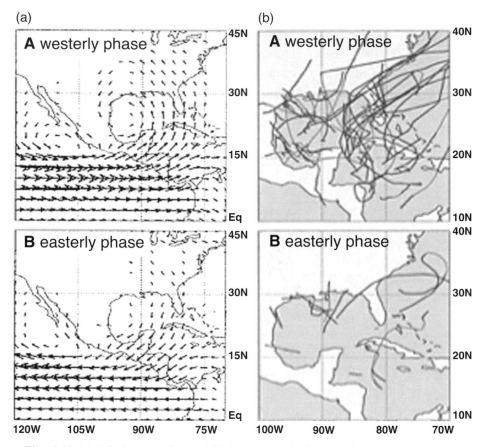

Fig. 4.43 (a) Wind anomalies at 850 hPa when the MJO index has a magnitude greater than 1 SD from zero during (A) positive phases and (B) negative phases. (b) As in left, but for TC tracks over the Gulf of Mexico, Caribbean Sea, and western Atlantic.

Source: Maloney and Hartmann, 2000b; ©Science. Used with permission.

O'Brien et al., 1996; Bove et al., 1998; Saunders et al., 2000; Kim et al., 2009), and reduced hurricane damage in the US (Pielke and Landsea, 1999; Klotzbach et al., 2018). Figure 4.44 displays composites anomalies of TC track density during the boreal summer during EP El Niño, CP El Niño, and La Niña (Kim et al., 2009). As in the WNP (Section 4.2), the two types of El Niño may have different effects on Atlantic TCs.

During the EP El Niño years, TC track density is reduced over most of the North Atlantic (Fig. 4.44a). Kim et al. (2009) found that during the CP El Niño years, TC track density increases across the Caribbean, the Gulf of Mexico, and the US East Coast, while it decreases in the central and eastern North Atlantic (Fig. 4.44b). In contrast, during La Niña years, large increases in TC track density are observed

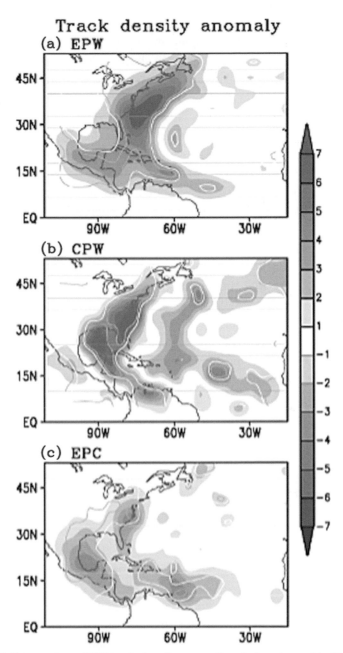

Fig. 4.44 Composites of TC track density anomalies during August to October for (a) canonical El Niño (EPW), (b) central Pacific El Niño (CPW), and (c) La Niña (EPC). Light (dark) contours show statistical significance at the 90% (95%) level. Units: number per 5° × 5° grid box per year multiplied by 0.1. A black and white version of this figure will appear in some formats. For the color version, refer to the plate section.

Source: Kim et al., 2009; ©Science. Used with permission.

across the entire North Atlantic (Fig. 4.44c). Larson et al. (2012) looked at the reason TC activity in the North Atlantic during CP El Niño years is relatively more active than during EP El Niño years. In short, the impact of a warming in the tropical central Pacific was not large enough to remotely influence the large-scale environment in the North Atlantic. This is mainly because the location of warming in the central Pacific was far from the Atlantic Ocean. In addition, the magnitude of warming during CP El Niño was relatively smaller than that during EP El Niño. It has been reported that, in recent decades, the CP El Niño has become more frequent while EP El Niño has become less frequent (e.g., Yeh et al., 2009; Lee and McPhaden, 2010; Lu et al., 2020). If this trend continues in the future, the suppressing effect of EP EL Niño on Atlantic TC activity may diminish. However, Patricola et al. (2016) found in model simulations that Atlantic TC activity was reduced in both EP El Niño and CP El Niño events. The reason for this difference in findings from the observational studies discussed previously may be due to the climate model's ability to control for Atlantic SST differences, while in the observed record, more CP El Niño events have occurred in recent years when tropical Atlantic SSTs are warmer (Lee et al., 2010).

Another significant mode of interannual natural variability is the Atlantic Meridional Mode (AMM; Chiang and Vimont, 2004). The AMM has a physical mechanism almost identical to PMM (Section 3.1.9) but occurs over the Atlantic Ocean. Previous studies (Kossin and Vimont, 2007; Vimont and Kossin, 2007) showed that the AMM significantly impacts TC activity in the North Atlantic. The spatial structure of the AMM is depicted in Fig. 4.45a and shows the composite difference between the five years with most positive and the five years with most negative AMM values. As in the PMM, the AMM is characterized by a meridional SST gradient near the equator, with winds blowing towards warmer surface water and veering to the right in the Northern Hemisphere in accordance with the Coriolis force. The ITCZ also shifts towards the warmer hemisphere. Although observations show that the maximum variance of the AMM occurs during the boreal spring, the AMM also influences hurricane variability during the boreal summer and autumn.

Figure 4.45b (left) and (right) show composite anomalies of the large-scale environment (shading and contours) superimposed on locations of tropical cyclone formations (cross and circle marks) for positive and negative AMM phases, respectively. During the five positive AMM years selected by Kossin and Vimont (2007), there were 84 named storms, 51 of which became hurricanes, and 29 of which became major hurricanes. In contrast, during the five negative AMM years, there were 45 named storms, 20 of which became hurricanes, and only 4 of which became major hurricanes. Therefore, the seasonal Atlantic hurricane frequency is related to the AMM phases. In addition, during the positive AMM phase, the mean

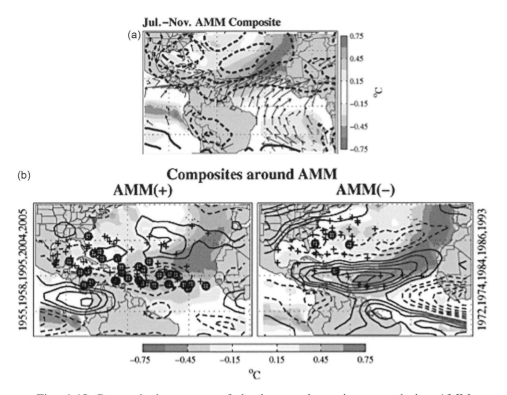

Fig. 4.45 Composited structure of the large-scale environment during AMM positive and negative events: (a) SST (shaded), 925-hPa height (contour, 2 m), and 925-hPa wind (vectors) anomalies around the 5-yr with the largest positive AMM index values minus the anomalies around the 5-yr with the largest negative AMM index values. For 925-hPa heights, solid contours denote positive anomalies, dashed contours denote negative anomalies. (b) Tropical cyclogenesis points for (left) the five strongest and (right) five weakest AMM years. Superimposed is the composites of SST (shaded) and vertical wind shear (contours) anomalies. Crosses show the genesis points for all storms that reached tropical storm strength. Storms that reached major hurricane strength ($\geq$ 49 m s^{-1}) also have a circle around their genesis point. Solid (dashed) shear contours denote positive (negative) values. The contour interval is 0.25 m s^{-1}, and the zero contour has been omitted. Shear was calculated every 6 h as the amplitude of the vector difference between the layer mean winds in the 300–150-hPA and 925–700-hPa layers, and means were formed around the hurricane season using monthly means. Number on the sides are the years during the AMM years. A black and white version of this figure will appear in some formats. For the color version, refer to the plate section.
Source: Kossin and Vimont, 2007; ©American Meteorological Society. Used with permission.

genesis location moves eastward and equatorward, or a southeastward shift, whereas a northwestward shift is noted in the negative AMM phase. In other words, the AMM variability modulates the northwest–southeast shift of the seasonal mean genesis location in the North Atlantic. The composite differences of

SST (shading) and vertical wind shear (contours) in Fig. 4.45b (left) and (right) reveal the causal evidence for the shift in TC genesis locations. During positive AMM phases, anomalously low vertical wind shear and high SST are observed in the main development region (MDR), favoring TC formation and intensification. In contrast, cooler SST and stronger wind shear over the MDR are unfavorable for TC development during negative AMM phases, although some storms form near the US East Coast. Under these conditions only a small percentage of TCs intensify into major hurricanes. Moreover, the positive phase of the AMM corresponds to an increase in storm duration, simply because storms formed farther away from landmass so they may persist longer over open ocean before making landfall.

   Although the variations of both ENSO and AMM are dominant at the interannual time scale, their effects on TC activity are spatially different. Figure 4.46 reveals

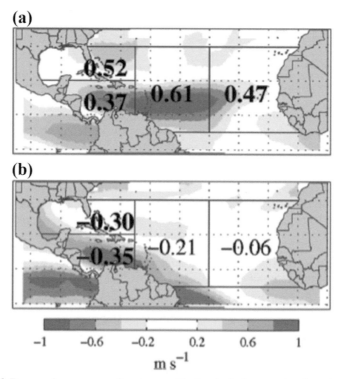

Fig. 4.46 Regression maps of mean vertical wind shear onto the standardized (a) AMM index and (b) Nino-3.4 index. Units: m s^{-1} per standard deviation of the respective time series. The listed number is the correlation between the number of storm days within each region and the respective index. Statistically significant correlations are listed in boldface. A black and white version of this figure will appear in some formats. For the color version, refer to the plate section.
Source: Kossin and Vimont, 2007; ©American Meteorological Society. Used with permission.

effects of the AMM and ENSO on vertical wind shear (shadings) and TC days (numbers) for each subregion. The positive AMM largely affects variations in vertical wind shear over the tropical Atlantic, which then affects TC activity specifically predominately over the tropical North Atlantic. On the other hand, positive ENSO (e.g., El Niño) significantly increases vertical wind shear in the western Atlantic, which then affects TC activity over the Caribbean and Gulf of Mexico. Given that most major hurricanes originate over the tropical Atlantic, there is a significant positive correlation between the AMM and the number of major hurricanes in the North Atlantic over the period 1950–2005 ($r = 0.61$; Vimont and Kossin, 2007).

The Atlantic hurricane activity is also influenced by "in-situ" sea surface temperatures (SSTs). For example, anomalously warm winter SSTs over the MDR accounted for up to 42% of the variance in regional SST variability during the subsequent hurricane season (Wang et al., 2017). Warm SSTs in winter can persist into the summer via a positive SST–water vapor feedback, as described in Section 3.1.2 (Wang et al., 2017). Other processes that may prolong warm winter SST anomalies include the WES feedback and SST–cloud feedback (Section 3.1.2). Anomalously warm SST from winter to summer increases atmospheric water vapor content, leading to increased occurrence of TC frequency and intensity during the hurricane season. The extremely active 2020 Atlantic hurricane season seemed to be fueled by a combination of several factors: anomalously warm SSTs over the MDR that helped precondition the Atlantic hurricane season along with a positive AMM as well as La Niña during the peak of the season.

Previous studies have found that relative SST anomalies over the MDR are more important than local SST anomalies for Atlantic TC activity (Vecchi and Soden, 2007a; Vecchi et al., 2013; Murakami et al., 2018; Yan et al., 2017). The relative MDR SST anomaly has been defined as the difference in SST between the mean SST over the MDR (i.e., 10°N–25°N, 80°W–20°W) and the mean SST over the global tropics (30°S–30°N). In other words, the key factor controlling Atlantic hurricane activity appears to be how much the tropical Atlantic warms relative to the rest of the global tropics. Murakami et al. (2018) found that the correlation coefficient between observed major hurricane frequency and relative SST anomaly was +0.61, while the correlation between observed major hurricane frequency and the MDR SST anomaly was +0.50, highlighting the improved relationship when relative SST is used. Murakami et al. (2018) also showed through the use of a high-resolution climate model that the relative SST anomaly is key for determining active or inactive hurricane season in a future climate.

Another potential factor influencing TC activity in the North Atlantic on an interannual time scale is the stratospheric quasi-biennial oscillation (QBO). The QBO represents interannual variability of the equatorial stratosphere, which

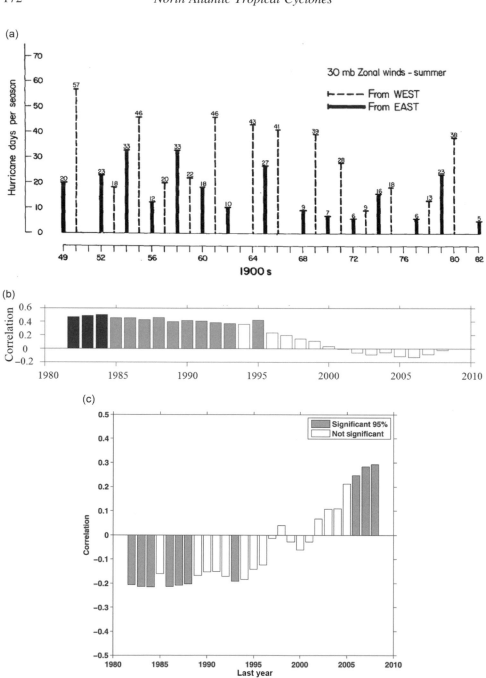

Fig. 4.47 Relationship between the QBO and storm days in the North Atlantic. (a) Relationship between 30 hPa stratospheric wind direction and the seasonal number of hurricane days from 1949 to 1982. (b) 30-yr correlations of Berlin 30 hPa QBO with the number of hurricane days in the North Atlantic. The first correlation (first bar) is calculated for the period 1953–1982, the subsequent one is calculated for

manifests itself with alternating periods of easterly and westerly zonal winds that repeat with a well-defined period of about 28 months. Gray (1984) first reported significant correlation between the QBO and TC activity in the North Atlantic. When the QBO was in its westerly wind phase, TC activity in the North Atlantic was greater than when the QBO was in the easterly wind phase, with more TCs and TC days occurring in the westerly QBO phase (Gray, 1984). Figure 4.47a shows interannual variations in the number of hurricane days and zonal wind direction in the stratosphere, highlighting a significant increase in hurricane days during the westerly phase of QBO compared with the easterly phase of QBO. However, the physical mechanism behind the relationship between the QBO and Atlantic TC activity is somewhat unclear.

Previous studies suggested that the QBO modified vertical wind shear, which in turn affected TC activity (Gray et al., 1992a, 1992b; Arpe and Leroy, 2009). However, changes in wind shear associated with QBO are restricted to the lower stratosphere, with a very weak extension into the uppermost troposphere where TCs have typically have their outflow. There are also a few studies that indicate that ENSO is modified by the QBO (Gray et al., 1992a, 1992b). Because ENSO is a large interannual driver of TC activity in the North Atlantic, these studies indicate an indirect effect of the QBO on TC activity in the North Atlantic via the QBO's modulation of ENSO.

Because most of the studies relating the QBO to Atlantic TC activity were published before the mid-1990s, Camargo and Sobel (2010) revisited the relationship between QBO and Atlantic TC activity by updating the analysis through 2008. They found that there was no significant correlation between QBO and TC activity in the North Atlantic since the mid-1990s. Figure 4.47b shows correlation coefficients between the QBO and the number of hurricane days in the North Atlantic during the previous 30 years. Although there were statistically significant correlations observed for the relationship through the 1980s, the correlation has been insignificant since that time. Figure 4.47c shows that the correlation between ENSO and the QBO reversed sign around that same time, indicating that the effect of the QBO on ENSO changed in the middle of 1990s, thereby reversing the relationship between the QBO and TC activity in the North Atlantic. Camargo and Sobel (2010) also examined the correlation between

Caption for fig. 4.47 (*cont.*). the next 30-yr period (1954–1983), until in the last bar the last 30-yr period (1979–2008) is considered. The *x*-axis shows the last year that was included in the calculation. Gray and black bars indicate statistical significance. (c) As in the middle panel, but for correlations between the Niño-3.4 index and Berlin 30-hPa QBO.
Sources: Gray, 1984; Camargo and Sobel, 2010; ©American Meteorological Society. Used with permission.

the QBO and TC activity after linearly removing the influence of ENSO and showed that the early record still retained a significant correlation between the QBO and TCs. This result implies that we cannot reject the hypothesis that the correlation in the early record is either a pure statistical fluke or a result of a physical connection between the QBO and TC activity in the North Atlantic. To date, it is still unclear if there is physical link between the QBO and TC activity in the North Atlantic.

Annual TC activity in the North Atlantic is also strongly correlated with convective activity over the African continent. Karnauskas (2006) and Karnauskas and Li (2016) used satellite observations to show that there is a statistically significant positive correlation between Atlantic seasonal hurricane activity and the meridional gradient of outgoing longwave radiation (OLR) across Africa during the period 1979–2015. The OLR gradient across Africa represents an integrated quantity reflecting the intensity of the ITCZ over the Sahel and the production of African easterly waves, as well as the meridional temperature gradient related to the African Easterly Jet (AEJ) via the thermal wind relationship. African easterly waves often initiate tropical cyclogenesis in both the North Atlantic and North Pacific Oceans. A large fraction of TC genesis in the North Atlantic is generated by TD-type disturbances in the form of easterly waves originating from Africa (Fig. 4.7). African easterly waves are well defined as wave perturbations with periods of 3–5 days and spatial scales of ~1,000 km. These waves occur with a maximum amplitude close to the level of the AEJ and a low-level maximum amplitude north of the AEJ. Consequently, the location of the AEJ and its intensity are important for determining easterly wave activity.

Previous studies have documented that Atlantic TCs and African easterly waves are linked on synoptic time scales, with ~85% of observed major hurricanes and ~60% of tropical storms originating from African easterly waves (Dunn, 1940; Frank, 1970; Landsea, 1993; Russell et al., 2017). Therefore, Atlantic TCs is supposed to be correlated with the OLR gradient across Africa via the variation of activity of the AEJ and African easterly waves. However, Patricola et al. (2018b) questioned that the interannual variation of TCs was dependent on the interannual variation of African easterly wave activity. They investigated the impact of suppression of African easterly waves on seasonal Atlantic TC activity through idealized dynamical model experiments. In these experiments, African easterly waves were artificially suppressed by removing the 2–10-day time scale from all dynamic and thermodynamic variables and all vertical levels at the boundary of the west coast of northern Africa during the model simulations. The idealized suppression of easterly wave experiment resulted in no significant change in seasonal Atlantic TC numbers relative to the original experiments that did not suppress easterly waves. These findings indicate that African easterly waves are not necessary for maintaining climatological basin-wide TC frequency and basin-wide seasonal Atlantic TC number, even though TCs readily originate from disturbances. Therefore, it is not clear if the correlation between the OLR meridional gradient across Africa and seasonal TC number in the North Atlantic at interannual time scale reflects physical causality with easterly waves or is more of an indicator of the large-scale environment (e.g., vertical wind shear, vertical motion).

4.6.4 Decadal to Multidecadal Tropical Cyclone Variations

There is also significant decadal variation in Atlantic TC activity associated with low frequency variability in the Atlantic Ocean that has been termed "the Atlantic Multidecadal Oscillation" (AMO; Goldenberg et al., 2001). The AMO index is defined as an average of low-pass filtered annual mean SST anomalies over the entire North Atlantic, after removing linear trends to remove the influence of greenhouse gas-induced global warming. The AMO has been hypothesized to be driven by changes in the Atlantic thermohaline circulation and associated ocean heat transport fluctuations. So far, there is no firm evidence of a unique periodicity for these multidecadal fluctuations in North Atlantic SST, therefore, the AMO has also been termed Atlantic Multidecadal Variability (AMV). For the remainder of this chapter, we will refer to this multidecadal variability in the Atlantic Ocean as the AMO.

Although the conventional AMO index is defined using SST over the North Atlantic Ocean, low-frequency multi-decadal ocean variability is also highly correlated with Atlantic sea level pressure (SLP) variations. Klotzbach and Gray (2008) developed a modified AMO index, which utilized low-frequency far North Atlantic SSTs (north of 50°N) and Atlantic SLP equatorward of 50°N. The modified AMO index revealed a marked multi-decadal oscillation with alternating warm phases (1878–1899; 1926–1969; 1995–2006) and cold phases (1900–1925; 1970–1994) in the Atlantic Ocean since the late nineteenth century (Fig. 4.48). Table 4.6 displays the average annual number of hurricanes (H), hurricane days (HD), major hurricanes (MH), and major hurricane days (MHD) during the five multidecadal periods in the North Atlantic Ocean. All metrics have higher values during positive phases of the AMO than in negative phases of the AMO, with the difference being greatest for MH and MHD. Klotzbach and Gray (2008) and Klotzbach et al. (2018) also found that landfalling hurricanes along the US coastline become more frequent during positive AMO phases, with the most dramatic increases during a positive AMO phase for hurricanes affecting the US East Coast and Florida Peninsula.

Zhang and Delworth (2006) utilized model simulations to highlight the substantial impact of the AMO on Atlantic TCs. They demonstrated that the positive AMO

Table 4.6. Observed annual averaged number of hurricanes (H), hurricane days (HD), major hurricane (MH), and major hurricane days (MHD) over the North Atlantic Ocean during the multidecadal periods of 1878–1899, 1900–1925, 1926–1969, 1970–1994, and 1995–2006. The bottom row provides the annual averaged ratio for the 77 positive AMO years and the 51 negative AMO years.

Period	AMO phase	H	HD	MH	MHD
1878–1899	Positive	5.9	29.0	1.6	4.4
1900–1925	Negative	3.6	15.4	1.2	3.2
1926–1969	Positive	5.6	24.8	2.6	6.5
1970–1994	Negative	5.0	16.0	1.5	2.5
1995–2006	Positive	8.2	35.3	3.9	10.1
Ratio (1878–1899/1900–1925)	Positive/negative	1.6	1.9	1.3	1.4
Ratio (1926–1969/1900–1925)	Positive/negative	1.6	1.6	2.2	2.0
Ratio (1926–1969/1970–1994)	Positive/negative	1.1	1.6	1.7	2.6
Ratio (1995–2006/1970–1994)	Positive/negative	1.6	2.2	2.6	4.0
All positive/all negative	Positive/negative	1.4	1.8	1.8	2.2

Source: Klotzbach and Gray, 2008; ©American Meteorological Society. Used with permission.

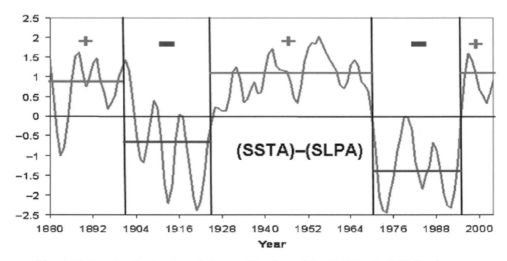

Fig. 4.48 Standardized value of the combination of North Atlantic SST for the area 50°N–60°N and 50°W–10°W minus the North Atlantic SLP anomaly for the area 0°–50°N and 70°W–10°W taken to be the strength of the AMO for the period 1880–2004. Horizontal lines indicate average values for the multidecadal period, while (+) and (–) symbols indicate that positive or negative values of the particular index predominated during that period.
Source: Klotzbach and Gray, 2008; ©American Meteorological Society. Used with permission.

phase leads to a northward shift of the ITCZ and an anomalous overturning circulation with rising motion north of the equator and descending motion south of the equator over the tropical Atlantic. Hence over the tropical North Atlantic, where most TCs generate (e.g., MDR), the surface easterly trade winds weaken and upper level zonal winds become less westerly, resulting in a reduction of the vertical wind shear of the zonal wind. This reduction in vertical wind shear leads to an increase in TCs. A negative AMO phase is a mirror image of a positive AMO phase – an increase in vertical wind shear suppresses TC activity in the North Atlantic.

Although the decade 1995–2005 was very active for TCs and especially major hurricanes in the North Atlantic along with a positive phase of AMO, the decade 2006–2016 had relatively lower major hurricane activity. Although the AMO index remained positive during this period, it had tended towards neutral during this time. Yan et al. (2017) examined this decline in major hurricane activity and associated it with a negative trend in the AMO index and a weakening of the Atlantic Meridional Overturning Circulation (AMOC). The AMOC is character-ized by a northward flow of warm, salty water in the upper layers of the Atlantic and southward flow of colder water in the deep Atlantic. The result by Yan et al. (2017) indicated a potentially important role for the extra-tropical North Atlantic Ocean in modulating Atlantic major hurricane frequency at the multidecadal time scale. Since the time that Yan et al. (2017) was published, the North Atlantic has experienced four above-average hurricane seasons in a row, including two extremely active seasons: 2017 and 2020.

Despite the substantial impact of AMO on TC activity in the North Atlantic, the origins of the observed AMO are still uncertain. Numerous studies based on global climate model simulations showed that the northward heat transport fluctuations associated with the AMOC variability are the main driver of the AMO (Kushnir, 1994; Delworth and Mann, 2000; Knight et al., 2005). However, during the last century, North Atlantic SST has also been affected by changes in both anthropogenic forcing such as aerosols (Booth et al., 2012), making the assessment of the contribution of the internal AMOC fluctuations to the observed AMO difficult. In addition, Mann et al. (2021) has recently hypothesized that the observed multi-decadal variability in the North Atlantic over the past millennium may be mostly due to volcanic eruptions. Although challenging, further studies are needed to distinguish between natural and anthropogenic elements of the AMO. This in turn will enable better predictions of long-term changes in TC activity in the North Atlantic.

4.7 Observed Tropical Cyclone Activity and Climate Change

We next examine observed changes in TC activity and its possible linkage to climate change. The term "climate change" is broadly used here and refers to both natural variability and anthropogenic warming effects. In the following sections, we will focus on several aspects of TC activity, including TC genesis frequency, motion, intensity, and poleward migration of the lifetime maximum intensity.

Review articles discussing TCs and climate change have been published by Walsh et al. (2015) and Knutson et al. (2019).

4.7.1 Tropical Cyclone Genesis

Historical time series of global TC frequency do not indicate a clear signal for either increasing or decreasing trends (e.g., Knutson et al., 2010, 2019; Elsner et al., 2008), despite an increase in global temperature. On the other hand, trends in TC frequency of occurrence (or TC density) at a global scale since 1980 exhibit a distinct spatial pattern, with significant decreases in the South Indian Ocean and Western North Pacific along with significant increases in the North Atlantic, Central Pacific, and Arabian Sea. Using a suite of high-resolution dynamical model experiments, Murakami et al. (2020) demonstrated that the observed spatial pattern in trends cannot be explained only by underlying multidecadal variability, but external forcing such as greenhouse gases, aerosols, and volcanic eruptions likely played an important role. In other words, the effect of climatic change on TC activity is more evident in the spatial distribution pattern, than the overall number of global TCs. Volcanic eruptions may also have exerted a great impact on where TCs have occurred (Pausata and Camargo, 2019). For example, the major eruptions of El Chichón in Mexico in 1982 and Pinatubo in the Philippines in 1991 caused the Northern Hemisphere atmosphere to cool, shifting ITCZ (and TC activity) southward for a few years. Particulate pollution and other aerosols can also influence TC activity. These particulates help create clouds and reflect sunlight away from the earth, causing cooling. The decline in particulate pollution due to pollution control measures since 1980 in the USA and Europe may have increased North Atlantic warming by allowing more sunlight to be absorbed by the ocean, consequently leading to increased SSTs in the North Atlantic that in turn led to increasing TCs there (Mann and Emanuel, 2006; Dunstone et al., 2013; Murakami et al., 2020).

Most climate models projected a decrease or an insignificant increase in global TC numbers in a warming scenario (Sugi et al., 2002; Bengtsson et al., 2007; Gualdi et al., 2008; Knutson et al., 2010). Knutson et al. (2010) examined a suite of climate models and found decrease that varied from 0% to 30% in the Northern Hemisphere and from 10% to 40% in the Southern Hemisphere. However, other climate models and related downscaling studies indicate an increase in global numbers (Haarsma et al., 1993; Emanuel, 2013). This diversity of future TC number scenarios may arise because of differences in model-generated parameters that specify TC formation and different ways in which the TC-vortex detection algorithms are applied. Zhao et al. (2013) hypothesized that changes in TC frequency were most strongly associated with 500-hPa vertical velocity.

4.7.2 Poleward Migration of the Latitude of Lifetime Maximum Intensity

Kossin et al. (2014) indicated that the latitude of lifetime maximum intensity (LMI) of TCs migrated poleward globally during 1980–2012 with significant poleward migration rates occurring in the WNP, South Pacific, and South Indian Ocean. In the meantime, the migration direction is equatorward in the North Atlantic and its rate is also significant. The predominantly poleward displacement is likely associated with the expansion of the Hadley circulation in a warmer climate. In the WNP, the LMI change is consistent with a significant poleward displacement of TC genesis latitude (Daloz and Camargo, 2018). Long-term changes in track density can also account for the shift of the LMI location. For example, Tu et al. (2009) showed that typhoon tracks since 2000 have shifted northward from the South China Sea (SCS) and the Philippine Sea towards the vicinity of Taiwan and the East China Sea. This shift may be caused by a weakening of easterly steering flows in the WNP and the SCS (Chu et al., 2012) and an eastward retreat of the WNP subtropical high (Tu et al., 2009; Liang et al., 2017).

4.7.3 Tropical Cyclone Motion

Kossin (2018) showed a significant decreasing trend in global TC translation speed (~10%) over 1949–2016. At regional scale, the slowdown was the largest for overland TCs in the Australian region (22%), followed by overland TCs in the WNP region (21%) and then overland TCs in the North Atlantic region (16%). The slowdown in TC motion is consistent with the weakening of the tropical circulation in a warmer climate (Vecchi and Soden, 2007b). Slower translation speeds have a direct and profound impact on TC-related rainfall because slower TCs impact the surrounding environment for longer periods, thus bringing more torrential rainfall and associated damage, including landslides. For example, the slow movement of Typhoon Morakot resulted in one of the highest recorded rainfall amounts in southern Taiwan. The four-day accumulated rainfall produced by this record-breaking storm in August 2009 exceeded 3,000 mm and killed more than 600 people (Ge et al., 2010; Chien and Kuo, 2011; Lin et al., 2011; Xie and Zhang, 2012; Yu and Cheng, 2013; Wu, 2013). In heavily populated coastal communities, the slow movement of TCs (such as Hurricane Harvey in Texas in 2017) lead to critical damage due to prolonged exposure to destructive winds, torrential rain, and storm surge (Emanuel, 2017).

The slowdown in TC translation speed reported by Kossin (2018) was challenged by Moon et al. (2019), who pointed out the under-sampling problems of TC detection when storms were weak or over open oceans during the pre-satellite era. This concern was also voiced by Chan (2019). Therefore, the long-term slowdown trend could be attributed to the temporal inhomogeneity of historical TC records caused by changes in observational practices before and after the satellite era rather than a climate signal. Lanzante (2019) also noted that the time series of TC translation speed exhibits a few abrupt, step-like changes, not a continuous slowing trend. Moreover, these step-like changes appear to be associated with the PDO phase shifts (Lanzante, 2019). Climate modeling studies also do not support the observed significant decreasing trends in TC translation speed over the historical period 1951–2010 due to significant unforced internal variability of the climate system (Yamaguchi et al., 2020; Zhang et al., 2020). However, Yamaguchi et al. (2020) and Zhang et al. (2020) showed possible impacts of anthropogenic on the slowdown of TC translation speed, specifically in extratropical regions. These findings indicate possible future change in TC translation speed when the magnitude of anthropogenic warming becomes more significant.

By analyzing global mean tropical cyclone translation speed (TCTS) in the post-satellite era (1982–2017), Kim et al. (2020) noted an increase of 0.3 km h^{-1} per decade over the last 36 years, but the steering flow controlling the local TCTS decreased by –0.15 km h^{-1} per decade in the major TC passage regions. The inconsistency between these two related variables (TCTS and steering flow) was caused by relative TC frequency changes by basin and latitude. The TCTS is closely related to the latitude of the TC position, and the mean TCTS is different in each TC basin. The increase in global-mean TCTS was mainly attributed to by two factors: (1) an increase (4.5% per decade) in the relative frequency of TC tracks in the North Atlantic which shows the fastest mean TCTS among all TC basins; and (2) poleward shift of TC activity. These two effects mainly account for the observed global-mean TCTS trend, thus overwhelming the impacts of a slowing steering flow related to a weakening of the large-scale tropical circulation.

4.7.4 Tropical Cyclone Intensity Change

An increasing trend in the global numbers of very intense TCs (category 4 and 5) derived from the best-track archives during 1970–2004 was reported by Webster et al. (2005), but the finding was subsequently questioned by several others on the grounds of data quality in the early decades and changing observational practices (e.g., Landsea et al., 2006; Klotzbach and Landsea, 2015). To reduce the effects of temporal inhomogeneity and uncertainty about intensity estimates in early years, Kossin et al. (2013) created a globally homogenized satellite data set during 1982–2009 called the Advanced Dvorak Technique-Hurricane-Satellite-B1 (ADT-HURSAT) and investigated TC activity for 1982–2009 using a quantile regression to estimate trends in TC intensity. Based on the corrected ADT-HURSAT, they found that stronger TCs, on a global scale, have become more intense at a rate of

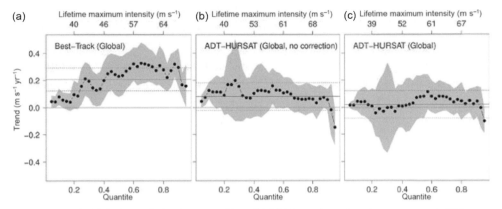

Fig. 4.49 Global trends in the quantiles of LMI for at least hurricane strength TCs (LMI $\geq$ 33 m s^{-1}) during 1982–2009. (a) trends in the best-track, (b) trends in the uncorrected ADT-HURSAT, and (c) trends with the additional homogenization correction. The dots denote trends in each quantile from 0.05 to 0.95 and the shading is the 95% confidence interval. The solid line denotes the trend in the mean, and the dashed lines denote the confidence interval.

Source: Kossin et al., 2013; ©American Meteorological Society. Used with permission.

about +1 m s^{-1} decade^{-1}, although this trend is only marginally significant (Fig. 4.49). Nevertheless, dramatic changes were observed in the North Atlantic, with a very strong positive trend in LMI for quantiles between 0.4 and 0.8 (moderate to strong TCs) based on the ADT-HURSAT and best-track data, while small positive trends were noted at very high quantiles in the South Pacific (Fig. 4.50). For the WNP, no clear trends were evident in any quantiles based on ADT-HURSAT although there was a weak positive trend for moderately strong TCs from the best-track data. For the eastern North Pacific, an increasing trend was noted for the strongest hurricanes from the best-track data, but this signal was absent in the ADT-HURSAT. The inconsistent trends of LMI between these two data sets leave doubt about best track intensity estimates for the WNP and eastern North Pacific. For the South Pacific, the agreement between these two data sets is good for the strongest storms, with both showing an upward trend, although the trend was weaker in the ADT-HURSAT.

Based on satellite-derived maximum wind speeds since 1981, an increase in the LMI of the strongest tropical cyclones was noted for all ocean basins, with the greatest increases over the North Atlantic and the north Indian Ocean (Elsner et al., 2008). The proportion of hurricanes reaching categories 4 and 5 has also increased significantly (Holland and Bruyere, 2014). Kossin et al. (2020) documented that from 1979 to 2017 there was a statistically significant increase in the ratio of category 3–5 to category 1–5 TCs in the ADT-HURSAT dataset. For the WNP, Park et al. (2014) found that the location of the maximum intensity of

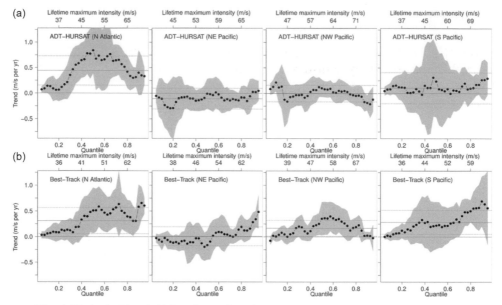

Fig. 4.50 As in Fig. 4.49 but for regional trends in the North Atlantic, eastern and western North Pacific, and South Pacific in the homogenized (a) ADT-HURSAT and (b) best-track data.

Source: Kossin et al., 2013; ©American Meteorological Society. Used with permission.

TCs moved closer to East Asia, resulting in increased TC landfalling intensity. Mei and Xie (2016) found that landfalling typhoons in East and Southeast Asia increased by 12–15% in the period 1977–2014, with the proportion of storms of category 4–5 doubling or even tripling. Mei and Xie (2016) attributed the increased intensity of landfalling typhoons to strengthening intensification rates, which were tied to locally enhanced ocean surface warming along the rim of East and Southeast Asia.

The intensification rate of TC (IR), which is defined as the 24-h wind speed change, is another metric of TC intensity. Specifically, over the period 1982–2009, statistically significant positive trends have occurred in the percentage of IR events exceeding 30 knots for both the globe and the North Atlantic basin individually (Bhatia et al., 2019). Bhatia et al. (2019) found, using a suite of climate model simulations, that the observed positive trends in frequency in IR in the North Atlantic were mainly due to increased anthropogenic forcing.

4.8 Summary

This chapter describes the impact of subseasonal, interannual, decadal and interdecadal climate variability on TC activity in the WNP, ENP, CNP, South

Pacific, and North Atlantic individually. Various kinds of equatorial waves are shown to be TC precursors. These wave types include the MJO, equatorial Rossby waves, tropical-depression (TD) type disturbances, mixed Rossby-gravity waves, and Kelvin waves. TD type disturbances and equatorial Rossby waves prevail in the WNP, ENP, South Pacific, and North Atlantic.

The dominant climate modes on subseasonal scales are the MJO and quasi-biweekly oscillation (QBWO), while ENSO, PMM, and AMM are major modes operating on interannual time scales. The El Niño phenomenon has been categorized as either an Eastern Pacific (EP) or Central Pacific (CP) event. On the interdecadal time scale, the PDO and AMO are two well-known modes and they modulate TC activity. These modes of climate variability can alter large-scale environmental conditions that influence various TC attributes such as genesis location, frequency of occurrence, intensity, track, lifespan, and landfall.

Five distinctive large-scale low-level flow patterns associated with tropical cyclogenesis in the WNP are presented. They include monsoon shear line, monsoon confluence zone, Rossby wave energy dispersion, easterly waves, and monsoon gyre. TC activity in the WNP is enhanced when the MJO is convectively active over the Pacific Ocean. Statistics reveal that 23% (20%) of TCs in the WNP are associated with the actively active MJO (QBWO) phase. In the convectively active MJO phase, TCs tend to track westward or northward and result in increased activity in Philippines, Vietnam, and Japan. The modulation of MJO on TC genesis is stronger in the early season (May–June), than in the peak season (July–September), and the peak season modulation is stronger than in the late season (October–December). The QBWO in the boreal summer appears to originate near the equatorial dateline and propagates northwestward through the tropical WNP and the South China Sea. The impact of the QBWO is more localized because of its smaller spatial extent, and it appears to modulate TC landfall for the Philippines and Japan. Recent studies have shown that the MJO determines primarily TC genesis frequency, whereas the QBWO affects TC genesis location.

For the WNP, an EP El Niño reduces vertical wind shear and increases mid-tropospheric relative humidity over the central equatorial Pacific. These changes are accompanied by a notable increase in TC genesis and frequency in the southeastern quadrant of the basin (150°E to the dateline) (Table 4.7). More intense typhoons and longer TC lifespans are expected given that storms forming in the far eastern portion of the WNP can traverse a larger region with warm SSTs. A concomitant decrease in TC frequency is noted in the northwestern quadrant of the basin in EP El Niño. During a La Niña, an opposite pattern of genesis and track to that of EP El Niño is observed. For CP El Niño, the overall pattern in genesis location and track is similar to that of the EP event, with the exception that TCs

Table 4.7. Tropical cyclone activity in the Pacific and Atlantic Oceans relative to climatology in Eastern Pacific (EP) El Niño events. Opposite-signed anomalies typically occur in La Niña events. NA denotes information not available.

Basin	Frequency	Genesis Location	Intensity	Track	Life Span
Western North Pacific	Significant increase (decrease) in the southeast (northwest) quadrant; no overall significant changes in developing year but significant reduction in year following El Niño	Farther east and south	More intense typhoons	More likely to recurve	Longer
Eastern North Pacific	Overall frequency unchanged but fewer storms off of the Central American coast	Farther west and south	More intense hurricanes in the western part and in the year following El Niño	Westward expansion to the central Pacific	Longer in the western portion of the basin
Central North Pacific (180°W–140°W)	More storms	More likely to form in the south	Likely more intense	More erratic	Longer
South Pacific (west of 160°E)	Fewer storms	Farther equatorward	NA	Likely to track westward in the tropics	NA
Southwest Pacific (160°E–160°W)	More storms	Farther east and north	Likely more intense equatorward of 15°S	Eastward and equatorward expansion	Probably longer
North Atlantic	Fewer storms	Farther north in the mid and late seasons	weaker	Fewer storms cross the US East Coast, the Gulf of Mexico, and the Caribbean	NA

generated in the southeast quadrant tend to track northwestward through the WNP, reaching Japan and the eastern coast of China and Taiwan (Table 4.8).

The PMM has been shown to also modulate WNP TC activity. For example, the mean genesis location (MGL) of TCs in summer of 2015 shifted farther eastward by ~10-degree longitude relative to that in 1997. Both 1997 and 2015 were very

Table 4.8. As in Table 4.7 but for Central Pacific (CP) El Niño events.

Basin	Frequency	Genesis Location	Intensity	Track	Life Span
Western North Pacific	Significant increase in a band from the southeast quadrant to southern Japan and east China	Farther east and south	Probably more intense typhoons	Likely to track northwestward to Japan, Korea, and China	Longer
Eastern North Pacific	Fewer storms off of the central American coast	Probably farther west	Probably more intense	Overall suppressed track density but increased westward track to the south of Hawaii	Likely longer
Central North Pacific (180°W–140°W)	Likely more storms	Likely shift to the south of Hawaii	Probably more intense	Likely to track westward	Likely longer
South Pacific (west of 160°E)	NA	NA	NA	NA	NA
Southwest Pacific (160°E–160°W)	Probably more storms	Probably farther east	NA	Likely increased southeastward	NA
North Atlantic	More storms over the US East Coast, the Gulf of Mexico, and the Caribbean	Shift westward	NA	Increased track across the US coastline	NA

strong El Niño years that had comparable Niño-3.4 index values. However, the PMM-like SST anomalies in summer 2015 were twice as large as those in the summer of 1997, and this extra subtropical warming likely caused the observed shift in MGL. On interdecadal time scale, based on peak season TC data from 1979 to 2012, an abrupt shift in WNP TC frequency appeared to occur in the late 1990s, when the PDO transitioned from a positive to a negative phase. Decadal changes in TC genesis frequency have also been linked to corresponding variations in tropical Indo-Pacific SST. An abrupt shift in October–December (late season) typhoon counts in 1995 was noted through a Bayesian change-point analysis.

For the eastern North Pacific (ENP), a composite study indicated that hurricanes were four times likely to occur when the MJO was in its westerly phase as opposed to its easterly phase, and VWS was a key reason why the MJO modulated TC genesis. The difference in contrasting ENSO states (El Niño versus La Niña) appears to be minimal for overall hurricane frequency in the entire basin (140°W to the Central American coast). However, when the ENP was divided into two sub-regions using 116°W as the dividing line, the western main development region

(MDR) experienced more frequent, longer TC duration, and more intense TCs during El Niño years relative to La Niña years. When a higher resolution (2° × 2°) data set was used, TC activity near the Pacific coast was reduced but more TCs were generated in the southern and western portion of the MDR during an EP El Niño hurricane season (Table 4.7). TC tracks also shift westward and lifespans become longer during EP years. For La Niña years, tropical cyclogenesis and tracks are greatly reduced. TC genesis to the east of 130°W was reduced in a CP event, with a large area of reduced track from the eastern to the central Pacific to the north of Hawaii (Table 4.8).

Post-El Niño effects also can significantly alter hurricane activity in the ENP because of its shallow thermocline. That is, the additional heat supply from the equatorial subsurface ocean can fuel TCs into major hurricanes in the summer and fall following an El Niño event. This effect appears to be stronger following an EP than a CP El Niño. TC activity after CP events are not statistically different from neutral or even La Niña conditions. A case study of the strong El Niño of 2015 found that the subtropical warming associated with the positive PMM was more important than the tropical warming for TC genesis over the ENP that year. The PMM-like SST anomalies appear to be conducive for significantly shifting the mean TC genesis location.

Hurricane activity over the ENP has undergone decadal variations. Using a Bayesian paradigm and examining TC data for 1972–2003, abrupt shifts in annual major hurricane counts occur around 1982 and 1999, with three periods identified spanning an inactive 1972–1981 epoch, active 1982–1998 epoch, and inactive 1999–2003 epoch. The identification of abrupt shifts in major hurricane activity provides a basis for studying decadal hurricane variations. The Bayesian model also can be used to forecast decadal hurricane fluctuations. In this case, a forecast issued in 2003 for ENP major hurricane activity for 2004–2013 called for low activity, which is consistent with actual observation.

The central North Pacific (CNP) is defined to be north of the equator and extends from 140°W to the date line. In this sub-basin, TC counts include storms generated in-situ and those that form in the ENP and subsequently track into the CNP. This region has low TC counts but large interannual variability. For instance, the mean annual number of CNP TCs is only 2.9 but the standard deviation is as large as the annual mean. This is particularly exemplified by the hyperactive 2015 CNP hurricane season when 14 TCs were reported. Of these 14 TCs, 9 reached hurricane strength. Other years, however, have had 0 or just one CNP TC.

TC activity in the CNP is modulated by the MJO, with increased TC activity observed in the CNP when the CNP and ENP are in the convectively enhanced phase of the MJO. This sub-basin also sees an elevated level of TC frequency during an EP El Niño (Table 4.7), and this increase is due to both an eastward excursion of the monsoon trough from the WNP and the westward expansion of genesis location and tracks from the ENP. During a CP event (Table 4.8), there is also an increase in TC genesis and track density, whereas TC activity is much reduced during a La Niña. Moreover, subtropical warming associated with a positive PMM phase is perhaps more important than tropical warming for modulating TC activity in this region. Decadal TC variations are also seen in this region with shifts occurring in 1982 and 1995. Diagnostic studies contrasting the active 1982–1994 and the inactive 1966–1981 epochs during the peak hurricane season reveal that warmer SST, lower SLP, reduced VWS, an increase in low-level cyclonic vorticity, and a deeper moisture layer are found over the MDR of the CNP in the active epoch. While being statistically significant based on a nonparametric test, these changes in environmental conditions favor more CNP TCs during the active 1982–1994 epoch.

The South Pacific (SP) TC genesis rate is also increased during the convectively active MJO phase in the central southwest Pacific. Based on the genesis potential index (GPI), mid-tropospheric relative humidity is the single most important parameter in the GPI. As in the WNP, TC frequency and formation locations in SP change substantially when El Niño years are compared with La Niña years (Table 4.7). While TC frequency in the Coral Sea is reduced during EP El Niño, increased cyclogenesis and tracks are noticed in the central southwest Pacific. This region includes Vanuatu, New Caledonia, Fiji, Samoa, and Tonga (Fig. 4.36). Islands that are normally not hurricane-prone such as French Polynesia are threatened more often by TCs during strong EP El Niño events. To the west of 170°E, TCs tend to track westward between 10°S and 15°S during EP years. Larger ACE values are noted equatorward of 15°S during El Niño than La Niña phase. The impact of CP El Niño on TC activity in the South Pacific has not yet been investigated.

While the North Atlantic is far removed from the typical MJO source region of the Indian Ocean, TC activity in this basin is still modulated by the MJO. For example, the number of major hurricanes in the North Atlantic during the convectively active MJO phase over Africa and the western Indian Ocean is three times higher than during the inactive phase. For the North Atlantic, fewer hurricanes and major hurricanes, fewer hurricane days and a lower probability of US hurricane landfalls occur in EP El Niño years relative to La Niña years (Table 4.7). The US coast faces a higher risk of hurricane landfalls in La Niña years, with more hurricanes also crossing the lower Caribbean basin. For CP El Niño (Table 4.8), TCs track more often across the US East Coast and the Gulf of Mexico, while they decrease in the central and eastern North Atlantic.

During the five strongest positive AMM years from 1950 to 2005, the average number of hurricanes (major hurricanes) is 10 (6), in contrast to 4 hurricanes and 1 major hurricane during the five strongest negative AMM. The AMM affects VWS in the MDR, thus affecting TC activity over the tropical North Atlantic, while El Niño primarily increases VWS closer to the coast of Central America, which in turn modulates TC activity over the Caribbean and Gulf of Mexico. TC formation locations shift equatorward and westward as the hurricane season progresses so the Caribbean Sea typically becomes more vulnerable to hurricane risks later in season. On a longer time scale, Atlantic TCs are also modulated by the Atlantic Multidecadal Oscillation (AMO). Storm activity is higher during positive phases of AMO than the negative phases of AMO, and this difference is most evident for major hurricanes and major hurricane days. Hurricane landfalls along the US East Coast are more frequent during positive AMO phases.

The last section of this chapter deals with the observed changes in TC activity. Historical TC frequency does not provide a clear indication for whether there is an increasing or decreasing trend on a global scale, with only the Atlantic basin seeing an increasing trend in recent decades. The latitude of lifetime maximum intensity of TCs appears to shift poleward globally during 1980–2012, with this poleward shift most evident in the WNP, South Pacific, and South Indian Ocean. For TC motion, a significant decreasing trend in translation speed was observed based on the record from 1949 to 2016, and this slowdown appears to coincide with the weakening of tropical circulation in a warmer climate. However, a recent study using post-satellite data (1982–2017) found an increase in global mean TC translation speed. This increase was attributed to two factors: an increase (4.5% per decade) in the relative frequency of TC passage points in the North Atlantic and a poleward shift of TC activities. The proportion of hurricanes reaching category 4 and 5 increased significantly from 1970 to 2004, although that proportion has leveled off in recent years. Landfalling typhoons in East Asia and Southeast Asia increased by 12–15% from 1977 to 2014.

Exercises

4.1 Describe the modulation of different MJO phases on TC activity over the western North Pacific according to the classification by Wheeler and Hendon (2004) of eight MJO phases. In your discussion, please include changes in named storms, hurricanes, major hurricanes, ACE, and rapid intensification during 24-h periods. (b) What environmental conditions are most important for these changes?

4.2 (a) Describe the modulation of different MJO phases on TC activity over the eastern and central North Pacific. (b) What environmental conditions are most important for the MJO phase changes on eastern North Pacific TC activity? You may use the figures, table, and eight MJO phases from Wheeler and Hendon (2004) for your discussion.

4.3 (a) Discuss the differences in typhoon landfalls on the East Asian coasts in JASO during the Eastern and Central Pacific El Niño events. (b) Justify your answer.

4.4 Discuss changes in tropical cyclone (TC) genesis location and tracks during El Niño, neutral, and La Niña phases over the southwest Pacific TC season and their possible causes.

4.5 (a) Explain the spatial patterns of sea surface temperature anomalies and anomalous low-level winds associated with the positive phase of the Atlantic Meridional Mode (AMM). (b) How might the AMM modulate the North Atlantic seasonal hurricane frequency, genesis location, and storm duration?

4.6 (a) Describe the Atlantic Multidecadal Oscillation (AMO) and its variations over the last 100 years or so. (b) How might the AMO influence the seasonal TC activity over the North Atlantic?

4.7 (a) What is the linear trend in TC frequency of occurrence in various ocean basins (e.g., North Atlantic, western North Pacific) since 1980 when satellite observations became available? (b) How would you attribute the observed spatial patterns in global TC trends (e.g., natural climate variability or external forcing)? Explain your reasoning.

4.8 Why there is a significant increase in TC frequency over the North Atlantic over the last 40 years?

4.9 (a) What are the trends in translation speed for the global, Northern Hemisphere, and Southern Hemisphere over the last 30–40 years? (b) What is the corresponding change in steering flow and is this change consistent with translation speed?

References

Aiyyer, A., and J. Molinari, 2008: MJO and tropical cyclogenesis in the Gulf of Mexico and eastern Pacific: Case study and idealized numerical modeling. *J. Atmos. Sci.*, **65**, 2691–2704.

Alexander, M. A., H. Seo, S.-P. Xie, and J. D. Scott, 2012: ENSO's impact on the gap wind regions of the eastern Tropical Pacific Ocean. *J. Climate*, **25**, 3549–3565.

Arpe, K., and S. A. G. Leroy, 2009: Atlantic hurricanes – Testing impacts of local SSTs, ENSO, stratospheric QBO: Implications for global warming. *Quat. Int.*, **195**, 4–14.

Avila, L. A., 2016: *2015 Eastern North Pacific Hurricane Season. National Hurricane Center Annual Summary.* National Weather Service, NOAA.

Barrett, B. S., and L. M. Leslie, 2009: Links between tropical cyclone activity and Madden–Julian Oscillation phase in the North Atlantic and Northeast Pacific basins. *Mon. Wea. Rev.*, **137**, 727–744.

Basher, R. E., and X. Zheng, 1995: Tropical cyclones in the southwest Pacific: Spatial patterns and relationships to Southern Oscillation and sea surface temperature. *J. Climate*, **8**, 1249–1260.

Bell, G. D., and Coauthors, 2000: Climate assessment for 1999. *Bull. Amer. Meteorol. Soc.*, **81**, S1–S50.

Bengtsson, L., and Coauthors, 2007: How may tropical cyclones change in a warmer climate? *Tellus*, **59A**, 539–561.

Bhatia, K. T., and Coauthors, 2019: Recent increases in tropical cyclone intensification rates. *Nat. Commun.*, **10**, 635.

Booth, B. B. B., and Coauthors, 2012: Aerosols implicated as a prime driver of twentieth-century North Atlantic climate variability. *Nature*, **484**, 228–232.

Boucharel, J., and Coauthors, 2016: Different controls of tropical cyclone activity in the eastern Pacific for two types of El Niño. *Geophys. Res. Lett.*, **43**, 1679–1686. doi:10 .1002/2016GL067728.

Bove, M. C., and Coauthors, 1998: Effect of El Niño on U.S. landfalling hurricanes, revisited. *Bull. Amer. Meteorol. Soc.*, **76**, 2477–2482.

Briegel, L. M., and W. M. Frank, 1997: Large-scale influences on tropical cyclogenesis in the western North Pacific. *Mon. Wea. Rev.*, **125**, 1397–1413.

Camargo, S. J., and A. H. Sobel, 2005: Western North Pacific tropical cyclone intensity and ENSO. *J. Climate*, **18**, 2996–3006.

Camargo, S. J., and A. H. Sobel, 2010: Revisiting the influence of the Quasi-Biennial Oscillation on tropical cyclone activity. *J. Climate*, **23**, 5810–5825.

Camargo, S. J., K. A. Emanuel, and A. H. Sobel, 2007a: Use of a genesis potential index to diagnose ENSO effects on tropical cyclone genesis. *J. Climate*, **20**, 4819–4834.

Camargo, S. J., A. W. Robertson, S. J. Gaffney, P. Smith, and M. Ghil, 2007b: Cluster analysis of typhoon tracks. Part I: General properties. *J. Climate*, **20**, 3635–3653.

Camargo, S. J., A. W. Robertson, S. J. Gaffney, P. Smith, and M. Ghil, 2007c: Cluster analysis of typhoon tracks. Part II: Large-scale circulation and ENSO. *J. Climate*, **20**, 3654–3676.

Camargo, S. J., M. C. Wheeler, and A. H. Sobel, 2009: Diagnosis of the MJO modulation of tropical cyclogenesis using an empirical index. *J. Atmos. Sci.*, **66**, 3061–3074.

Chan, J. C. L., 1995: Tropical cyclone activity in the western North Pacific in relation to the stratospheric quasi-biennial oscillation. *Mon. Wea. Rev.*, **123**, 2567–2571.

Chan, J. C. L., 2000: Tropical cyclone activity over the western North Pacific associated with El Niño and La Niña events. *J. Climate*, **13**, 2960–2972.

Chan, J. C. L., and W. M. Gray, 1982: Tropical cyclone movement and surrounding flow relationships. *Mon. Wea. Rev.*, **110**, 1354–1374.

Chan, J. C. L., and K. S. Liu, 2004: Global warming and western North Pacific typhoon activity from an observational perspective. *J. Climate*, **17**, 4590–4602.

Chan, K. T. F., 2019: Are global tropical cyclones moving slower in a warming climate? *Environ. Res. Lett.*, **14**(10), 104014.

Chand, S. S., and K. J. E. Walsh, 2009: Tropical cyclone activity in the Fiji region: Spatial patterns and relationship to large-scale circulation. *J. Climate*, **22**, 3877–3893.

Chand, S. S., and K. J. E. Walsh, 2010: The influence of the Madden–Julian Oscillation on tropical cyclone activity in the Fiji region. *J. Climate*, **23**, 868–886.

Chand, S. S., and K. J. E. Walsh, 2011: Influence of ENSO on tropical cyclone intensity in the Fiji region. *J. Climate*, **24**, 4096–4108.

Chand, S. S., et al., 2013: Impact of different ENSO regimes on Southwest Pacific tropical cyclones. *J. Climate*, **26**, 600–608.

Chelton, D. B., M. H. Freilich, and S. K. Esbensen, 2000: Satellite observations of wind jets off the Pacific coast of Central America. Part I: Case studies and statistical characteristics. *Mon. Wea. Rev.*, **128**, 1993–2018.

Chen, G., 2011: How does shifting Pacific Ocean warming modulate on tropical cyclone frequency over the South China Sea? *J. Climate*, **24**, 4695–4700.

Chen, G., and C. Chou, 2014: Joint contribution of multiple equatorial waves to tropical cyclogenesis over the western North Pacific. *Mon. Wea. Rev.*, **142**, 79–93.

Chen, G., and C.-Y. Tam, 2010: Different impacts of two kinds of Pacific warming on tropical cyclone frequency over the western North Pacific. *Geophy. Res. Lett.*, **37**, L01803.

Chen, J.-M., C.-H. Wu, P.-H. Chung, and C.-H. Sui, 2018: Influence of intraseasonal-interannual oscillations on tropical cyclone genesis in the western North Pacific. *J. Climate*, **31**, 4949–4961.

Chen, T.-C., S.-P. Weng, N. Yamazaki, and S. Kiehne, 1998: Interannual variation in the tropical cyclone formation over the western North Pacific. *Mon. Wea. Rev.*, **126**, 1080–1090.

Chen, T.-C., S.-Y. Wang, and M.-C. Yen, 2006: Interannual variation of the tropical cyclone activity over the western North Pacific. *J. Climate*, **19**, 5709–5720.

Chia, H. H., and C. F. Ropelewski, 2002: The interannual variability in the genesis location of tropical cyclones in the northwest Pacific. *J. Climate*, **15**, 2934–2944.

Chiang, J. C. H., and D. J. Vimont, 2004: Analogous Pacific and Atlantic Meridional Modes of tropical atmosphere-ocean variability. *J. Climate*, **17**, 4143–4158.

Chien, F. C., and H. C. Kuo, 2011: On the extreme rainfall of Typhoon Morakot (2009). *J. Geophys. Res.*, **116**, D05104.

Chowdhury, M. R., and P.-S. Chu, 2015: Sea level forecasts and early-warning application: Expanding cooperation in the South Pacific. *Bull. Amer. Meteorol. Soc.*, **96**, 381–386.

Chowdhury, M. R., P.-S. Chu, and C. Guard, 2014: An improved sea level forecasting scheme for hazards management in the U.S. affiliated Pacific islands. *Int. J. Climatol.*, **34**, 2320–2329.

Chu, P.-S., 2002: Large-scale circulation features associated with decadal variations of tropical cyclone activity over the central North Pacific. *J. Climate*, **15**, 2678–2689.

Chu, P.-S., 2004: ENSO and tropical cyclone activity. In *Hurricanes and Typhoons: Past, Present and Future*. R. J. Murnane and K. B. Liu, Eds. Columbia University Press, 297–332.

Chu, P.-S., and J. Wang, 1997: Tropical cyclone occurrences in the vicinity of Hawaii: Are the differences between El Niño and non-El Niño years significant? *J. Climate*, **10**, 2683–2689.

Chu, P.-S., and X. Zhao, 2004: Bayesian change-point analysis of tropical cyclone activity: The central North Pacific case. *J. Climate*, **17**, 4893–4901.

Chu, P.-S., and X. Zhao, 2011: Bayesian analysis for extreme climatic events: A review. *Atmos. Res.*, **102**, 243–262.

Chu, P.-S., X. Zhao, and J.-H. Kim, 2010: Regional typhoon activity as revealed by track patterns and climate change. In *Hurricanes and Climate Change*, Vol. 2. J. Elsner et al., Eds. Springer, 137–148.

Chu, P.-S., J.-H. Kim, and Y. R. Chen, 2012: Have steering flows in the western North Pacific and the South China Sea changed over the last 50 years? *Geophys. Res. Lett.*, **39**, L10704, doi:10.1029/2012GL051709.

Church, J. A., N. J. White, and J. R. Hunter, 2006: Sea-level rise at tropical Pacific and Indian Ocean islands. *Glob. Planet. Change*, **53**, 155–168.

Church, J. A., P. L. Woodworth, T. Aarup, and S. Wilson, 2010: *Sea Level Rise and Vulnerability.* Wiley.

Clark, J. D., and P.-S. Chu, 2002: Interannual variation of tropical cyclone activity over the central North Pacific. *J. Meteorol. Soc. Japan,* **80,** 403–418.

Collimore, C. C., et al., 2003: On the relationship between the QBO and tropical dep convection. *J. Climate,* **16,** 2552–2568.

Collins, J. M., 2010: Contrasting high North East Pacific tropical cyclone activity. *Southeast Geogr.,* **50,** 83–98, doi:10.1353/sgo.0.0069.

Collins, J. M., and I. M. Mason, 2000: Local environmental conditions related to seasonal tropical cyclone activity in the Northeast Pacific basin. *Geophys. Res. Lett.,* **27,** 3881–3884. doi:10.1029/2000GL011614.

Collins. J. M., and Coauthors, 2016: The record-breaking 2015 hurricane season in the eastern North Pacific: An analysis of environmental conditions. *Geophys. Res. Lett.,* **43,** 9217–9224.

Daloz, A. S., and S. J. Camargo, 2018: Is the poleward migration of tropical cyclone maximum intensity associated with a poleward migration of tropical cyclone genesis? *Clim. Dyn.,* **50,** 705–715.

Delworth, T. L., and M. E. Mann, 2000: Observed and simulated multidecadal variability in the Northern Hemisphere. *Clim. Dyn.,* **16,** 661–676.

Diamond, H. J., A. M. Lorrey, and J. A. Renwick, 2013: A Southwest Pacific tropical cyclone climatology and linkages to the El Niño-Southern Oscillation. *J. Climate,* **26,** 3–25.

Dowdy, A. J., and Coauthors, 2012: Tropical cyclone climatology of the South Pacific Ocean and its relationship to El Niño-Southern Oscillation. *J. Climate,* **25,** 6108–6122.

Du, Y., L. Yang, and S.-P. Xie, 2011: Tropical Indian Ocean influence on northwest Pacific tropical cyclones in summer following strong El Niño. *J. Climate,* **24,** 315–322.

Dunn, G. E., 1940: Cyclogenesis in the tropical Atlantic. *Bull. Amer. Meteorol. Soc.,* **21,** 215–229.

Dunstone, N. J., and Coauthors, 2013: Anthropogenic aerosol forcing of Atlantic tropical storms. *Nat. Geosci.,* **6,** 534–539, doi:10.1038/ngeo1854.

Elsner, J. B., and A. B. Kara, 1999: *Hurricanes of the North Atlantic.* Oxford University Press.

Elsner, J. B., T. Jagger, and X.-F. Niu, 2000: Changes in the rates of North Atlantic major hurricane activity during the 20th century. *Geophys. Res. Lett.,* **27,** 1743–1746.

Elsner, J. B., J. P. Kossin, and T. H. Jagger, 2008: The increasing intensity of the strongest tropical cyclones. *Nature,* **455**(7209), 92–95.

Emanuel, K. A., 1988: The maximum intensity of hurricanes. *J. Atmos. Sci.,* **45,** 1143–1155.

Emanuel, K. A., 2013: Downscaling CMIP5 climate models shows increased tropical cyclone activity over the 21st century. *Proc. Natl. Acad. Sci.,* **110**(30), 12219–12224, doi:10.1073/pnas.1301293110.

Emanuel, K. A., 2017: Assessing the present and future probability of Hurricane Harvey's rainfall. *Proc. Natl. Acad. Sci., USA,* **114,** 12681–12684.

Emanuel, K. A., and D. S. Nolan, 2004: Tropical cyclone activity and global climate. Preprint, *26th Conference on Hurricanes and Tropical Meteorology, Miami, FL.* Amer. Meteorol. Soc., 240–241.

Evans, J. L., and R. Allan, 1992: El Niño/Southern modification to the structure of the monsoon and tropical cyclone activity in the Australian region. *Int. J. Climatol.,* **12,** 611–623.

Folland, C. K., J. A. Renwick, M. J. Salinger, and A. B. Mullan, 2002: Relative influences of the interdecadal Pacific Oscillation and ENSO on the South Pacific Convergence Zone. *Geophys. Res. Lett.,* **29**(13), 1643, doi:10.1029/2001GL014201.

Frank, N. L., 1970: Atlantic tropical systems of 1969. *Mon. Wea. Rev.*, **98**, 307–314.

Frank, W. M., 1982: Large-scale characteristics of tropical cyclones. *Mon. Wea. Rev.*, **110**, 577–586.

Frank, W. M., and P. E. Roundy, 2006: The role of tropical waves in tropical cyclogenesis. *Mon. Wea. Rev.*, **134**, 2397–2417.

Fu, B., T. Li, M. S. Peng, and F. Weng, 2007: Analysis of tropical cyclogenesis in the western North Pacific for 2000 and 2001. *Wea. Forecasting*, **22**, 763–780.

Fu, D., P. Chang, and C. M. Patricola, 2017: Intrabasin variability of East Pacific tropical cyclones during ENSO regulated by central American gap winds. *Sci. Rep.*, **7**, 1658, doi:10.1038/s41598-017-01962-3.

Gao, S., L. Zhu, W. Zhang, and Z. Chen, 2018: Strong modulation of the Pacific Meridional Mode on the occurrence of intense tropical cyclones over the western North Pacific. *J. Climate*, **31**, 7739–7749.

Ge, X., T. Li, S. Zhang, and M. S. Peng, 2010: What causes the extremely heavy rainfall in Taiwan during Typhoon Morakot (2009)? *Atmos. Sci. Lett.*, **11**, 46–50.

Gill, A. E., 1980: Some simple solutions for heat-induced tropical circulation. *Quart. J. Roy. Meteorol. Soc.*, **106**, 447–462.

Goldenberg, S. B., C. W. Landsea, A. M. Mestas-Nuñez, and W. M. Gray, 2001: The recent increase in Atlantic hurricane activity: Causes and implications. *Science*, **293**, 474–479.

Gray, W. M., 1979: Hurricanes: Their formation, structure and likely role in the tropical circulation. In *Meteorology over Tropical Oceans*, D. B. Shaw, Ed. Royal Meteorological Society, 155–218.

Gray, W. M., 1984: Atlantic seasonal hurricane frequency: Part I: El Nino and 30 mb quasi-biennial oscillation influences. *Mon. Wea. Rev.*, **112**, 1649–1668.

Gray, W. M., 1993: Seasonal forecasting. Chapter 5 in *Global Guide to Tropical Cyclone Forecasting*. World Meteorological Organization, WMO/TD-No. 560.

Gray, W. M., J. D. Sheaffer, and J. A. Knaff, 1992a: Hypothesized mechanism for stratospheric QBO influence on ENSO variability. *Geophys. Res. Lett.*, **19**(2), 107–110.

Gray, W. M., J. D. Sheaffer, and J. A. Knaff, 1992b: Influence of the stratospheric QBO on ENSO variability. *J. Meteorol. Soc. Japan*, **70**, 975–995.

Gualdi, S., E. Scoccimarro, and A. Navarra, 2008: Changes in tropical cyclone activity due to global warming: Results from a high-resolution coupled general circulation model. *J. Climate*, **21**, 5204–5228.

Haarsma, R. J., J. F. B. Mitchell, and C. A. Senior, 1993: Tropical disturbances in a GCM. *Clim. Dyn.*, **8**, 247–257.

Hall, J. D., A. J. Matthews, and D. J. Karoly, 2001: The modulation of tropical cyclone activity in the Australian region by the Madden–Julian Oscillation. *Mon. Wea. Rev.*, **129**, 2970–2982.

Harr, P. A., and R. L. Elsberry, 1995a: Large-scale circulation variability over the tropical western North Pacific. Part I: Spatial patterns and tropical cyclone characteristics. *Mon. Wea. Rev.*, **123**, 1225–1246.

Harr, P. A., and R. L. Elsberry, 1995b: Large-scale circulation variability over the tropical western North Pacific. Part II: Persistence and transition characteristics. *Mon. Wea. Rev.*, **123**, 1247–1268.

Hastings, P. A., 1990: Southern Oscillation influences on tropical cyclone activity in the Australian/Southwest Pacific region. *Int. J. Climatol.*, **10**, 291–298.

He, H., and Coauthors, 2015: Decadal changes in tropical cyclone activity over the western North Pacific in the late 1990s. *Clim. Dyn.*, **45**, 3317–3329.

Ho, C.-H., H. S. Kim, J. H. Jeong, and S. W. Son, 2009: Influence of stratospheric quasi-biennial oscillation on tropical cyclone tracks in the western North Pacific. *Geophys. Res. Lett.*, **36**, L06702.

Holbach, H. M., and M. A. Bourassa, 2014: The effects of gap-wind-induced vorticity, the monsoon trough, and the ITCZ on east Pacific tropical cyclogenesis. *Mon. Wea. Rev.*, **142**, 1312–1325.

Holland, G. J., 1995: Scale interaction in the western Pacific monsoon. *Meteorol. Atmos. Phys.*, **56**, 57–79.

Holland, G. J., and C. Bruyere, 2014: Recent intense hurricane response to global climate change. *Clim. Dyn.*, **42**, 617–627.

Hong, C.-C., M.-Y. Lee, H.-H. Hsu, and W.-L. Tseng, 2018: Distinct influences of the ENSO-like and PMM-like SST anomalies on the mean TC genesis location in the western North Pacific: The 2015 summer as an extreme example. *J. Climate*, **31**, 3049–3059.

Huang, P., C. Chou, and R. Huang, 2011: Seasonal modulation of tropical intraseasonal oscillations on tropical cyclone geneses in the western North Pacific. *J. Climate*, **24**, 6339–6352.

Hsu, P.-C., P.-S. Chu, H. Murakami, and X. Zhao, 2014: An abrupt decrease in the late-season typhoon activity over the western North Pacific. *J. Climate*, **27**, 4296–4312.

Huo, L., P. Guo, S. N. Hameed, and D. Jin, 2015: The role of tropical Atlantic SST anomalies in modulating western North Pacific tropical cyclone genesis. *Geophys. Res. Lett.*, **42**, 2378–2384.

Irwin, R. P. III, and R. E. Davis, 1999: The relationship between the Southern Oscillation index and tropical cyclone tracks in the eastern North Pacific. *Geophys. Res. Lett.*, **26**, 2251–2254, doi:10.1029/1999GL900533.

Jien, J. Y., W. A. Gough, and K. Butler, 2015: The influence of El Niño-Southern Oscillation on tropical cyclone activity in the eastern North Pacific basin. *J. Climate*, **28**, 2459–2474.

Jin, C.-S., and Coauthors, 2013: Critical role of northern off-equatorial sea surface temperature forcing associated with central Pacific El Niño in more frequent tropical cyclone movements toward East Asia. *J. Climate*, **26**, 1534–2545.

Jin, F.-F., J. Boucharel, and I.-I. Lin, 2014: Eastern Pacific tropical cyclones intensified by El Niño delivery of subsurface ocean heat. *Nature*, **516**, 82–85.

Karnauskas, K. B., 2006: The African meridional OLR contrast as a diagnostic for Atlantic tropical cyclone activity and implications for predictability. *Geophys. Res. Lett.*, **33**, L06809.

Karnauskas, K. B., and L. Li, 2016: Predicting Atlantic seasonal hurricane activity using outgoing longwave radiation over Africa. *Geophys. Res. Lett.*, **43**, 7152–7159.

Kim, H.-M., P. J. Webster, and J. A. Curry, 2009: Impact of shifting patterns of Pacific Ocean warming on North Atlantic tropical cyclones. *Science*, **325**, 77–80.

Kim, H.-M., P. J. Webster, and J. A. Curry, 2011a: Modulation of North Pacific tropical cyclone activity by three phases of ENSO. *J. Climate*, **24**, 1839–1849.

Kim, H.-S., J.-H., Kim, C.-H. Ho, and P.-S. Chu, 2011b: Pattern classification of typhoon tracks using the fuzzy c-Means clustering method. *J. Climate*, **24**, 488–508.

Kim, J.-H., C.-H. Ho, H.-S. Kim, C.-H. Sui, and S. K. Park, 2008: Systematic variation of summertime tropical cyclone activity in the western North Pacific in relation to the Madden-Julian Oscillation. *J. Climate*, **21**, 1171–1191.

Kim, S.-H., I.-J. Moon, and P.-S. Chu, 2020: An increase in global trends of tropical cyclone translation speed since 1982 and its physical causes. *Environ. Res. Lett.*, **15**, 094084.

Klotzbach, P. J. 2010: On the Madden–Julian Oscillation-Atlantic hurricane relationship. *J. Climate*, **23**, 282–293.

Klotzbach, P. J., 2014: The Madden–Julian Oscillation's impacts on worldwide tropical cyclone activity. *J. Climate*, **27**, 2317–2330.

Klotzbach, P. J., and E. S. Blake, 2013: North-central Pacific tropical cyclones: Impact of El Niño-Southern Oscillation and the Madden–Julian oscillation. *J. Climate*, **26**, 7720–7733.

Klotzbach, P. J., and W. M. Gray, 2008: Multidecadal variability in North Atlantic tropical cyclone activity. *J. Climate*, **21**, 3929–3935.

Klotzbach, P. J., and C. Landsea, 2015: Extremely intense hurricanes: Revisiting Webster et al. (2005) after 10 years. *J. Climate*, **28**, 7621–7629.

Klotzbach, P. J., S. G. Bowen, R. Pielke Jr., and M. Bell, 2018: Continental U.S. hurricane landfall frequency and associated damage: Observations and future risks. *Bull. Amer. Meteorol. Soc.*, **99**, 1359–1376.

Knaff, J. A., 1997: Implications of summertime sea level pressure anomalies in the tropical Atlantic region. *J. Climate*, **10**, 789–804.

Knight, J. R., R. J. Allan, C. K. Folland, M. Vellinga, and M. E. Mann, 2005: A signature of persistent natural thermohaline circulation cycles in observed climate. *Geophys. Res. Lett.*, **32**, L20708.

Knutson, T., and Coauthors, 2010: Tropical cyclones and climate change. *Nat. Geosci.*, **3**, 157–163.

Knutson, T., and Coauthors, 2019: Tropical cyclones and climate change assessment. Part I: Detection and attribution. *Bull. Amer. Meteorol. Soc.*, **100**, 1987–2007.

Kossin, J. P., 2018: A global slowdown of tropical cyclone translation speed. *Nature*, **558**, 104–108.

Kossin, J. P., and D. J. Vimont, 2007: A more general framework for understanding Atlantic hurricane variability and trends. *Bull. Amer. Meteorol. Soc.*, **88**, 1767–1782.

Kossin, J. P., T. L. Olander, and K. R. Knapp, 2013: Trend analysis with a new global record of tropical cyclone intensity. *J. Climate*, **26**, 9960–9976.

Kossin, J. P., K. A. Emanuel, and G. A. Vecchi, 2014: The poleward migration of the location of tropical cyclone maximum intensity. *Nature*, **509**, 349–352.

Kossin, J. P., K. Knapp, T. Olander, and C. Velden, 2020: Global Increase in major tropical cyclone exceedance probability over the past four decades. *Proc. Natl. Acad. Sci.*, **117**, 11975–11980.

Kuleshov, Y., L. Qi, R. Fawcett, and D. Jones, 2008: On tropical cyclone activity in the Southern Hemisphere: Trends and the ENSO connection. *Geophys. Res. Lett.*, **35**, L14S08, doi:10.1029/2007GL032983.

Kug, J. S., F.-F. Jin, and S.-I. An, 2009: Two types of El Niño events: Cold tongue El Niño and warm pool El Niño. *J. Climate*, **22**, 1499–1515.

Kushnir, Y., 1994: Interdecadal variations in North Atlantic sea surface temperature and associated atmospheric conditions. *J. Climate*, **7**, 141–157.

Lander, M. A., 1994: An exploratory analysis of the relationship between tropical storm formation in the western North Pacific and ENSO. *Mon. Wea. Rev.*, **122**, 636–651.

Landsea, C. W., 1993: A climatology of intense (or major) Atlantic hurricanes. *Mon. Wea. Rev.*, **121**, 1703–1713.

Landsea, C. W., and J. L. Franklin, 2013: Atlantic hurricane data-base uncertainty and presentation of a new database format. *Mon. Wea. Rev.*, **141**, 3576–3592.

Landsea, C. W., B. A. Harper, K. Horaru, and J. A. Knaff, 2006: Can we detect trends in extreme tropical cyclones? *Science*, **313**, 452–454.

Lanzante, J. R., 2019: Uncertainties in tropical-cyclone translation speed. *Nature*, **570**, E6–E15.

Larson, S., and Coauthors, 2012: Impacts of non-canonical El Niño patterns on Atlantic hurricane activity. *Geophys. Res. Lett.*, **39**, L14706.

Lau, K.-M., and N.-C. Lau, 1992: The energetics and propagation dynamics of tropical summertime synoptic-scale disturbances. *Mon. Wea. Rev.*, **120**, 2523–2539.

Lee, S.-K., C. Wang, and D. B. Enfield, 2010: On the impact of central Pacific warming events on Atlantic tropical cyclone activity. *Geophys. Res. Lett.*, **37**, L17702.

Lee, T., and M. J. McPhaden, 2010: Increasing intensity of El Niño in the central equatorial Pacific. *Geophys. Res. Lett.*, **37**, L14603.

Li, R. C. Y., and W. Zhou, 2012: Changes in Western Pacific tropical cyclones associated with the El Niño-Southern Oscillation cycle. *J. Climate*, **25**, 5864–5878.

Li, R. C. Y., and W. Zhou, 2013a: Modulation of western North Pacific tropical cyclone activity by the ISO. Part I: Genesis and intensity. *J. Climate*, **26**, 2904–2918.

Li, R. C. Y., and W. Zhou, 2013b: Modulation of western North Pacific tropical cyclone activity by the ISO. Part II: Tracks and landfalls. *J. Climate*, **26**, 2919–2930.

Liang, A., L. Oey, S. Huang, and S. Chou, 2017: Long-term trends of typhoon-induced rainfall over Taiwan: In-situ evidence of poleward shift of typhoons in western North Pacific in recent decades. *J. Geophys. Res. Atmos.*, **122**, 2750–2765.

Liebmann, B., H. H. Hendon, and J. D. Glick, 1994: The relationship between tropical cyclones of the western Pacific and Indian Oceans and the Madden–Julian Oscillation. *J. Meteorol. Soc. Japan*, **72**, 401–412.

Lin, C. Y., et al., 2011: Mesoscale processes for super heavy rainfall of Typhoon Morakot (2009) over Southern Taiwan. *Atmos. Chem. Phys.*, **11**, 345–361.

Lin, I.-I., I.-F. Pun, and C.-C. Lien, 2014: "Category-6" supertyphoon Haiyan in global warming hiatus: Contribution from subsurface ocean warming. *Geophys. Res. Lett.*, **41**(23), 8547–8553, doi:10.1002/2014GL061281.

Liu, K. S., and J. C. L. Chan, 2008: Interdecadal variability of western North Pacific tropical cyclone tracks. *J. Climate*, **21**, 4464–4476.

Liu, K. S., and J. C. L. Chan, 2020: Interdecadal variation of frequencies of tropical cyclones, intense typhoons and their ratio over the western North Pacific. *Int. J. Climatol.*, **40**, 3954–3970.

Lu, B., P.-S. Chu, S.-H. Kim, and C. Karamperidou, 2020: Hawaiian regional climate variability during two types of El Niño. *J. Climate*, **33**, 9929–9943.

Maloney, E. D., and D. L. Hartmann, 2000a: Modulation of Eastern North Pacific hurricanes by the Madden–Julian Oscillation. *J. Climate*, **13**, 1451–1460.

Maloney, E. D., and D. L. Hartmann, 2000b: Modulation of hurricane activity in the Gulf of Mexico by the Madden–Julian Oscillation. *Science*, **287**, 2002–2004.

Maloney, E. D., and D. L. Hartmann, 2001: The Madden–Julian Oscillation, barotropic dynamics, and North Pacific tropical cyclone formation. Part I: Observations. *J. Atmos. Sci.*, **58**, 2545–2558.

Mann, M. E., K. A. Emanuel, 2006: Atlantic hurricane trends linked to climate change. *Eos, Trans. Amer. Geophys. Union*, **87**, 233–244.

Mann, M. E., B. A. Steinman, D. J. Brouilette, and S. K. Miller, 2021: Multidecadal climate oscillations during the past millennium driven by volcanic forcing. *Science*, **371**, 1014–1019.

Mao, J. Y., and G. Wu, 2010: Intraseasonal modulation of tropical cyclogenesis in the western North Pacific: A case study. *Theor. Appl. Climatol.*, **100**, 397–411.

McBride, J. L., 1995: Tropical cyclone formation. In *Global Perspectives on Tropical Cyclones*. R. L. Elsberry, Ed. World Meteorological Organization, WMO/TD-No. 693, 63–105.

Mei, W., and S.-P. Xie, 2016: Intensification of land falling typhoons over the northwest Pacific since the late 1970s. *Nat. Geosci.*, **9**, 753–757.

Molinari, J., D. Vollaro, S. Skubis, and M. Dickinson, 2000: Origins and mechanisms of Eastern North Pacific tropical cyclogenesis: A case study. *Mon. Wea. Rev.*, **128**, 125–139.

Moon, I.-J., S. H. Kim, and J. C. L. Chan, 2019: Climate change and tropical cyclone trend. *Nature*, **570**, E3–E5.

Murakami, H., and Coauthors, 2017: Dominant role of subtropical Pacific warming in extreme eastern Pacific hurricane seasons: 2015 and the future. *J. Climate*, **30**, 243–264.

Murakami, H., and Coauthors, 2018: Dominant effect of relative tropical Atlantic warming on major hurricane occurrence. *Science*, **362**, 794–799.

Murakami, H., and Coauthors, 2020: Detected climatic change in global distribution of tropical cyclones. *Proc. Natl. Acad. Sci.*, **117**, 10706–10714.

Nicholls, N., 1979: A possible method for predicting seasonal tropical cyclone activity in the Australian region. *Mon. Wea. Rev.*, **107**, 1221–1224.

Nicholls, N., 1985: Predictability of interannual variations of Australian seasonal tropical cyclone activity. *Mon. Wea. Rev.*, **113**, 1144–1149.

Nugent, A., and Coauthors, 2020: Fire and rain: The legacy of Hurricane Lane in Hawaii. *Bull. Amer. Meteorol. Soc.*, **101**, 954–967.

O'Brien, J. J., T. S. Richards, and A. C. Davis, 1996: The effect of El Niño on U.S. landfalling hurricanes. *Bull. Amer. Meteorol. Soc.*, **77**, 773–774.

Okun, H., 2021: Cluster analysis of eastern and central North Pacific tropical cyclones and the influences of ENSO and MJO. M.S. thesis, Department of Atmospheric Sciences, University of Hawaii, 45 pp.

Park, D.-S., C.-H. Ho, and J.-H. Kim, 2014: Growing threat of intense tropical cyclones to East Asia over the period 1977-2010. *Environ. Res. Lett.*, **9**, 014008, doi:10.1088/1748.9326/9/1/014008.

Patricola, C. M., P. Chang, and R. Saravanan, 2016: Degree of simulated suppression of Atlantic tropical cyclones modulated by flavour of El Niño. *Nat. Geosci.*, **9**, 155–160.

Patricola, C. M., and Coauthors, 2018a: The influence of ENSO flavors on western North Pacific tropical cyclone activity. *J. Climate*, **31**, 5395–5416.

Patricola, C. M., R. Saravanan, and P. Chang, 2018b: The response of Atlantic tropical cyclones to suppression of African easterly waves. *Geophys. Res. Lett.*, **45**, 471–479.

Pausata, F. S. R., and S. J. Camargo, 2019: Tropical cyclone activity affected by volcanically induced ITCZ shifts. *Proc. Natl. Acad. Sci.*, **116**, 7732–7737.

Pielke, R. A., Jr., and C. W. Landsea, 1999: La Nina, El Niño, and Atlantic hurricane damages in the United States. *Bull. Amer. Meteorol. Soc.*, **80**, 2027–2033.

Ralph, T. U., and W. A. Gough, 2009: The influence of sea surface temperature on Eastern North Pacific tropical cyclone activity. *Theor. Appl. Climatol.*, **95**, 257–264, doi:10.1007/s00704-008-0004-x.

Ramsay, H. S., S. J. Camargo, and D. Kim, 2012: Cluster analysis of tropical cyclone tracks in the Southern Hemisphere. *Clim. Dyn.*, **39**, 897–917.

Rappaport, E. N., and Coauthors, 1998: Eastern North Pacific hurricane season of 1995. *Mon. Wea. Rev.*, **126**, 1152–1162.

Revell, C. G., and S. W. Goulter, 1986: South Pacific tropical cyclones and the Southern Oscillation. *Mon. Wea. Rev.*, **114**, 1138–1145.

Ritchie, E. A., and G. J. Holland, 1999: Large-scale patterns associated with tropical cyclogenesis in the western Pacific. *Mon. Wea. Rev.*, **127**, 2027–2043.

Rodionov, S. N., 2004: A sequential algorithm for testing climate regime shifts. *Geophys. Res. Lett.*, **31**, L09204, doi:10.1029/2004GL019448.

Romero-Centeno, R., J. Zavala-Hidalgo, A. Gallegos, and J. J. O'Brien, 2003: Isthmus of Tehuantepec wind climatology and ENSO signal. *J. Climate*, **16**, 2628–2639.

Russell, J. O., A. Aiyyer, J. D. White, and W. Hannah, 2017: Revisiting the connection between African easterly waves and Atlantic tropical cyclogenesis. *Geophys. Res. Lett.*, **44**, 587–595.

Sadler, J., 1983: Tropical Pacific atmospheric anomalies during 1982–83. In *Proceedings of the 1982/83 El Niño/Southern Oscillation Workshop*. Miami, NOAA Atlantic Oceanographic and Meteorological Laboratory, 1–10.

Saunders, M. A., R. E. Chandler, C. J. Merchant, and F. P. Robert, 2000: Atlantic hurricanes and NW Pacific typhoons: ENSO spatial impacts on occurrence and landfall. *Geophys. Res. Lett.*, **27**, 1147.

Schreck, C. J., and J. Molinari, 2009: A case study of an outbreak of twin tropical cyclones. *Mon. Wea. Rev.*, **137**, 863–875.

Schreck, C. J., J. Molinari, and K. I. Mohr, 2011: Attributing tropical cyclogenesis to equatorial waves in the western North Pacific. *J. Atmos. Sci.*, **68**, 195–209.

Schreck, C. J., J. Molinari, and A. Aiyyer, 2012: A global view of equatorial waves and tropical cyclogenesis. *Mon. Wea. Rev.*, **140**, 774–788.

Schreck, C. J., K. R. Knapp, and J. P. Kossin, 2014: The impact of best track discrepancies and on global tropical cyclone climatologies using IBTrACS. *Mon. Wea. Rev.*, **142**, 3881–3899.

Shan, K., and X. Yu, 2020: Interdecadal variability of tropical cyclone genesis frequency in western North Pacific and South Pacific Ocean basins. *Environ. Res. Lett.*, **15**, 064030.

Shan, K., and X. Yu, 2021: Variability of tropical cyclone landfalls in China. *J. Climate*, **34**, 9235–9247.

Sobel, A. H., and C. S. Bretherton, 1999: Development of synoptic-scale disturbances over the summertime tropical northwest Pacific. *J. Atmos. Sci.*, **56**, 3106–3127.

Song, J., P. J. Klotzbach, and Y. Duan, 2020: Differences in western North Pacific tropical cyclone activity among three El Niño phases. *J. Climate*, **33**, 7983–8002.

Sugi, M., A. Noda, and N. Sato, 2002: Influence of the global warming on tropical cyclone climatology: An experiment with the JMA global model. *J. Meteorol. Soc. Japan*, **80**, 249–272.

Takahashi, C., M. Watanabe, and M. Mori, 2017: Significant aerosol influence on the recent decadal decrease in tropical cyclone activity over the western North Pacific. *Geophys. Res. Lett.*, **44**, 9496–9504.

Timmermann, A., and Coauthors, 2018: El Niño-Southern Oscillation complexity. *Nature*, **559**, 535–545.

Tu, J.-Y., C. Chou, and P.-S. Chu, 2009: The abrupt shift of typhoon activity in the vicinity of Taiwan and its association with western North Pacific-East Asian climate change. *J. Climate*, **22**, 3617–3628.

Vecchi, G. A., and B. J. Soden, 2007a: Effect of remote sea surface temperature change on tropical cyclone potential intensity. *Nature*, **450**, 1066–1071.

Vecchi, G. A., and B. J. Soden, 2007b: Global warming and the weakening of the tropical circulation. *J. Climate*, **20**, 4316–4340.

Vecchi, G. A., et al., 2013: Multiyear predictions of North Atlantic hurricane frequency: Promise and limitations. *J. Climate*, **26**, 5537–5557.

Ventrice, M. J., C. D. Thorncroft, and C. J. Schreck, 2012: Impacts of convectively coupled Kelvin waves on environmental conditions for Atlantic tropical cyclogenesis. *Mon. Wea. Rev.*, **140**, 2198–2214.

Vimont, D. J., and J. P. Kossin, 2007: The Atlantic meridional mode and hurricane activity. *Geophys. Res. Lett.*, **34**, L07709.

Vincent, D. G., 1994: The South Pacific convergence zone (SPCZ): A review. *Mon. Wea. Rev.*, **122**, 1949–1970.

Walsh, K. J. E., K. McInnes, and J. L. McBride, 2012: Climate change impacts on tropical cyclones and extreme sea levels in the South Pacific – A regional assessment. *Glob. Planet. Change*, **80–81**, 149–164.

Walsh, K. J. E., and Coauthors, 2015: Hurricanes and climate: The U.S. CLIVAR working group on hurricanes. *Bull. Amer. Meteorol. Soc.*, **96**, 997–1017.

Wang, B., and J. C. L. Chan, 2002: How strong ENSO events affect tropical storm activity over the western North Pacific? *J. Climate*, **15**, 1643–1658.

Wang, C., F. Kucharski, R. Barimalala, and A. Braccco, 2009: Teleconnections of the tropical Atlantic to the tropical Indian and Pacific Oceans: A review of recent findings. *Meteorol. Z.*, **18**, 445–454.

Wang, X., H. Liu, and G. R. Foltz, 2017: Persistent influence of tropical North Atlantic wintertime sea surface temperature on the subsequent Atlantic hurricane season. *Geophys. Res. Lett.*, **44**, 7927–7935.

Webster, P. J., G. J. Holland, J. A. Curry, and H. R. Chang, 2005: Changes in tropical cyclone number, duration, and intensity in a warming environment. *Science*, **309**, 1844–1846.

Weinkle, J., and Coauthors, 2018: Normalized hurricane damage in the continental United States 1900-2017. *Nat. Sustain.*, **1**, 808–813.

Wheeler, M. C., and H. H. Hendon, 2004: An all-season real-time multivariate MJO index: Development of an index for monitoring and prediction. *Mon. Wea. Rev.*, **132**, 1917–1932.

Whitney, L. D., and J. Hobgood, 1997: The relationship between sea surface temperatures and maximum intensities of tropical cyclones in the eastern North Pacific Ocean. *J. Climate*, **10**, 2921–2930.

Wu, C. C., 2013: Typhoon Morakot (2009): Key findings from the journal TAO for improving prediction of extreme rains at landfall. *Bull. Amer. Meteorol. Soc.*, **94**, 155–160.

Wu, L., Z. Wen, R. Huang, and R. Wu, 2012: Possible linkage between the monsoon trough variability and the tropical cyclone activity over the western North Pacific. *Mon. Wea. Rev.*, **140**, 140–150.

Wu, L., H. Zhang, J.-M. Chen, and T. Feng, 2018: Impact of two types of El Niño on tropical cyclones over the western North Pacific: Sensitivity to location and intensity of Pacific warming. *J. Climate*, **31**, 1725–1742.

Wu, L.-G., and B. Wang, 2004: Assessing impacts of global warming on tropical cyclone tracks. *J. Climate*, **17**, 1686–1698.

Wu, M. C., W. L. Chang, and W. M. Leung, 2004: Impacts of El Niño–Southern Oscillation events on tropical cyclone landfalling activity in the western North Pacific. *J. Climate*, **17**, 1419–1428.

Wu, P., and P.-S. Chu, 2007: Characteristics of tropical cyclone activity over the eastern North Pacific: The extremely active 1992 and the inactive 1977. *Tellus*, **59A**, 444–454.

Xie, B., and F. Zhang, 2012: Impacts of typhoon track and island topography on the heavy rainfalls in Taiwan associated with Morakot (2009). *Mon. Wea. Rev.*, **140**, 3379–3394.

Xie, S.-P., and Coauthors, 2009: Indian Ocean capacitor effect on Indo-Western Pacific climate during the summer following El Niño. *J. Climate*, **22**, 730–747.

Yamaguchi, M., and Coauthors, 2020: Global warming changes tropical cyclone translation speed. *Nat. Commun.*, **11**, 47.

Yan, X., R. Zhang, and T. R. Knutson, 2017: The role of Atlantic overturning circulation in the recent decline of Atlantic major hurricane frequency. *Nat. Commun.*, **8**, 1695.

Yeh, S.-W., and Coauthors, 2009: El Niño in a changing climate, *Nature*, **461**, 511–514, doi:10.1038/nature08316.

Yoshida, R., and H. Ishikawa, 2013: Environmental factors contributing to tropical cyclone genesis in the Western North Pacific. *Mon. Wea. Rev.*, **141**, 451–467.

You, L., J. Gao, H. Lin, and S. Chen, 2019: Impact of the intra-seasonal oscillation on tropical cyclone genesis over the western North Pacific. *Int. J. Climatol.*, **39**, 1969–1984.

Yu, C. K., and L.-W. Cheng, 2013: Distribution and mechanisms of orographic precipitation associated with Typhoon Morakot (2009). *J. Atmos. Sci.*, **70**, 2894–2915.

Yu, J., T. Li, Z. Tan, and Z. Zhu, 2016: Effects of tropical North Atlantic SST on tropical cyclone genesis in the western North Pacific. *Clim. Dyn.*, **46**, 865–877.

Yu, J.-Y., and S. T. Kim, 2013: Identifying the types of major El Niño events since 1870. *Int. J. Climatol.*, **33**, 2105–2112.

Zhan, R., Y. Wang, and X. Lei, 2011: Contribution of ENSO and East Indian Ocean SSTA to the interannual variability of Northwest Pacific tropical cyclone frequency. *J. Climate*, **24**, 509–521.

Zhan, R., Y. Wang, and M. Wen, 2013: The SST gradient between the southwestern Pacific and the western Pacific warm pool: A new factor controlling the norwestern Pacific tropical cyclone genesis frequency. *J. Climate*, **26**, 2408–2415.

Zhan, R., Y. Wang, and J. Zhao, 2017: Intensified mega-ENSO has increased the proportion of intense tropical cyclones over the western Northwest Pacific since the late 1970s. *Geophys. Res. Lett.*, **44**, 11959–11966.

Zhang, G., H. Murakami, T. R. Knutson, R. Mizuta, and K. Yoshida, 2020: Tropical cyclone motion in a changing climate. *Sci. Adv.*, **6**(17), eaaz7610.

Zhang, R., and T. L. Delworth, 2006: Impact of Atlantic multidecadal oscillation on India/Sahel rainfall and Atlantic hurricanes. *Geophys. Res. Lett.*, **33**, L17712.

Zhang, W., H.-F. Graf, Y. Leung, and M. Herzog, 2012: Different El Niño types and tropical cyclone landfall in East Asia. *J. Climate*, **25**, 6510–6523.

Zhang, W., and Coauthors, 2016: The Pacific meridional mode and the occurrence of tropical cyclones in the western North Pacific. *J. Climate*, **29**, 381–398.

Zhang, W., and Coauthors, 2018: Dominant role of Atlantic multidecadal oscillation in the recent decadal changes in western North pacific tropical cyclone activity. *Geophys. Res. Lett.*, **45**, 354–362.

Zhao, H., P. J. Klotzbach, and S. Chen, 2020: Dominant influence of ENSO-like and global sea surface temperature patterns on changes in prevailing boreal summer tropical cyclone tracks over the western North Pacific. *J. Climate*, **33**, 9551–9565.

Zhao, J., R. Zhan, Y. Wang, and H, Xu, 2018: Contribution of the interdecadal Pacific oscillation to the recent abrupt decrease in tropical cyclone frequency over the western North Pacific since 1998. *J. Climate*, **31**, 8211–8224.

Zhao, M., and Coauthors, 2013: Response of global tropical cyclone frequency to a doubling of CO_2 and a uniform SST warming: A multi-model intercomparison. US CLIVAR Hurricane Workshop, Princeton, NJ, Geophysical Fluid Dynamics Laboratory [Online]. Available at www.usclivar.org/sites/default/files/meetings/Zhao_Ming_Hurricane2013.pdf

Zhao, X., and P.-S. Chu, 2006: Bayesian multiple changepoint analysis of hurricane activity in the eastern North Pacific: A Markov Chain Monte Carlo approach. *J. Climate*, **19**, 564–578.

Zwiers, F. W., 1990: The effect of serial correlation on statistical inferences made with resampling procedures. *J. Climate*, **3**, 1452–1461.

5

Subseasonal to Seasonal Tropical Cyclone Prediction

5.1 Introduction

This chapter provides a current knowledge regarding the state of TC prediction on subseasonal to seasonal time scales. The TC forecasting methods can be generally classified into statistical, dynamical, and statistical-dynamical approaches. Statistical methods are generally based on logistic regression, multiple linear regression, Poisson regression, and the Poisson regression cast in the Bayesian paradigm. Dynamical forecast methods rely on dynamical climate models using either atmospheric general circulation models (GCMs) forced by observed or predicted SST anomalies, or coupled atmosphere–ocean models. Statistical-dynamical methods are a hybrid approach by combining forecast information from dynamical models and the observed statistical relationship between TC and environmental conditions to forecast TC changes.

Recent thrusts for dynamical subseasonal forecasts use multi-model datasets from the Subseasonal to Seasonal (S2S) project to evaluate the predictability of TCs and results are promising. A brief review of the current real-time subseasonal to seasonal TC forecasts at operational centers is provided. Recommendations for future development of TC forecasts are summarized at the end of this chapter.

5.2 Subseasonal Tropical Cyclone Prediction

Statistical methods based on logistic regression have been used to predict the probability of TC genesis with leads out to seven weeks (Leroy and Wheeler, 2008; Slade and Maloney, 2013). Logistic regression is used to model the probability of an event. The basic form of the logistic regression is a logistic function that models a binary dependent (or response) variable with two possible values such as TC genesis or no TC genesis for a specific time period. These two values are represented by an indicator variable, labelled as 1 or 0.

In the intraseasonal prediction of weekly TC formation for the Southern Hemisphere centered on Australia, Leroy and Wheeler (2008) employed five predictors: two representing the eastward propagation of the MJO; one for the time-varying climatology of TC frequency; and two representing the leading two modes of interannual SST variations in the Indo-Pacific Oceans. Cross-validated hindcasts are conducted from 1969/1970 to 2003/2004. The Brier skill score (BSS) and reliability diagrams are used to assess the hindcast performance. When the BSS is positive, it indicates the percentage improvement over reference climatology.

Leroy and Wheeler (2008) found that the inclusion of the MJO indices increases the forecast skill to about the third week. Beyond this point, there is no significant overall improvement derived from the MJO signal. Following the same methodology and statistical model developed by Leroy and Wheeler (2008), Slade and Maloney (2013) predicted weekly TC activity over the North Atlantic and East Pacific basins using predictors from the multivariate MJO indices, the climatology of TC genesis for the Atlantic and east Pacific separately, an ENSO index, and the Atlantic SST in the MDR for the Atlantic model. Figure 5.1 displays reliability curves for the Atlantic out to a week 2 forecast lead during 1975–2009. The hindcast probability is very close to the observed probability at almost all bins from lead 0 to 2 weeks in advance, indicating a fairly reliable hindcast from the logistic model at short leads. There appears to be a slight overestimation of cyclone genesis at higher probabilities but they lie within the 10% interval about the perfect forecast (diagonal solid line).

Using a dynamical atmospheric model (ECMWF Cy32r3), Vitart (2009) showed good skill in subseasonal prediction of tropical storms and storm landfall. Because this dynamical model is capable of maintaining the amplitude of MJO for approximately 30 days, it has skill in predicting the risks of storm landfall over

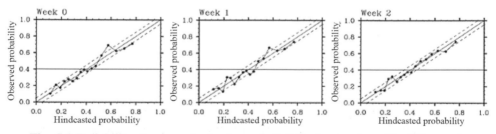

Fig. 5.1 Reliability diagrams for leads week 0–2 hindcasts of weekly TC genesis probabilities over the North Atlantic during 1975 and 2009. Observations and hindcasts are binned into 17 groups and each group is denoted as a solid circle. The perfect forecast and a 10% interval around the perfect forecast is shown by a diagonal line and dashed lines, respectively. The solid horizontal line indicates the average observed probability.

Source: Slade and Maloney, 2013; ©American Meteorological Society. Used with permission.

Australia and North America by about 20 days in advance. A comparison between the logistic regression and aforementioned dynamical model for the Southern Hemisphere indicates that the statistical model is more reliable than the dynamical model for week-1 to week-3 forecast leads, although the former has relatively low resolution (Vitart et al., 2010). Low resolution implies that the forecasts resolve the observed events into different groups poorly, leading to low BSS. No resolution means that the outcomes following different forecasts are nearly the same on average.

After a simple calibration, the dynamical model shows considerable improvement in the reliability curves and higher BSS than the statistical model for the first three weeks, although the calibrated ECMWF model still shows less reliability than the statistical model. One interesting aspect of that study is the use of a simple equally weighted average of forecasts from the calibrated dynamical model and statistical model. The multi-model mean has higher BSS and better reliability than the calibrated model, but at the expense of less sharpness. Sharpness is the tendency to forecast extreme values (e.g., probabilities near 0 or 100%). Although a combination of only two model forecasts are used (Vitart et al., 2010), the multi-model approach seems to be able to yield better hindcast skills than the individual models, regardless of whether purely statistical model, uncalibrated dynamical model, or calibrated dynamical model is employed.

Because of the increasing demand for subseasonal prediction, an international effort was initiated to develop the Subseasonal to Seasonal (S2S) prediction project (Vitart et al., 2017). The multi-model (models from 11 centers) dataset from the S2S Project contains dynamical subseasonal forecasts and reforecasts with leads up to 60 days. For the subseasonal to seasonal range, the model error is usually large so that reforecasts are created to remove systematic bias. This extensive data set offers excellent opportunities to better explore the MJO predictability from a suite of dynamical models and may improve our knowledge concerning MJO-TC modulation.

Taking advantage of this relatively new data set, Lee et al. (2018) investigated TC genesis prediction and MJO at a weekly time scale using coupled, global GCMs from six operational centers (Australian Bureau of Meteorology, China Meteorological Administration, ECMWF, Japan Meteorological Agency, Meteo-France, and NCEP). Forecast skill (probability of weekly storm occurrence) is measured using the BSS. Relative to the climatology, most models show skill for week 1, the period when model initialization is important because of the influence from pre-existing TCs. Without including those storms, the BSS drops considerably. From week 2 to 5, the BSS in most models decreases and levels off, although they still exhibit the eastward propagation of TC genesis anomalies for week 2 reforecasts, particularly evident in the Southern Hemisphere (Fig. 5.2). This is consistent with the observations. Among the six S2S models examined,

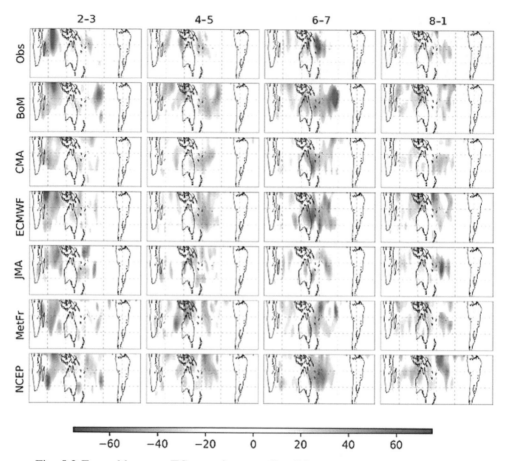

Fig. 5.2 Ensemble mean TC genesis anomalies (%) at every two MJO phases for 2–3, 4–5, 6–7, and 8–1 in the Southern Hemisphere from the observations (top row) and the week 2 reforecasts from six S2S models. A black and white version of this figure will appear in some formats. For the color version, refer to the plate section.
Source: Lee et al., 2018; ©American Meteorological Society. Used with permission.

models from the ECMWF, Australian Bureau of Meteorology, and Meteo-France appear to have skill in predicting TC occurrence four weeks in advance (Lee et al., 2020).

The state of TC prediction on subseasonal time scales at various operational centers and private sectors around the world was reviewed by Camargo et al. (2019). The source of TC subseasonal predictability appears to be the MJO, various kinds of equatorial waves, and possibly extratropical processes via Rossby wave breaking. Also presented in Camargo et al. (2019) is the useful guidance regarding the probability of TC development out to three weeks in advance by the ECMWF and Australian Bureau of Meteorology among six centers. As an example, for cyclone Hilda which made landfall in northwest Australia in late December 2017 (Gregory et al., 2019), the landfall location and shape of TC tracks were correctly predicted out to two weeks ahead. Although improvements in subseasonal TC prediction has been made during the last few years, some issues remain which impede a direct comparison of forecast skill among operational centers and private enterprises. Central to the issues is the lack of uniformity in forecast verification measures. For example, forecast lead times, periods of forecast, and variables used are quite different from one center to another. In this regard, it is urged by Camargo et al. (2019) to have "common standards and verification metrics for subseasonal forecasts."

5.3 Tropical Cyclone Prediction on Seasonal Time Scales

5.3.1 Statistical Methods

Gray et al. (1992, 1993, 1994) pioneered the seasonal hurricane prediction enterprise using regression-based statistical models. The model they used is called the least absolute deviation (LAD). The least square errors (LSE) method is perhaps the best known way of fitting a linear regression model and by far the most widely used. The LSE is not necessarily the optimal fitting method if the regression error is not Gaussian distributed. Moreover, the key statistic in LSE is sample mean, whereas in LAD is the sample median. Because the sample median is a more robust estimator for a location parameter than sample mean, LAD regression estimates are less sensitive to sample outliers (e.g., extreme values) than the LSE method. In addition, the former is adequate when working with a small sample size because it does not assume that the residuals follow a Gaussian distribution. A LAD fitting problem is to find a minimizer of the distance function (e.g., absolute deviation) but regression parameters cannot be solved in an analytical way because of the absolute operation used. The Bloomfield and Steiger (1980) algorithm efficiently finds the normalized steepest direction in each iteration. A detailed explanation and pseudo code for this algorithm are given in Chu et al. (2007).

Gray et al. (1992, 1993, 1994) showed that nearly half of the interannual variability of hurricane activity in the North Atlantic could be predicted in advance. Klotzbach and Gray (2004) continued to revise their forecasts as peak seasons approach. Seven parameters are routinely predicted at long lead times, including the number of hurricanes, number of named storms, number of hurricane days, number of named storm days, intense hurricanes, intense hurricane days, and net tropical cyclone activity (NTC). The NTC is a combined measure of the aforementioned six parameters normalized by their climatological averages. When verified for a 52-yr record (1950–2001), hindcast skill is lowest for named storms but is better for intense hurricanes and NTC when the forecast is issued in early December, 6–11 months prior to the Atlantic hurricane season. Using only July wind anomalies between 925 and 400 hPa, Saunders and Lea (2005) developed a linear regression model to successfully predict the US ACE for the following hurricane season (August–October).

Elsner and Schmertmann (1993) considered a different approach to predict intense annual Atlantic hurricane counts. Specifically, the annual hurricane occurrence is modeled as a Poisson process, which is governed by a single parameter, the Poisson intensity. The intensity of the Poisson process is then made to depend upon a set of covariates such as the west Sahel rainfall and the stratospheric zonal winds via a multiple linear regression model. Compared to the LAD model, the Poisson regression model exhibits higher hindcast skill. Many studies indicated that the Poisson distribution is appropriate for modeling the occurrence of rare, discrete events such as the occurrences of TCs. This distribution also restricts to the non-negative integer values, making it ideal for TC counts. Parameters of the regression are estimated by maximum likelihood. Hess et al. (1995) further improved the hurricane forecasting scheme of Elsner and Schmertmann (1993) by separating tropical-only hurricanes from those influenced by extra-tropical factors.

McDonnell and Holbrook (2004) found that a Poisson regression model was skillful in predicting tropical cyclogenesis in the Australian-Southwest Pacific Ocean region. They also advocated the use of this method for subregional and subseasonal TC frequency forecasts. For the western North Pacific, Chan et al. (1998) developed a scheme to predict annual and seasonal TC numbers using a project pursuit regression technique, which reduces high-dimensional data to a lower dimensional subspace before the regression is performed. Chan et al. (2001) further improved their operational statistical prediction model by including new predictors and predictors in the preceding April and May.

In addition to the prediction of basin-total TC frequency, efforts have also been made for the seasonal prediction of regional TC activity using statistical methods, e.g., the number of TCs influencing a specific domain in a basin (Liu and Chan, 2003; Elsner and Jagger, 2006; Chu et al., 2007, 2010a; Chu and Zhao, 2007; Lu et al., 2010; Kim et al., 2012). Compared with the prediction of basin-wide seasonal TC activity, the prediction for a specific region is more challenging because the predictive skill is affected not only by TC genesis but also by TC tracks. Chu et al. (2007) used the LAD regression technique to predict the seasonal (June to October) TC counts in the vicinity of Taiwan using large-scale climate variables from the antecedent May. Through lagged correlation analysis, five variables (sea surface temperature, sea-level pressure, precipitable water, low-level relative vorticity, and vertical wind shear) in key locations of the WNP are identified as predictor datasets. Results from the leave-one-out cross-validation

(LOOCV) suggest that the multivariate LAD regression model is skillful, with a correlation coefficient of 0.63 during 1970–2003. To predict higher than normal seasonal TC frequency near Taiwan, warmer SSTs, a moist troposphere, and the presence of a low-level cyclonic circulation coupled with low-latitude westerlies in the Philippine Sea in the preceding May appear to be important.

Kim et al. (2010a) predicted summertime (July–September) TC frequency over the East China Sea (25°N–35°N, 120°E–130°E) using the LAD and the Poisson regression methods. The Poisson regression model is described in the following equation. Given the Poisson intensity parameter λ (i.e., the mean seasonal TC rates), the probability mass function of h TCs occurring in T years is

$$P(h|\lambda, T) = \exp\left(-\lambda T\right)\frac{(\lambda T)^h}{h!}, \text{ where } h = 0, 1, 2, \ldots \text{ and } \lambda > 0, T > 0. \quad (5.1)$$

The Poisson mean is simply the product of λ and T, so is its variance. When the sample TC count is annual or seasonal values, the unit observation period is one year or season, thus $T = 1$. Assume there are N observations and for each observation there are K relative predictors. We define a latent random N-vector $\mathbf{Z}$, such that for each observation h_i, $i = 1, 2, \ldots, N$, $Z_i = \log \lambda_i$, where λ_i is the relative Poisson intensity. The link between this latent variable and the predictors is expressed as the Poisson regression model

$$Z_i = \mathbf{X}_i\boldsymbol{\beta} + \varepsilon_i, \quad (5.2)$$

where $\boldsymbol{\beta} = [\beta_0, \beta_1, \beta_2, \ldots, \beta_K]'$ is a set of regression parameters, ε_i is assumed to be independent and identically distributed (IID) and is normally distributed with zero mean and σ^2 variance, and $\mathbf{X}_i = [1, X_{i1}, X_{i2}, \ldots, X_{iK}]$ is the predictor vector for h_i. From Eq. 5.2, the model uses the logarithm of the TC rate as the response variable but is linear in the regression structure.

Many TCs, forming in the Philippine Sea, pass across the East China Sea, recurve near 25°N, and then approach the East Asian coasts. Among the 114 TCs passing through this region during July–September of 1979–2007, 82% strike either China, Taiwan, Korea, or Japan (Kim et al., 2010a). Physically interpretable and statistically significant pre-season environmental parameters are first identified as potential predictors and then subjected to a screening method based on the stepwise regression to avoid overfitting issues. When tested through cross-validation, the correlation coefficient between the predicted and observed frequency is 0.75 for the LAD model and 0.78 for the Poisson regression model (Fig. 5.3). To further examine the predictive skill, the seasonal TC counts are classified into three categories: above normal; normal; and below normal. The skill of categorical forecasts is then assessed using the Gerrity skill score (WMO, 2002). A Gerrity skill score of one represents a perfect forecast, zero is the same as the

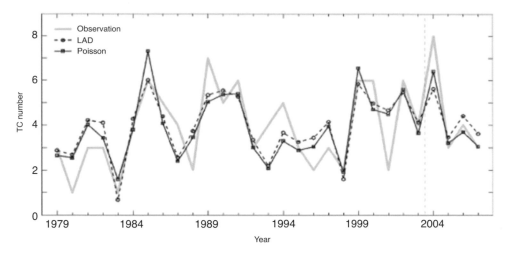

Fig. 5.3 Time series of cross-validated hindcasts (1978–2003) and the independent forecasts (2003–2007) of seasonal TC frequency over the East China Sea based on the LAD model (dashed lines) and the Poisson regression model (solid line). The observation is denoted by a gray solid line.
Source: Kim et al., 2010a; ©RightsLink/John Wiley & Sons. Used with permission.

reference forecasts based on the climatology, and negative if the forecasts are worse than the climatology. For this tercile forecast, the Gerrity skill score is 0.593 in the LAD and 0.693 in the Poisson model. Thus, the predictive skills of the two selected statistical models are much higher than the reference forecasts. Moreover, the Poisson regression model is more successful in the prediction of regional and seasonal TC frequency than the LAD model.

The Poisson regression model is also used to forecast seasonal (July–October) typhoon genesis frequency over the WNP by Zhang et al. (2018). The two predictors chosen by Zhang et al. (2018) are the tri-polar pattern of April SSTA in the North Pacific and SSTA off east coast of Australia. The former is conducive to the formation of a large anomalous cyclonic circulation over the North Pacific while the latter helps maintain the cyclonic circulation through weakening of northeast trade winds over the North Pacific. For the basin-total frequency, the correlation coefficient between the prediction and observation is 0.65 during the training period (1965–1998) and this value drops to 0.52 for the independent testing period (1996–2015).

Extending from their earlier approach, which yields a deterministic forecast, Elsner and Jagger (2004) introduced a Bayesian method to the Poisson regression model so that the predicted annual (or seasonal) hurricane numbers could be cast in terms of probability distributions. This is an advantage over the deterministic forecasts because the uncertainty inherent in forecasts is quantitatively expressed in the probability statement and probability is the universal language of uncertainty. The conventional statistical or dynamical models provide deterministic forecasts, but not information on the likelihood of the range of TC metrics. Because probabilistic forecasts yield the uncertainty of the prediction, forecasts are better presented in a probabilistic format. The probabilistic forecasts are deemed as useful for risk and emergency managers in estimating a possible range of potential disaster losses before the target forecast period.

Elsner and Jagger (2004) also addressed the issue regarding unreliable records by introducing an informative prior for the coefficient parameters of the model via a bootstrap procedure. The bootstrap operates by generating artificial data batches from the existing sample *with replacement*. Because of this, repetitions of the same observations and absence of others may occur simultaneously in a particular bootstrap batch. With a similar Bayesian regression model, Elsner and Jagger (2006) attempted to predict annual US hurricane counts. The model includes predictors representing the North Atlantic Oscillation (NAO), the Southern Oscillation, the Atlantic multidecadal oscillation, as well as an indicator variable that is either 0 or 1, depending on the time period specified. As a baseline for comparison, a climatology model is used, which contains only the regression constant (i.e., intercept) and the indicator variable. In an out-of-sample cross-validation test, the Bayesian model appears to have a lower mean-squared error relative to climatology.

For the central North Pacific, a Poisson regression model cast in the Bayesian framework is applied to forecast the TC frequency during the peak season (July–September) using large-scale environmental variables of the antecedent May and June (Chu and Zhao, 2007). Specifically, the seasonal TC counts are modeled by a Poisson process, where the Poisson intensity is governed by a multiple linear regression model via a latent variable. The critical regions for which the correlation between the pre-season environmental predictors and the seasonal TC counts is statistically significant are determined. With a non-informative prior assumption for the model parameters, a Bayesian inference is derived in detail. A Gibbs sampler based on the Markov Chain Monte Carlo (MCMC) method is designed to integrate the desired posterior predictive distribution, which involves complex integrations of high-dimensional functions. The MCMC methods are used for numerical approximations of multi-dimensional integrals when direct sampling is difficult. Gibbs sampling solves complex integrals by expressing them as expectations for some specified multivariate probability distribution and then estimating the expectations by drawing samples from that distribution. The proposed hierarchical model is physically based and yields a probabilistic prediction for seasonal TC frequency. A cross-validation method is applied to predict TC counts within the period 1966–2003 and satisfactory results are obtained (e.g., the correlation

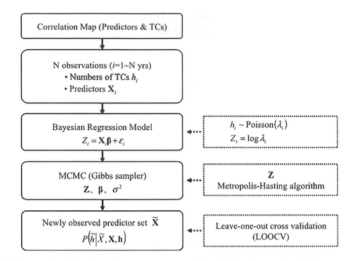

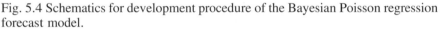

Fig. 5.4 Schematics for development procedure of the Bayesian Poisson regression forecast model.
Source: Lu et al., 2010; ©American Meteorological Society. Used with permission.

coefficient between the median of predictive rate and independent observations is 0.63). Note that a similar approach for predicting seasonal TC activity was also adopted by Ho et al. (2009) for the East China Sea, Lu et al. (2010) for Taiwan, and Chand et al. (2010) for Fiji.

Figure 5.4 illustrates the procedure of the Bayesian regression forecast system (Lu et al., 2010). The Metropolis–Hasting algorithm operates through generation of random samples and it is necessary to know only the mathematical form of the target probability density function (PDF), not its cumulative distribution function (CDF) so the PDF need not be analytically integrable (e.g., Wilks, 2011). This algorithm is commonly applied to 1-parameter problems. When two or more parameters need to be computed, the Gibbs sampler is the typical choice. Out of several potential predictors, the low-level relative vorticity extending from the southern Philippine to the WNP in the preceding May is identified as the primary one for predicting the seasonal TC counts in the vicinity of Taiwan through the stepwise screening procedure. This reflects the variation of the ridge position of the westward extension of the WNP subtropical high, which is considered as a precursor signaling how the high will evolve in the following months.

In Lu et al. (2010), the probabilistic forecasts are calibrated and then converted to categorical forecasts because the latter are easier to understand for most users. A common practice in issuing the seasonal outlook of TC counts in an operational center is to categorize them as above normal, normal, or below normal. In principle, the CDF corresponding to 33% and 67% is used as the cutoff for categorizing the forecast results when the outcome is expressed in a tercile.

Table 5.1. Contingency table for the tercile category of TCs near Taiwan predicted by the Bayesian Poisson regression model after calibration using the LOOCV results. BN denotes below normal, N for near normal, and AN for above normal.

		Prediction			
		BN	N	AN	Total
Obs	BN	5	4	0	9
	N	4	5	4	13
	AN	0	2	5	7
	Total	9	11	9	29

Source: Lu et al., 2010; ©American Meteorological Society. Used with permission.

Because there is a negative bias in the forecast, the probabilistic forecast results need to be calibrated. The contingency table for the categorical forecast after calibration is displayed in Table 5.1. The highest numbers are along the main diagonal and the Heidke skill score, which is based on the hit rate as the accuracy measure, is 0.27. Note that forecasts equivalent to the climatology forecasts receive a Heidke score of zero. The zero number of false forecasts for the opposite category clearly indicates the capability of the Bayesian regression forecasts to capture the categorical forecasts correctly.

Also cast in the Bayesian regression paradigm, Chu et al. (2010a) attempted to forecast regional typhoon activity using a track clustering approach. That is, TC tracks over the entire WNP basin are clustered before building the prediction model (Chu et al., 2010a, 2010b; Kim et al., 2011). Previous studies noted that an effective way to elucidate the characteristics of various TC tracks is to classify TC trajectories into a finite number of patterns (e.g., Harr and Elsberry, 1991, 1995a, 1995b; Hodanish and Gray, 1993; Lander, 1996; Elsner and Liu, 2003; Camargo et al., 2007a, 2007b). As a result, TC tracks are objectively classified using both the similarity of track shape and contiguity of geographical path. Indeed, the TC clustering is particularly important if the focus is confined to a limited region where TC risks are mainly determined by track patterns. Moreover, a categorization of the historical TC tracks and forecasting of each individual track type may also yield a better physical understanding of the forecast skills because some track types are inherently related to certain climate modes.

In Chu et al. (2010a), for each cluster, a Poisson or probit regression model is applied individually to forecast the seasonal TC activity for that particular type. As mentioned previously, the Poisson regression model is appropriate for modeling rare event count series. However, if the underlying TC rate is significantly below one, this model may introduce significant bias. This may occur when TC tracks are clustered into several types, the analysis is limited to a smaller region, or the sample size is extremely small. Therefore, if the mean of the historical seasonal occurrence for a cluster is less than 0.5, then a binary classification model is adopted. In essence, the response variable is a binary class label, which is termed "Y." For each observation period, a class "$Y = 1$" is defined if one or more TC are observed, and "$Y = 0$" otherwise. The structure of the probit regression model is very similar to the Poisson regression model, and a Gibbs sampler based on the MCMC method is designed to integrate the posterior predictive distribution. A LOOCV procedure is applied to one particular type, which affects Taiwan the most, and the correlation skill is found to be 0.76 during 1979–2006.

For the WNP, there are two major methods used in probabilistic clustering of TC tracks. The first one is based on a mixture regression model by which each track path is modeled as a second-order polynomial regression function of the lifetime of the TC, provided that the number of clusters is given (Camargo et al., 2007a, 2007b; Chu et al., 2010b; Chu and Zhao, 2011). The first-order term of this regression model features the characteristic direction of the path type and the second-order polynomial determines the recurving shape of the typical path of a storm type. In short, this curve clustering method considers both track shape and position. Unlike the k-means clustering that requires data vectors with equal and fixed length, which is sometimes unrealistic (e.g., always set TC lifetime at five days), the mixture regression model can accommodate tracks of varying lengths in the analysis, which is very useful because each individual TC is characterized by a different lifetime (three or seven days). The expectation-maximization algorithm is used to derive the maximum likelihood estimation of all model parameters. On the basis of this clustering method, Camargo et al. (2007a, 2007b) classified historical TC tracks from 1950 to 2002 into seven types such that it is assumed that each individual TC track is generated by one of these seven clusters. Chu and Zhao (2011) used the curve-based mixture model to classify TC tracks into eight groups and examined their long-term changes in terms of TC frequency, lifespan, intensity, and ACE.

The second clustering method is the fuzzy c-means clustering (Chu et al., 2010a; Kim et al., 2011, 2012). Unlike the common k-means clustering or mixture regression method, where the data objects are allocated to one cluster, the fuzzy method allows each TC track to belong to all cluster types with different membership degrees. This property is more amenable when attempting to classify widespread data with some ambiguity in its cluster boundaries, such as the TC track data. Details of the fuzzy clustering methods are given in Kim et al. (2011).

As an example of the fuzzy clustering, Fig. 5.5 shows the seven patterns or types (C1–C7) of TC tracks during JJASO of 1965–2006. C1 features storms that developed in the northern Philippine Sea with a northward track that eventually hits the East Asian region, including Korea and Japan. Out of all 855 TCs, 16% of them (133/855) have the largest membership coefficient in C1. C2 storms

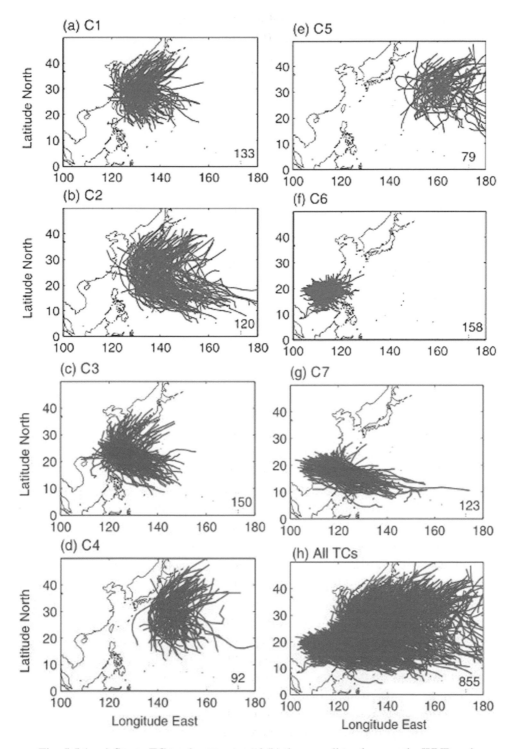

Fig. 5.5 (a–g) Seven TC track patterns and (h) the overall tracks over the WNP and their frequencies during the TC season for the period 1965–2006.
Source: Kim et al., 2012; ©American Meteorological Society. Used with permission.

developed in the far southeastern portion of the WNP and moved northwestward before recurving northeastward to the south of Japan. C2 represents 14% (120/855) of all storms and has the largest traveling distance and lifetime (seven days) among all seven types (Kim et al., 2011). C3 storms formed in the Philippine Sea, follow a west-northwestward path, and many of them made landfall on Taiwan and the eastern China coast. Some of them also affected Japan and Korea. C3 includes 18% (150/855) of all storms and is the second largest cluster after C6. Storms associated with C4 and C5 are relatively few (11% and 9%, respectively) and they form away from Asia. C6 storms developed in the South China Sea and are landlocked by the Indochina peninsula and China's southern coast, with very short paths and lifespans (three days), although they have the highest frequency of occurrence (19%). C7 storms originated in low latitudes, traverse straight westward through the Philippines, and made landfall on Vietnam and southeast China. This type accounts for 14% of all storms.

Note that C1, C2, and C3 reflect the influences of La Niña, EP El Niño, and CP El Niño, respectively (Kim et al., 2011, 2012). Based on the mixture regression model, Camargo et al. (2007a) found that two of the seven clusters are typical of El Niño events and one cluster type occurs more often during La Niña events. C4 is possibly related to the stratospheric quasi-biennial oscillation (Ho et al., 2009; Kim et al., 2010a, 2011). The other three patterns (C5–C7) are probably influenced by local environments such as SST and steering flow (Kim et al., 2011).

The clustering method also can be applied to improve short-term typhoon intensity forecasts. Recently, statistical-dynamical typhoon intensity predictions were made using track pattern clustering and ocean-coupled predictors in the WNP. Based on multiple linear regression models, prediction errors of track-based patterns are reduced by about 12–25% between 24- and 96-h lead, relative to the non-clustering approach (Kim et al., 2018). The regression model is also compared to four operational dynamical forecast models. At short leads (up to 24-h), the statistical-dynamical model based on track clustering has the smallest mean absolute errors. After a 24-h lead-time, this model still shows skill that is comparable with the best operational dynamical models.

TC genesis in the vast WNP experiences regional characteristics (Wang and Chan, 2002; Kim et al., 2010b). Accordingly, Wang et al. (2019) divided the entire WNP into four subregions and identified predictors for each subregion plus the South China Sea. For the WNP, the dividing lines separating these four quadrants are 17°N and 140°E. While TC genesis frequency in the northwest and southeast quadrants of the WNP are sensitive to the ENSO phenomenon (Chapter 4), TC genesis in the northeast quadrant is related to the preseason Pacific dipole SST and the west Indian Ocean surface temperature tendency predictors. An enhanced TC genesis in this quadrant is associated with an anomalous cyclonic circulation to the north and west of the anomalously warm SST in the subtropical central North Pacific as a Rossby wave response. The surface westerly and southwesterly anomalies around the eastern and southern flanks of the cyclonic circulation reduce the climatological easterlies associated with the subtropical high pressure. This reduces evaporative and entrainment cooling, and results in prolonged SST warming (Wang et al., 2000; Chiang and Vimont, 2004). As a result, the anomalous cyclonic circulation can be maintained and conditions are favorable for TC formation. Through a cross-validation, the predictive skill of seasonal TC genesis in terms of the correlation coefficient is 0.65, 0.76, 0.62, 0.61, and 0.60 for the northwest, southeast, northeast, and southwest WNP and the South China Sea respectively during 1965–2016. Apparently, the best skill is found for the southeast subregion, where TC variations are highly coupled to ENSO events.

5.3.2 Dynamical Methods

Parallel to the statistical-based forecasting enterprise, dynamical climate models for forecasting seasonal TC activity have also been explored by many researchers and organizations in the past three to four decades (e.g., Wu and Lau, 1992; Bengtsson et al., 1995; Vitart et al., 1997, 2007; Vitart and Stockdale, 2001; Camargo and Barnston, 2009; Wang et al., 2009; LaRow et al., 2010; Chen and Lin, 2013; LaRow, 2013; Manganello et al., 2016; Murakami et al., 2016a, 2016b; Zhang et al., 2017). In early years, the low horizontal resolution of atmospheric general circulation models (AGCM) forced by observed SSTA were used. Although these models were able to simulate some aspects of TC activity (mainly TC frequency) and their year-to-year variations, they could not realistically reproduce the structure of an individual storm such as eyewall and rainbands. The horizontal resolution in the AGCM has improved remarkably over the last 20 years and the forecast SST has also been used to make a genuine TC forecast. The conventional metrics used in dynamical forecasting are the basin-wide frequency of TCs, and/or basin-total values of ACE. With the advances in the coupled atmosphere–ocean models, they have also been used in seasonal TC forecasts and climate drift is calibrated against model climatology in forecasting experiments.

Dynamical seasonal TC forecasts have been issued operationally since the early 2000s by the European Center for Medium-Range Weather Forecasts (ECMWF) (Vitart and Stockdale, 2001; Vitart et al., 2007) and the International Research Institute for Climate and Society (IRI) (Camargo et al., 2005; Camargo and Barnston, 2009). The ECMWF forecast system is based on a coupled model, without flux correction, and has been updated constantly with increasingly higher horizontal resolution and multi-model ensemble techniques. The IRI system relies on a two-tier approach, in which a range of SSTs are predicted first and then they are used to force an AGCM (ECHAM 4.5). Camargo and Barnston (2009) found that the hindcast skill for the dynamical and simple statistical models is approximately the same but fairly modest. The dynamical forecasts require calibration (statistical post-processing) to be competitive with, or sometimes better than, the statistical models. Along with the ECMWF and IRI, the Florida State University (FSU) issued its operational forecast of Atlantic hurricane activity around June 1 based on an AGCM that has a grid spacing of approximately 1° latitude by longitude (LaRow et al., 2010; LaRow, 2013). The retrospective forecast SST dataset from CFSv1 is either corrected or uncorrected for bias to test its impact on seasonal North Atlantic hurricane forecasts. A linear method is used to correct systematic biases in the model simulation and few ensemble members are developed. When tested for a 28-yr period (1982–2009), much greater skill is achieved when using the bias-corrected SSTs (Fig. 5.6). The correlation coefficient between the simulated and observed hurricane counts is 0.74 in the bias correction compared to 0.42 when no correction is applied. However, the observed counts clearly lie outside the ensemble spread for many years such as 1984, 1989, 1993, 1999, 2002, and 2005. In terms of forecasting basin-wide seasonal TC frequency from the AGCM, the skill is usually higher in the Atlantic and lower in the WNP because of the well-known ENSO-TC relationship through modulation of

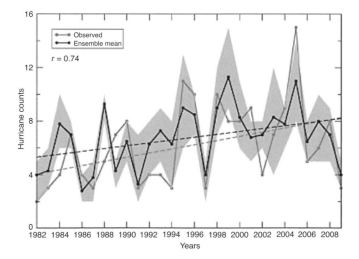

Fig. 5.6 Time series of observed (red solid line) and the ensemble mean (black solid line) of hurricane counts over the North Atlantic from 1982–2009. The shading region denotes the ensemble spread using the bias correction. The dashed lines are linear trends. A black and white version of this figure will appear in some formats. For the color version, refer to the plate section.
Source: LaRow, 2013; ©American Meteorological Society. Used with permission.

local vertical wind shear over the North Atlantic (Gray, 1984; Goldenberg and Shapiro, 1996; Aiyyer and Thorncroft, 2006; Shaman et al., 2009; LaRow, 2013). Generally, coupled atmosphere–ocean models exhibit lower skill in simulating TC frequency than the AGCM (Vitart and Stockdale, 2001; Vitart et al., 2007; LaRow, 2013).

Using a higher-resolution coupled climate model with a grid spacing of 25-km for its atmospheric component (HiFLOR), Murakami et al. (2016a) showed improved skill in predicting the frequency of major hurricanes (category 3, 4, and 5 where wind speed $\geq$49.4 m s^{-1}) and most intense hurricanes (category 4 and 5 where wind speed $\geq$58.1 m s^{-1}) in the North Atlantic, as well as landfalling TCs over the USA and Caribbean Islands a few months in advance, relative to a 50-km resolution atmospheric model (FLOR) for different lead months. The zero lead month means for retrospective forecasts initialized in July, the target season is July–November. Even for a lead-time of three months (forecasts from April) and six months (forecasts from January), HiFLOR still yields statistically significant rank correlation (0.28 to 0.50), indicative of skillful forecasts of major and most intense hurricanes in the North Atlantic several months in advance. Shown in Fig. 5.7 is the prediction of landfalling TC frequency at zero month lead. Both FLOR and HiFLOR show good skill in predicting landfalling TCs for the USA, the Caribbean islands, and the Hawaiian Islands.

Do forecast skills always improve when a higher-resolution dynamical model is used? Manganello et al. (2016) used a high-atmospheric resolution coupled long-range prediction system (ECMWF System 4) with different horizontal resolutions to examine whether the increased resolution from T319 (62 km), T639 (31 km), to T1279 (16 km) would improve the hindcast skills of seasonal TC activity. Retrospective forecasts initialized from May 1 are used for a seven-month run during 1980–2011. Ensemble members range from 15 for the two higher-resolution models (T639 and T1279) to 51 members for the lower-resolution model (T319).

Table 5.2 lists the linear correlations between the ensemble mean and observed TC frequency (top) and ACE (bottom) from three models with different resolutions for the three ocean basins in the Northern Hemisphere. For TC frequency, the highest correlations are 0.51 in the North Atlantic, 0.58 in the ENP, and 0.52 in the WNP. For the former two basins, the best skill comes from T1279, but for the WNP the best skill is from T319, the model with the lowest resolution. The correlations for the ACE are usually higher than that of TC frequency, but the highest correlations are found in T639 in the North Atlantic and WNP, and T319 in the ENP. Therefore, the best forecast skill is not necessarily associated with the model of finest resolution. For ACE, increases in correlations are largest from T319 to T639, with a slight drop in correlations from T639 to T1279. Note that although T319 has 51 ensemble members, which is three time more than T639 and

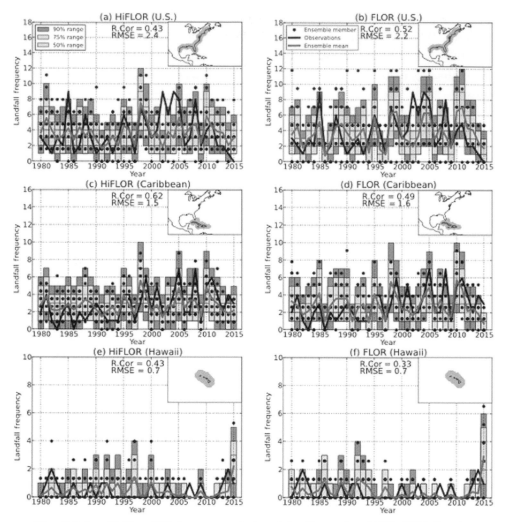

Fig. 5.7 Landfalling TC frequency during July–November 1980–2015 for the retrospective forecasts initialized in July using HiFLOR (left) and FLOR (right) for (a), (b) United States, (c), (d) Caribbean islands, and (e), (f) Hawaiian Islands. The black lines denote the observed statistics, the green lines are the mean forecast, and shading indicates the confidence intervals computed by convolving interensemble spread based on the Poisson distribution. The black dots denote the forecast value from each ensemble member. R.Cor and RMSE indicate the rank correlation coefficient and root-mean-square error between the black and green lines. A black and white version of this figure will appear in some formats. For the color version, refer to the plate section.

Source: Murakami et al., 2016a; ©American Meteorological Society. Used with permission.

Table 5.2. Correlation coefficients between the ensemble mean predicted and observed TC frequency and ACE for MJJASO of 1980–2011. Correlation coefficients for the detrended time series are given in parentheses. NA denotes the North Atlantic, EP the eastern North Pacific, and WP the western North Pacific. Boldface values indicated statistical significance at the 5% level using a one-sided Student's t test after taking into account of serial correlation in the time series.

	NA	EP	WP
		TC frequency	
T1279	**0.51 (0.48)**	**0.58 (0.45)**	0.44
T639	**0.48 (0.57)**	**0.52 (0.38)**	0.46
T319	0.28 **(0.34)**	**0.56 (0.42)**	0.52
		ACE	
T1279	**0.61 (0.59)**	**0.70 (0.64)**	0.72
T639	**0.64 (0.69)**	**0.71 (0.65)**	0.76
T319	**0.48 (0.50)**	**0.72 (0.66)**	0.68

Source: Manganello et al., 2016; ©American Meteorological Society. Used with permission.

T1279, it does not lead to more skillful forecasts. Retrospective forecasts of seasonal TC track density are also performed using the above three models and their skill shows some similarity but also differences in three basins. For example, in the WNP, the rank correlation is highest among all three basins but the area of highest correlation is confined to the southeast quadrant of the basin, indicative of the profound influence of ENSO. Besides this quadrant, there is essentially no skill elsewhere in the WNP and along the coastline of east and southeast Asia. No improvement in track density forecasts is found when the resolution is increased.

However, increasing horizontal resolution of atmospheric models certainly helps to improve predictions of intense tropical cyclones relative to the low resolution. This is because the low-resolution models cannot resolve intense TCs due to the limitation in the model resolution. Therefore, high-resolution models are better in terms of ACE predictions than the low resolutions without calibration because ACE partially depends on TC intensity. There is also model dependency on the results. Unlike Manganello et al. (2016), another model showed an improved skill in TC predictions by increasing horizontal resolution. For example, Murakami et al. (2016a) showed improved skill in TC predictions in terms of number of named storms, hurricanes, major hurricanes, and ACE over the North Atlantic by the HiFLOR model (25-km mesh) than the FLOR model (50-km mesh). Murakami et al. (2016a) investigated if the improved skill in predicting Atlantic TCs are related to the relevant large-scale parameters to TC activity such as vertical wind shear and SST. They found that the improvements in TC predictions by HiFLOR relative to FLOR is not directly related to the improvements in prediction of the large-scale parameters. Probably, the improvement may be due to the differences in the simulation of TCs themselves and the response of TC activity to the same large-scale conditions between HiFLOR and FLOR. In any case, incorporating a high-resolution model is preferable for dynamical seasonal TC prediction despite the huge computational costs.

5.3.3 Statistical-Dynamical Methods

Some of the limitations of dynamical methods for forecasting TCs can be alleviated using so-called "statistical-dynamical predictions." In the statistical-dynamical predictions, a statistical model is constructed using the empirical relationship between observed TC activity and predicted large-scale parameters simulated by a dynamical model. This can be achieved by using forecast products from coupled dynamical models such as the NCEP Climate Forecast System (CFS) retrospective forecasts, which provide nine-month forecast fields for every month since 1981 (Saha et al., 2006), or other agencies. The CFS forecasts include 15 ensemble members, which were run using different initial conditions. Because the statistical model (e.g., Poisson regression) is used to establish the relationship between TC activity and large-scale environments, while the latter are predictions from the dynamical model, this scheme is called a hybrid statistical-dynamical prediction (Wang et al., 2009; Vecchi et al., 2011; Kim et al., 2012; Murakami et al., 2016a, 2016b; Zhang et al., 2017). Wang et al. (2009) first used this approach to forecast Atlantic seasonal hurricane activity based on the CFS retrospective forecasts. Vecchi et al. (2011) also applied the statistical-dynamical approach but with a high-resolution atmospheric model. They suggested that skillful forecasts for the Atlantic hurricane frequency during August–October could be made as early as November of the previous year.

For the WNP, another important advance in the hybrid statistical-dynamical method is the production of spatial maps of predicted track density over the WNP. This represents a milestone for seasonal TC forecasting enterprises because this gives the spatial distribution of anomalous track density over the entire basin so that government agencies in east and southeast Asia affected by typhoons can use that information for better decision making (Kim et al., 2012). For the seasonal TC prediction using a hybrid model, the TC track patterns are identified as a first step (Fig. 5.5). For the second step, the seasonal TC frequencies of each type are separately correlated with environmental fields predicted from a dynamical model to find potential predictors. This step also involves a separate forecast for each track type using the Poisson regression model (Fig. 5.8).

Because there are 15 ensemble members from the CFS, 15 forecasts can be generated for each type. Figure 5.9 shows the hindcasts for the seven types from the LOOCV using 15 ensemble members during 1981–2006. Predictions

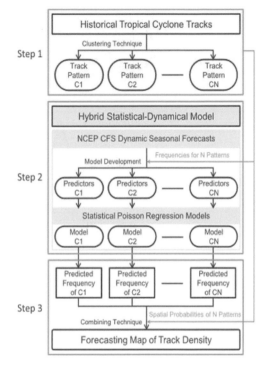

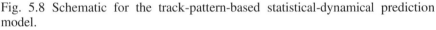

Fig. 5.8 Schematic for the track-pattern-based statistical-dynamical prediction model.
Source: Kim et al., 2012; ©American Meteorological Society. Used with permission.

from each member have larger errors in some years but their ensemble mean tends to smooth out variations in individual members and correspond well to the observations. The correlation coefficient between the ensemble mean and observations for each cluster is rather high, from 0.71 to 0.81, and the root-mean-squared error is around one (0.85 to 1.44). These verification statistics suggest that the hybrid statistical-dynamical model has good skill. The final step in Kim et al. (2012) is the construction of the forecasting map of TC track density over the entire WNP by combining the prediction results from each cluster. In this step, the seasonal TC track density is obtained by assembling the climatological probability of the TC tracks for the seven patterns weighted by the predicted number of TCs for each pattern. Biases in the mean and standard deviation of forecast TC track density are then corrected to yield more reliable forecasts.

The hindcasts of TC track density constructed by the hybrid model using 15 ensemble members for five best cases are shown in Fig. 5.10. The contours are the seasonal means of the probability of the TC tracks for each year with their anomalies indicated by shading. The model reasonably reproduces the seasonal distribution of the observed TC tracks, pointing to the feasibility of the track-pattern-based scheme

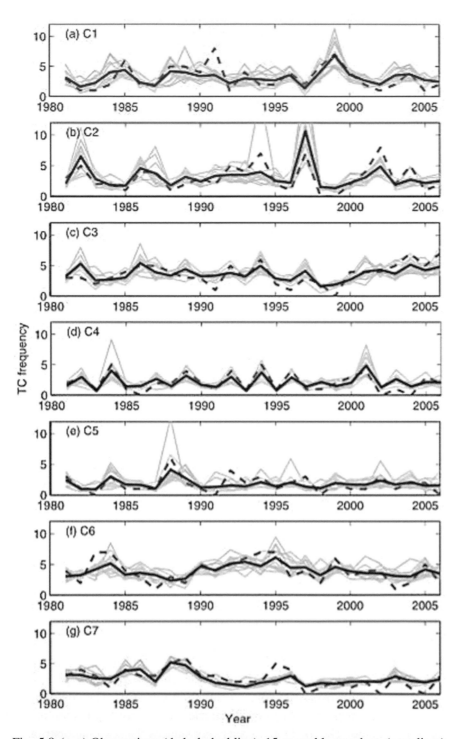

Fig. 5.9 (a–g) Observations (dark dashed line), 15 ensemble members (gray lines), and the ensemble mean (dark solid line) of the hindcasts for each TC track cluster over the WNP from the hybrid statistical-dynamical models for the period 1981–2006.
Source: Kim et al., 2012; ©American Meteorological Society. Used with permission.

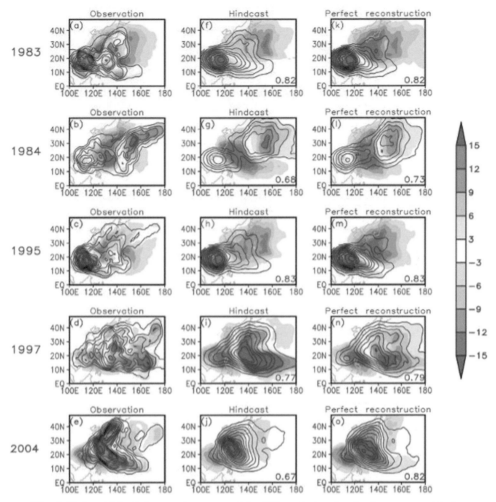

Fig. 5.10 (a–o) Final forecasting map of the TC track density for five years with the most skillful hindcasts: (left column) observation, (middle column) maps constructed from the hindcasts of the hybrid model, and (right column) maps constructed using the observed TC frequency in the seven patterns. The seasonal means of the probability of the TC tracks are in contours and their anomalies in shading. The contour interval is 5% and the zero line is omitted. The pattern correlation coefficient with the observation is shown in the bottom-right corner. A black and white version of this figure will appear in some formats. For the color version, refer to the plate section.

Source: Kim et al., 2012; ©American Meteorological Society. Used with permission.

to predict the anomalous spatial map of the seasonal TC tracks over the entire WNP. However, there are also years when the model shows poor performance in predicting the anomalous TC track density (not shown). One possible reason for this is that poor forecasts may occur when TC occurrence is not evenly distributed in time (i.e., clustered together in a short time span), which would be against the independence assumption of the Poisson distribution (Lu et al., 2010). Another possibility is due to the finite number of the chosen cluster types (i.e., seven) as they are not always sufficient to characterize all TC tracks and outliers do exist. The success of the predicted TC track maps also depends heavily on the forecast fields by the CFS, which are by no means perfect (e.g., Wang et al., 2009). By assuming that the hybrid model perfectly predicts the observed TC frequency of the seven patterns, a perfect reconstruction is also shown in Fig. 5.10. The perfect reconstruction provides an estimate of the upper limit of the model's skill in forecasting TC track density. Generally, it shows better pattern correlation than the hindcasts. The reproduction of the seasonal TC track density is more skillful during the ENSO episodes because the identified track types capture the ENSO-related anomalous track patterns well (Kim et al., 2012).

Statistical-dynamical retrospective forecasts of North Atlantic seasonal TC activity over the period 1980–2014 are conducted using a GFDL coupled climate model (FLOR) by Murakami et al. (2016b). The statistical component is the Poisson regression model. Observed and predicted TC tracks are classified into four types using fuzzy c-means clustering. Compared to the dynamical model (FLOR), the hybrid approach exhibits noticeable improvement in predicting TC frequency for each cluster at all lead times (0–7 months) in terms of RMSE. As a result, prediction of total TC frequency over the North Atlantic is more skillful using the hybrid model for lead 0–7 months when TC tracks are first classified into four distinct groups relative to the dynamical prediction without track clustering. For example, the rank correlation can reach as high as 0.75 for lead 0 and 1 month and remains 0.6 even at the 7-month lead forecast using the hybrid model. In comparison, under FLOR the correlation is 0.62 at lead 0 and drops to 0.4 at lead 7-month. Seasonal prediction of US landfalling TCs prove to be more challenging than the basin total TC frequency, but comparable or higher skill is found in the hybrid model relative to the FLOR.

Zhang et al. (2017) also applied a statistical-dynamical approach to forecast the seasonal WNP TC frequency and East Asia landfalling TCs for 1980–2015. The statistical model used is the Poisson regression and the dynamical model is the GFDL FLOR with flux adjustment (FLOR-FA) coupled system. For predicting the WNP TC frequency, the predictors chosen include the PMM, AMM, North Atlantic SSTA, and Niño-3 index. Shown in Fig. 5.11, the hybrid model outperforms FLOR-FA in predicting basin-total TC frequency for all lead months (0–6 months). Interestingly, this hybrid model also has a good skill in predicting landfalling TCs for forecasts initialized in June for the target season of June–October. One important aspect of the study by Zhang et al. (2017) is the adoption of a strategy called "closer-gets-richer." That is, all available dynamical forecasts

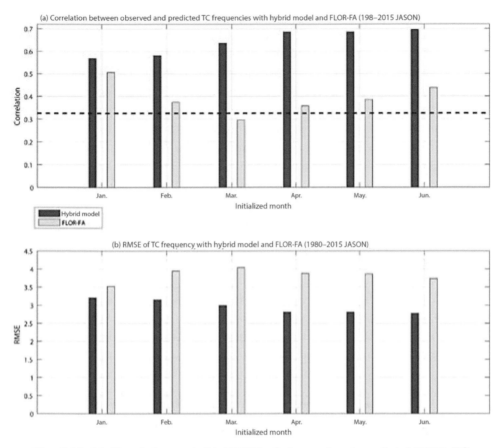

Fig. 5.11 (a) Correlation and (b) RMSE of observed and predicted WNP TC frequencies using the hybrid model (blue) and FLOR-FA (yellow) for initialization months from January to June after LOOCV. The black dashed line in (a) denotes the value for which the correlation coefficients are significantly different from zero at the 5% level. A black and white version of this figure will appear in some formats. For the color version, refer to the plate section.
Source: Zhang et al., 2017; ©American Meteorological Society. Used with permission.

in and prior to the initialization month are considered if they can improve the hybrid model even well before the initialization month relative to prediction from the dynamical model. This is justified by the fact that zero-lead-month forecasts are not necessarily better than those obtained from longer leads, probably due to initial shock in the dynamical model. Employing this strategy in the statistical-dynamical method also helps improving the forecasts.

Statistical-dynamical seasonal prediction has also been attempted based on a multi-model ensemble (MME) approach. Kim et al. (2017) used the MME from five different coupled models available at the APEC Climate Center (APCC) to forecast seasonal typhoon frequency (JASO) in the WNP for 1982 2008, where the LAD regression is the statistical model. The correlation between the hindcasts and observations through LOOCV is 0.67 for a one-month lead, 0.64 for a two-month lead, and 0.60 for a three-month lead. The hybrid model based on the MME also consistently outperforms the five single models in the prediction. Although the performance of the forecast skill is not particularly high, the concept of MME is also worth considering because no single model can be regarded as best for all circumstances. Each model has strengths and weaknesses. Therefore, in principle, the use of multi-model averaging may produce more skillful predictions than a forecast derived from a single model.

Apart from predicting seasonal TC frequency, Zhan and Wang (2016) also used the hybrid approach to predict seasonal (JJASO) ACE over the WNP. Several predictors with various lead times are chosen, including the vertical zonal wind shear in the equatorial central Pacific, the SST gradient between the southwestern Pacific and the western Pacific warm pool, the Niño 3.4 index, and the SST in the southwestern Pacific. Among these four predictors, the vertical wind shear is the best. The ensemble mean of the CFSv2 hindcasts is used. Through LOOCV, the predictive skills in terms of correlation coefficients vary from 0.51 to 0.81 for forecasts initiated from January to May when the vertical wind shear and the SST gradient predictors are employed.

5.4 Real-Time Seasonal Tropical Cyclone Forecasts

Zhan et al. (2012) and Klotzbach et al. (2019) reviewed real-time seasonal TC forecasts conducted by various universities, private sectors, and operational agencies around the world. The entities include, but not limited to, Colorado State University; NOAA; Tropical Storm Risk; UK Met Office; Australian Bureau of Meteorology; City University of Hong Kong; ECMWF; Geophysical Fluid Dynamics Laboratory; Shanghai Typhoon Institute; and National Typhoon Center of Korea. Some of the agencies provide publicly available forecasts while others do not. The forecast methods vary from statistical, dynamical, and statistical-dynamical. Currently, the majority of the agencies use statistical-dynamical methods. Real-time good skill is achieved for both the North Atlantic and the WNP when the forecast is issued at the commencement of an active TC season (by August for the former and July for the latter) (Klotzbach et al., 2019). In addition to predicting storm frequency and ACE, current attempts also include predicting seasonal TC track density and landfall likelihood.

Besides from the above agencies, the Central Weather Bureau (CWB) in Taiwan also routinely issued real-time seasonal TC forecasts using a statistical approach. The forecast is verified every year from 2009 to 2019. The forecast variables include the total number of TCs over the WNP during June–December, the TC frequency expected to affect Taiwan during June–November, and ACE in the vicinity of Taiwan during June–September. When verified during the last 11 years, the ranked probability skill score (RPSS) for the basin-wide TC frequency was negative from 2012 to 2016 but was positive and varied from 0 to 0.60 during 2009–2011 and 2017–2019. The ranked probability score is the sum of squared differences between the components of the cumulative forecasts and observation vectors. Note that if the forecasts being evaluated are inferior to the climatological forecasts, the RPSS will be negative, and vice versa. The trend in CWB's forecasts from negative to positive RPSS in recent years is encouraging. For 2020, their forecast issued in late June calls for a higher chance of near normal (50%) and below normal (40%) TC frequency over the WNP, whereas the observed counts during June–December was 22, which is slightly below the mean number. Therefore, the 2020 forecast was basically consistent with the observations.

5.5 Recommendations for Future Development

Several recommendations are suggested here with the aim of potentially improving subseasonal to seasonal TC forecasts.

- The track-pattern-based approach of TCs provides a better understanding of the effect of the large-scale environment on TC activity and the potential to isolate predictable aspects of landfall risks (Camargo et al., 2007b; Kim et al., 2011, 2012). As mentioned previously, a statistical-dynamical approach is the major tool used in the seasonal (and possibly subseasonal) TC forecasting enterprise. Given the relative simplicity and good results, this approach is highly recommended but there are ways in which further improvement can be made. First, historical TC tracks can be clustered into distinct groups and prediction of each group and a combination of group forecasts for a basin can be made. The track-pattern-based method yields higher forecast skill than the non-clustering approach (Kim et al., 2012, 2018; Murakami et al., 2016b). Another suggestion is to consider using all available dynamical forecasts prior to the initialization month, not just forecasts in the initialization month ("closer-gets-richer" strategy) (Zhang et al., 2017).
- New predictors should be explored. Most statistical or statistical-dynamical methods use SST as one of the predictors. However, the subsurface ocean heat content is also deemed as a potential important predictor to improve ENSO predictability (Barnston et al., 2012) and short-term TC intensity prediction (Kim et al., 2018). It would be interesting to see if this new predictor can improve subseasonal to seasonal TC prediction. For the subseasonal TC prediction over the North Atlantic and the Gulf of Mexico, Slade and Maloney (2013) suggested

some additional predictors such as the West African intraseasonal index, NAO, and PDO. One cautionary note is that increasing the number of predictors in a statistical model may result in overfitting problems. As such, predictor screening is suggested and this can be performed using a stepwise regression approach.

- For seasonal or subseasonal TC forecasts, the common approach is to find a statistical relationship between historical storm activity and large-scale environments and then use this empirical relation to forecast (or hindcast) unobserved TC activity. While in principle this is a valid approach, forecast values based on a regression model may be biased because of trend, decadal variations, or abrupt shifts in climate systems. That is, the regression coefficients do not vary with time and cannot capture the nonstationary relationship between the predictand and predictors as time evolves. This is also echoed by Wang et al. (2019), who noticed a weakening relationship between the PMM and TC frequency in the northeast subregion of the WNP, giving rise to a lower predictive skill during 2000–2016. To deal with this problem, the Bayesian dynamic linear model in which regression coefficients are updated at each time step, conditional on available data and use of the Bayesian theorem, is proposed (West and Harrison, 1997; Petris et al., 2009). Recently, Xing et al. (2020) used this method to improve summer precipitation prediction over the WNP.

- Dynamical downscaling with the use of high-resolution regional climate models (RCMs) aimed for seasonal TC forecasts is another avenue of potential research. The RCM is driven by outputs from the reanalysis or forecasts from global climate models (e.g., Walsh et al., 2004). Because of the higher resolution, representations of TC characteristics can be better resolved by RCMs. Published literature concerning dynamical downscaling with applications to seasonal TC prediction is scarce. Huang and Chan (2014) used a RCM with a horizontal resolution of 50 km to downscale retrospective forecasts from a global model to estimate WNP TC genesis and landfall during 2000–2010. The correlation coefficient in TC genesis number between model and observation during May–October is 0.63.

- While forecasts from some statistical models (e.g., LAD) are deterministic, Kwon et al. (2007) suggested the use of an ensemble method in the statistical prediction of seasonal TC frequency over the WNP. Thirty ensemble members are generated based the performance of the predictors, closeness of the lag month to a target season (e.g., June–August), and some subjective judgements. As a result, probabilistic forecasts can be made using relatively large ensemble members from a statistical model.

- The current practice used in statistical subseasonal TC forecast is the logistic regression model, which predicts the probability of at least one storm incidence on a weekly basis. It is also possible to use the Poisson regression model, as demonstrated in seasonal TC prediction, to forecast the probability of weekly TC frequency (one, two, or more), as suggested by McDonnell and Holbrook (2004).

- Statistical post-processing of forecasts (or reforecasts) from multi-models is expected to lead to better results. Numerous studies showed that the multi-model ensemble concept is superior to the average single-model prediction (e.g., Hagedorn et al., 2005). This was also demonstrated by Vitart et al. (2010) and Kim et al. (2017) in TC forecasts from subseasonal to seasonal scales. For multi-model ensembles, a common approach is to take a simple average to synthesize their outputs. However, because some models are better than others, taking equal weights from all models may degrade the overall skill. The Bayesian model averaging (BMA) can be used to estimate the relative importance of each model and hence used as a basis for model selection (e.g., Raftery et al., 2005; Zhang et al., 2019). In other words, the BMA treats each model weight differently and the weights can be considered as their relative contribution to predictive skill over the training period.
- Machine learning techniques also look promising for improving subseasonal to seasonal TC forecasts. For example, the support vector machine (SVM) has been used to demonstrate a better seasonal prediction of TC genesis over the North Atlantic and northwest Australian region (Richman and Leslie, 2012; Richman et al., 2017). In SVM, the input data are mapped onto a high-dimensional feature space using some nonlinear mapping and trained with a learning algorithm from optimization theory. The SVM can be applied to the case of regression in which a nonlinear function is learned by a linear learning machine in a kernel-induced feature space.

5.6 Summary

This chapter provides a review of the current status regarding subseasonal to seasonal TC prediction and the corresponding methods used by various operational centers and researchers around the world. Broadly speaking, TC prediction methods are classified into three approaches: purely statistical; dynamical; and statistical-dynamical. Statistical methods use a range of regression techniques such as logistic regression, least-absolute-deviation regression, Poisson regression and multiple linear regression. While the logistic regression is used to model the probability of the storm events (e.g., the likelihood of TC formation weeks in advance), the other three methods focus on deterministic forecasts of TC frequency. In addition, a Bayesian paradigm is also introduced so that the predicted TC counts could be cast in terms of probabilistic distributions. To solve complex integrations of high-dimensional functions, a Gibbs sampler based on the Markov

Chain Monte Carlo method is used to integrate the desired posterior predictive distribution.

Traditionally, attempts have been made to either forecast TC activity for an entire basin or for a specific region within a basin. In this regard, forecasts are categorized by their geographic location without considering the nature and variability of TC tracks. It is worth noting that the origin of each TC and its attendant tracks within a season are not the same from year to year. Some TCs are straight movers while others are prone to recurve to higher latitudes or even make loops. Moreover, some track types are linked to certain climate modes. Therefore, a categorization of the historical TC tracks into several distinct types and forecasting of each individual track type may result in a better physical understanding of the overall forecast skills than the traditional attempts. To help improve the forecast skills, a new approach based on TC track types using the fuzzy clustering method or the mixture regression model, is advocated in this chapter.

Dynamical climate models are also used for forecasting seasonal TC or landfall activity at various centers. The models are either atmospheric GCMs forced by observed or predicted SST anomalies, or coupled atmosphere–ocean models. The dynamical forecasts require statistical post-processing to remove systematic biases. The statistical-dynamical method is perhaps the most prevailing approach currently employed in TC forecasting communities. This hybrid scheme takes advantage of the statistical relationship between TC activity and large-scale environments, and dynamical forecasts (after calibrations) from climate models. This approach can also be extended to create spatial maps of predicted TC track density over an entire ocean basin based on the classified track types so the major operational centers can use that information to better inform countries potentially affected by TCs in the coming season.

For the subseasonal TC forecast, a reliable hindcast out to week 2 is shown for the North Atlantic based on the logistic regression. Most dynamical models exhibit the eastward propagation of TC genesis anomalies in the Southern Hemisphere for week 2 reforecasts based on the S2S data set. Some examples are also shown to illustrate the success of the probabilistic forecasts of TC formation near Australia out to three weeks in advance by the ECMWF and Australian Bureau of Meteorology.

For the seasonal TC forecast, many studies based on statistical approaches show good predictive skills for the western North Pacific, Atlantic, and southwest Pacific using antecedent predictors. The retrospective forecast skill from the dynamical models in early years is fairly modest. More recently, improved forecast skill in predicting the frequency of major hurricanes and most intense hurricanes over the North Atlantic, as well as landfalling TCs over the USA and Caribbean Islands a few months in advance, is shown based on a high-resolution coupled climate models. The track-pattern-based statistical-dynamical approach using 15 ensemble members from the NCEP CFS forecasts also yields good prediction skill for the western North Pacific. The LOOCV correlation coefficient between hindcasts and observations varies from 0.71 to 0.81 for the period 1981–2006. For real-time seasonal TC forecasts performed at various operational centers and private sectors, good skill can only be achieved when the forecast is issued at the beginning of an active TC season for both the North Atlantic and western North Pacific (e.g., August for the Atlantic and July for the western North Pacific). This lack of useful predictive information well in advance of the TC season at operational centers leads to recommendations for future development such as the exploration of new predictors (e.g., subsurface ocean heat content), new methods dealing with the nonstationary relationship between the predicted TC metrics and the chosen predictors, track-pattern-based forecasting, machine learning, and others.

Exercises

5.1 (a) What is logistic regression? (b) How this method can be used in TC prediction?

5.2 (a) What is Poisson regression? (b) How this method can be used in TC prediction?

5.3 (a) What is skill score? (b) What is the Brier skill score?

5.4 (a) What is cross validation? (b) Why is this method commonly used in forecast verification?

5.5 (a) Describe how the statistical-dynamical approach is used in TC forecasting. (b) Are there any advantages of this approach?

References

Aiyyer, A. R., and C. Thorncroft, 2006: Climatology of vertical wind shear over the tropical Atlantic. *J. Climate*, **19**, 2969–2983.

Barnston, A. G., and Coauthors, 2012: Skill of real-time seasonal ENSO model predictions during 2002–2011. *Bull. Amer. Meteorol. Soc.*, **93**, 631–651.

Bengtsson, L., M. Botzet, and M. Esch, 1995: Hurricane-type vortices in a general circulation model. *Tellus*, **47A**, 175–196.

Bloomfield, P., and W. L. Steiger, 1980: Least absolute deviations curve-fitting. *J. Sci. Stat. Comput.*, **1**, 290–301.

Camargo, S. J., and A. G. Barnston, 2009: Experimental dynamical seasonal forecasts of tropical cyclone activity at IRI. *Wea. Forecasting*, **24**, 472–491.

Camargo, S. J., A. G. Barnston, and S. E. Zebiak, 2005: A statistical assessment of tropical cyclones in atmospheric general circulation models. *Tellus*, **57A**, 589–604.

Camargo, S. J., and Coauthors, 2007a: Cluster analysis of typhoon tracks. Part I: General properties. *J. Climate*, **20**, 3635–3653.

Camargo, S. J., and Coauthors, 2007b: Cluster analysis of typhoon tracks. Part II: Large-scale circulation and ENSO. *J. Climate*, **20**, 3654–3676.

Camargo, S. J., and Coauthors, 2019: Tropical cyclone prediction on subseasonal time-scales. *Trop. Cyclone Res. Rev.*, **8**, 150–165.

Chan, J. C. L., J. E. Shi, and C. M. Lam, 1998: Seasonal forecasting of tropical cyclone activity over the western North Pacific and the South China Sea. *Wea. Forecasting*, **13**, 997–1004.

Chan, J. C. L., J. E. Shi, and K. S. Liu, 2001: Improvements in the seasonal forecasting of tropical cyclone activity over the western North Pacific. *Wea. Forecasting*, **16**, 491–498.

Chand, S. S., K. J. E. Walsh, and J. C. L. Chan, 2010: A Bayesian regression approach to seasonal prediction of tropical cyclones affecting the Fiji region. *J. Climate*, **23**, 3425–3445.

Chen, J.-H., and S.-J. Lin, 2013: Seasonal predictions of tropical cyclones using a 25-km resolution general circulation model. *J. Climate*, **26**, 380–398.

Chiang, J. C. H., and D. J. Vimont, 2004: Analogous Pacific and Atlantic meridional modes of tropical atmosphere-ocean variability. *J. Climate*, **17**, 4143–4158.

Chu, P.-S., and X. Zhao, 2007: A Bayesian regression approach for predicting seasonal tropical cyclone activity over the central North Pacific. *J. Climate*, **20**, 4002–4012.

Chu, P.-S., and X. Zhao, 2011: Bayesian analysis for extreme climatic events: A review. *Atmos. Res.*, **102**, 243–262.

Chu, P.-S., X. Zhao, C.-T. Lee, and M.-M. Lu, 2007: Climate prediction of tropical cyclone activity in the vicinity of Taiwan using the multivariate least absolute deviation regression method. *Terr. Atmos. Ocean. Sci.*, **18**, 805–825.

Chu, P.-S., and Coauthors, 2010a: Bayesian forecasting of seasonal typhoon activity: A track-pattern-oriented categorization approach for Taiwan. *J. Climate*, **23**, 6654–6668.

Chu, P.-S., X. Zhao, and J.-H. Kim, 2010b: Regional typhoon activity as revealed by track patterns and climate change. In *Hurricanes and Climate Change*, 2. J. B. Elsner, R. E. Hodges, J. C. Malmstadt, and K. N. Scheitlin, Eds. Springer, 137–148.

Elsner, J. B., and T. H. Jagger, 2004: A hierarchical Bayesian approach to seasonal hurricane modeling. *J. Climate*, **17**, 2813–2827.

Elsner, J. B., and T. H. Jagger, 2006: Prediction models for annual U.S. hurricane counts. *J. Climate*, **19**, 2935–2952.

Elsner, J. B., and K. B. Liu, 2003: Examining the ENSO-typhoon hypothesis. *Clim. Res.*, **25**, 43–54.

Elsner, J. B., and C. P. Schmertmann, 1993: Improving extended-range seasonal predictions of intense Atlantic hurricane activity. *Wea. Forecasting*, **8**, 345–351.

Goldenberg, S. B., and L. J. Shapiro, 1996: Physical mechanisms for the association of El Niño and West African rainfall with Atlantic major hurricane activity. *J. Climate*, **9**, 1169–1187.

Gray, W. M., 1984: Atlantic seasonal hurricane frequency: Part I: El Niño and 30 mb quasi-biennial oscillation influences. *Mon. Wea. Rev.*, **112**, 1649–1668.

Gray, W. M., C. W. Landsea, P. W. Pielke, and K. J. Berry, 1992: Predicting Atlantic basin seasonal hurricane activity 6–11 months in advance. *Wea. Forecasting*, **7**, 440–455.

Gray, W. M., C. W. Landsea, P. W. Pielke, and K. J. Berry, 1993: Predicting Atlantic basin seasonal hurricane activity by 1 August. *Wea. Forecasting*, **8**, 73–86.

Gray, W. M., C. W. Landsea, P. W. Pielke, and K. J. Berry, 1994: Predicting Atlantic basin seasonal hurricane activity by 1 June. *Wea. Forecasting*, **9**, 103–115.

Gregory, P., J. Camp, K. Bigelow, and A. Brown, 2019: Sub-seasonal predictability of the 2017–18 Southern Hemisphere tropical cyclone season. *Atmos. Sci. Lett.*, **20**(4), e886 doi:10.1002/asl.886.

Hagedorn, R., F. J. Doblas-Reyes, and T. N. Palmer, 2005: The rationale behind the success of multi-model ensembles in seasonal forecasting – I. Basic concept. *Tellus*, **57A**, 219–233.

Harr, P. A., and R. L. Elsberry, 1991: Tropical cyclone track characteristics as a function of large-scale circulation anomalies. *Mon. Wea. Rev.*, **119**, 1448–1468.

Harr, P. A., and R. L. Elsberry, 1995a: Large-scale circulation variability over the tropical western North Pacific. Part I: Spatial patterns and tropical cyclone characteristics. *Mon. Wea. Rev.*, **123**, 1225–1246.

Harr, P. A., and R. L. Elsberry, 1995b: Large-scale circulation variability over the tropical western North Pacific. Part II: Persistence and transition characteristics. *Mon. Wea. Rev.*, **123**, 1247–1268.

Hess, J. C., J. B. Elsner, and N. E. LaSeur, 1995: Improving seasonal hurricane predictions for the Atlantic basin. *Wea. Forecasting*, **10**, 425–432.

Ho, C.-H., H.-S. Kim, and P.-S. Chu, 2009: Seasonal prediction of tropical cyclone frequency over the East China Sea through a Bayesian Poisson regression method. *Asia-Pac. J. Atmos. Sci.*, **45**, 45–54.

Hodanish, S., and W. M. Gray, 1993: An observational analysis of tropical cyclone recurvature. *Mon. Wea. Rev.*, **121**, 2665–2689.

Huang, W.-R., and J. C. L. Chan, 2014: Dynamical downscaling forecasts of western North Pacific tropical cyclone genesis and landfall. *Clim. Dyn.*, **42**, 2227–2237.

Kim, H.-S., C.-H. Ho, P.-S. Chu, and J.-H. Kim, 2010a: Seasonal prediction of summer-time tropical cyclone activity over the East China Sea using the least absolute deviation regression and the Poisson regression. *Int. J. Climatol.*, **30**, 210–219.

Kim, H.-S., J.-H. Kim, C.-H. Ho, and P.-S. Chu, 2011: Pattern classification of typhoon tracks using the fuzzy c-means clustering method. *J. Climate*, **24**, 488–508.

Kim, H.-S., C.-H. Ho, J.-H. Kim, and P.-S. Chu, 2012: Track-pattern-based model for seasonal prediction of tropical cyclone activity in the western North Pacific. *J. Climate*, **25**, 4660–4678.

Kim, J.-H., C.-H. Ho, and P.-S. Chu, 2010b: Dipolar redistribution of summertime tropical cyclone genesis between the Philippine Sea and the northern South China Sea and its possible mechanism. *J. Geophys. Res.*, **115**, D06104.

Kim, O.-Y., H.-M. Kim, M.-I. Lee, and Y.-M. Min, 2017: Dynamical-statistical seasonal predictions for western North Pacific typhoons based on APCC multi-models. *Clim. Dyn.*, **48**, 71–88.

Kim, S.-H., I.-J. Moon, and P.-S. Chu, 2018: Statistical-dynamical typhoon intensity predictions in the western North Pacific using track pattern clustering and ocean coupling predictors. *Wea. Forecasting*, **33**, 347–365.

Klotzbach, P. J., and W. M. Gray, 2004: Updated 6–11 month prediction of Atlantic basin seasonal hurricane activity. *Wea. Forecasting*, **19**, 917–934.

Klotzbach, P. J., et al., 2019: Seasonal tropical cyclone forecasting. *Trop. Cyclone Res. Rev.*, **8**, 134–149.

Kwon, H. J., W.-J. Lee, S.-H. Won, and E.-J. Cha, 2007: Statistical ensemble prediction of the tropical cyclone activity over the western North Pacific. *Geophys. Res. Lett.*, **34**, L24805.

Lander, M. A., 1996: Specific tropical cyclone track types and unusual tropical cyclone motions associated with a reverse-oriented monsoon trough in the western North Pacific. *Wea. Forecasting*, **11**, 170–186.

LaRow, T. E., 2013: The impact of SST bias correction on North Atlantic hurricane retrospective forecasts. *Mon. Wea. Rev.*, **141**, 490–498.

LaRow, T. E., L. Stefanova, D. W. Shin, and S. Cocke, 2010: Seasonal Atlantic tropical cyclone hindcasting/forecasting using two sea surface temperature datasets. *Geophys. Res. Lett.*, **37**, L02804.

Lee, C.-Y., and Coauthors, 2018: Subseasonal tropical cyclone genesis prediction and MJO in the S2S dataset. *Wea. Forecasting*, **33**, 967–988.

Lee, C.-Y., and Coauthors, 2020: Subseasonal prediction of tropical cyclone occurrence and ACE in the S2S dataset. *Wea. Forecasting*, **35**, 921–938.

Leroy, A., and M. C. Wheeler, 2008: Statistical prediction of weekly tropical cyclone activity in the Southern Hemisphere. *Mon. Wea. Rev.*, **136**, 3637–3654.

Liu, K. S., and J. C. L. Chan, 2003: Climatological characteristics and seasonal forecasting of tropical cyclones making landfall along the South China coast. *Mon. Wea. Rev.*, **131**, 1650–1662.

Lu, M.-M., P.-S. Chu, and Y.-C. Lin, 2010: Seasonal prediction of tropical cyclone activity near Taiwan using the Bayesian multivariate regression method. *Wea. Forecasting*, **25**, 1780–1795.

Manganello, J. V., et al., 2016: Seasonal forecasts of tropical cyclone activity in a high-atmospheric-resolution coupled prediction system. *J. Climate*, **29**, 1179–1200.

McDonnell, K. A., and N. J. Holbrook, 2004: A Poisson regression model of tropical cyclogenesis for the Australian-Southwest Pacific Ocean region. *Wea. Forecasting*, **19**, 440–455.

Murakami, H., and Coauthors, 2016a: Seasonal forecasts of major hurricanes and land-falling tropical cyclones using a high-resolution GFDL coupled climate model. *J. Climate*, **29**, 7977–7989.

Murakami, H., and Coauthors, 2016b: Statistical-dynamical seasonal forecasts of North Atlantic and U.S. landfalling tropical cyclones using the high-resolution GFDL FLOR coupled model. *Mon. Wea. Rev.*, **144**, 2101–2123.

Petris, G., S. Petrone, and P. Campagnoli, 2009: *Dynamic Linear Models with R*. Springer, 31–84.

Raftery, A. E., T. Gneiting, T. F. Balabdaoui, and M. Polakowski, 2005: Using Bayesian model averaging to calibrate forecast ensembles. *Mon. Wea. Rev.*, **133**, 1155–11744.

Richman, M. B., and L. M. Leslie, 2012: Adaptive machine learning approaches to seasonal prediction of tropical cyclones. *Procedia Comput. Sci.*, **12**, 276–281.

Richman, M. B., L. M. Leslie, H. A. Ramsay, and P. J. Klotzbach, 2017: Reducing tropical cyclone prediction errors using machine learning approaches. *Procedia Comput. Sci.*, **114**, 314–323.

Saha, S., and Coauthors, 2006: The NCEP climate forecast system. *J. Climate*, **19**, 3483–3517.

Saunders, M. A., and A. S. Lea, 2005: Seasonal prediction of hurricane activity reaching the coast of the United States. *Nature*, **434**, 1005–1008.

Shaman, J., S. K. Esbensen, and E. D. Maloney, 2009: The dynamics of the ENSO-Atlantic hurricane teleconnection: ENSO-related changes to the North African-Asian jet affect Atlantic basin tropical cyclogenesis. *J. Climate*, **22**, 2458–2482.

Slade, S. A., and E. D. Maloney, 2013: An intraseasonal prediction model of Atlantic and East Pacific tropical cyclone genesis. *Mon. Wea. Rev.*, **141**, 1923–1942.

Vecchi, G .A., et al., 2011: Statistical-dynamical predictions of seasonal North Atlantic hurricane activity. *Mon. Wea. Rev.*, **139**, 1070–1082.

Vitart, F., 2009: Impact of the Madden Julian Oscillation on tropical storms and risk of landfall in the ECMWF forecast system. *Geophys. Res., Lett.*, **36**, L15802.

Vitart, F., and T. N. Stockdale, 2001: Seasonal forecasting of tropical storms using coupled GCM integrations. *Mon. Wea. Rev.*, **129**, 2521–2537.

Vitart, F., J. L. Anderson, and W. F. Stern, 1997: Simulation of interannual variability of tropical storm frequency in an ensemble of GCM integrations. *J. Climate*, **10**, 745–760.

Vitart, F., et al.,2007: Dynamically-based seasonal forecasting of Atlantic tropical storm activity issued in June by EUROSIP. *Geophys. Res. Lett.*, **34**, L16815.

Vitart, F., A. Leroy, and M. C. Wheeler, 2010: A comparison of dynamical and statistical predictions of weekly tropical cyclone activity in the Southern Hemisphere. *Mon. Wea. Rev.*, **138**, 3671–3682.

Vitart, F., and Coauthors, 2017: The subseasonal to seasonal (S2S) prediction project database. *Bull. Amer. Meteorol. Soc.*, **98**, 163–173.

Walsh, K. J. E., K.-C. Nguyen, and J. L. McGregor, 2004: Fine-resolution regional climate model simulations of the impact of climate change on tropical cyclones near Australia. *Clim. Dyn.*, **22**, 47–56.

Wang, B., and J. C. L. Chan, 2002: How strong ENSO events affect tropical storm activity over the western North Pacific. *J. Climate*, **15**, 1643–1658.

Wang, B., R. Wu, and X. Fu, 2000: Pacific-East Asian teleconnection: How does ENSO affect East Asian climate? *J. Climate*, **13**, 1517–1536.

Wang, C., B. Wang, and L. Wu, 2019: A region-dependent seasonal forecasting framework for tropical cyclone genesis frequency in the western North Pacific. *J. Climate*, **32**, 8415–8435.

Wang, H., and Coauthors, 2009: A statistical forecast model for Atlantic seasonal hurricane activity based on the NCEP dynamical seasonal forecast. *J. Climate*, **22**, 4481–4500.

West, M., and P. J. Harrison, 1997: *Bayesian Forecasting and Dynamic Models*. Springer-Verlag.

Wilks, D. S., 2011: *Statistical Methods in the Atmospheric Sciences*, 3rd ed. Academic Press.

Wu, G., and N. Lau, 1992: A GCM simulation of the relationship between tropical-storm formation and ENSO. *Mon. Wea. Rev.*, **120**, 958–977.

WMO, 2002: *Standardized Verification System for Long-Range Forecasts*. New Attachment 11-0 to the Manual on the GDPFS (WMO-No. 485). World Meteorological Organization.

Xing, W., W. Han, and L. Zhang, 2020: Improving the prediction of western North Pacific summer precipitation using a Bayesian dynamic linear model. *Clim. Dyn.*, **55**, 831–842.

Zhan, R., and Y. Wang, 2016: CFSv2-based statistical prediction for seasonal accumulated cyclone energy (ACE) over the western North Pacific. *J. Climate*, **29**, 525–541.

Zhan, R., and Coauthors, 2012: Seasonal forecasts of tropical cyclone activity over the western North Pacific: A review. *Trop. Cyclone Res. Rev.*, **1**, 307–324.

Zhang, H., P.-S. Chu, L. He, and D. Unger, 2019: Improving the CPC's ENSO forecasts using Bayesian model averaging. *Clim. Dyn.*, **53**, 3373–3385.

Zhang, W., and Coauthors, 2017: Statistical-dynamical seasonal forecast of western North Pacific and East Asia landfalling tropical cyclones using the GFDL FLOR coupled climate model. *J. Climate*, **30**, 2209–2232.

Zhang, X., S. Zhong, Z. Wu, and Y. Li, 2018: Seasonal prediction of the typhoon genesis frequency over the western North Pacific with a Poisson regression model. *Clim. Dyn.*, **51**, 4585–4600.

6

Extreme Typhoon Rainfall Under Changing Climate

6.1 Introduction

Extreme events, such as heavy precipitation and flooding associated with tropical cyclones, are a major concern because of their devastating consequences on society and economics. For instance, a heavy downpour in a short time span resulting from TCs can trigger landslides and mudslides in mountainous regions. This may lead to property and infrastructure damage, loss of human and animal life, along with severe environmental degradation. Densely populated areas such as major cities are also at a high risk for flash floods from TCs because the construction of buildings and roads in cities reduces the absorption capacity of the ground and, as a result, runoff of floodwater is increased. Torrential rainfall can also leave storm drains suddenly overwhelmed, resulting in floodwater spilling onto roads, buildings, and residential dwellings. Low-lying areas such as parking lots, building basements, underpasses, and subway stations are particularly vulnerable to such rapidly rising water. A short burst of intense rainfall, not necessarily related to the TC categorization according to the common sustained wind strength, often causes significant damage.

Observationally, there is no strong evidence for the effects of anthropogenic warming on TC rain rate (e.g., Lau and Zhou, 2012; Knutson et al., 2019). However, based on a suite of climate model simulations, a robust signal emerges to show a projected increase in TC rain rates on a global scale, with a median of +14% (Knutson et al., 2020). The increasing rain rates are dependent on the basin considered and appear to be modulated by the relative stronger SST warming in a given basin, where "relative" refers to a basin in comparison to the tropical average SST warming (e.g., Knutson et al., 2020). Changes in atmospheric water vapor content roughly follow the Clausius–Clapeyron scaling at constant humidity with a rate of about 7% per 1°C SST warming (Allen and Ingram, 2002; Trenberth et al., 2003; Chou and Chen, 2010). This relationship provides a clue about how extreme precipitation might scale with SST as the Earth warms. Higher relative SST warming corresponds to greater than 7% per 1°C scaling of TC rain rates and vice versa.

Apart from the increase in tropospheric water vapor capacity due to anthropogenic warming, moisture flux convergence, tightly linked to TC rainfall, is also enhanced according to a cloud-resolving modeling study applied to two landfalling typhoons in Taiwan (Wang et al., 2015). For both typhoons (Sinlaku and Jangmi in September 2008), the moisture convergence of the storm overwhelmed the evaporation term for the mean TC rainfall in the water budget analysis (Wang et al., 2015). As a consequence of these two processes (an increment in both water vapor and moisture convergence), TC rainfall increased by about 5% at 200–500 km radii from the typhoon center in a modern-day climate (1990–2009) relative to the past climate (1950–1969). Although the sample size is extremely small ($n = 2$), results from their cloud-resolving model also indicate an increase in the frequency of more intense typhoon rainfall (20 to $\geq$50 mm h^{-1}) by about 5–25%.

In this section, Taiwan is chosen as an example to illustrate how extreme typhoon rainfall is changing under a warmer climate. Taiwan possess reliable and long-term rainfall records, has a unique position in relation to the prevailing typhoon path, and also has abundance of literature on this topic. Taiwan is situated in one of the major regions where the WNP typhoons track through. On average, four or five typhoons are close to the vicinity of Taiwan every year with various degrees of rainfall-related damage (Tu et al., 2009). Climatologically, the typhoon-related rainfall accounts for 47.5% of the total annual rainfall in Taiwan (Chen et al., 2010). Modeling studies suggest an increase in the average TC-related rainfall rates with anthropogenic warming but there is a lack of agreement on the amount of the projected rainfall rates (e.g., Knutson et al., 2010, 2019, 2020; Liu et al., 2018). Under global warming, the water vapor capacity in the atmosphere should increase, as expected from the Clausius–Clapeyron relation. Because of the dominance of typhoons, long-term TC rainfall in Taiwan is likely influenced by global warming-induced effects and also by changes in typhoon characteristics such as prevailing tracks and translation speed, which impact storm duration and ultimately rainfall.

6.2 Long-Term Changes in Precipitation Extremes during the Typhoon Season

The prevailing typhoon tracks from the WNP bifurcate into two major groups just before reaching Taiwan (Fig. 6.1). These include (1) westward straight passing through southern Taiwan and the Bashi Channel onto the South China Sea; and

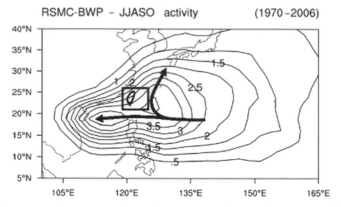

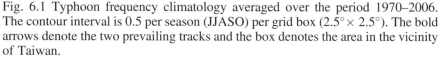

Fig. 6.1 Typhoon frequency climatology averaged over the period 1970–2006. The contour interval is 0.5 per season (JJASO) per grid box ($2.5° \times 2.5°$). The bold arrows denote the two prevailing tracks and the box denotes the area in the vicinity of Taiwan.

Source: Tu et al., 2009; ©American Meteorological Society. Used with permission.

(2) northward recurving. For the latter, the tracks either recurve to the east coast of Taiwan and/or China or recurve and veer east over the open ocean, eventually slamming into Japan and/or Korea (Harr and Elsberry, 1991; Tu et al., 2009; Liang et al., 2017). The island of Taiwan is dominated by the Central Mountain Range (CMR) with many peaks exceeding 3,000 m and the terrain–TC interaction plays an important role in TC structure and related rainfall (e.g., Wu et al., 2002, 2011; Yu and Cheng, 2008, 2013, 2014; Chang et al., 2013).

Precipitation extremes during the typhoon season in Taiwan have undergone a dramatic change over the last 60 years. This is evident from the analysis of a suite of climate change indices recommended by the joint World Meteorological Organization (WMO) Commission on Climatology/Climate Variability and Predictability (CLIVAR)/Joint Technical Commission for Oceanography and Marine Meteorology Expert Team on Climate Change Detection and Indices (ETCCDI) (Zhang et al., 2011). Because climate change indices defined by WMO/CLIVAR/ETCCDI are widely accepted tools, it is meaningful to investigate changes in precipitation extremes in Taiwan using these indices so that the results can be compared with other parts of the world (e.g., Frich et al., 2002; Alexander et al., 2006). The trends of four commonly used rainfall indices in Taiwan are investigated using daily observational records for the typhoon season (July–October) during 1950–2010 (Chu et al., 2013).

The trends of the indices are estimated by the nonparametric rank-based Mann–Kendall test and Sen's method (e.g., Chu et al., 2010). The former tests whether the trend is increasing or decreasing and estimates the significance of the trend, whereas the latter quantifies the slope of the trend. The advantage in combining these two methods is that missing values are allowed and the data do not need to conform to any parametric distribution. Moreover, Sen's method is robust against skewed distributions and outliers because the slopes of all data pairs are calculated and the median of these statistics is Sen's estimator of slope.

The first index is the simple daily "intensity" index (SDII), which is the average of wet-day rainfall intensity during the typhoon season, with unit of mm d^{-1}. Here the typhoon season runs from July to October. The second index (R50) is the seasonal number of days with daily rainfall $\geq$50 mm with unit in days. This definition is consistent with the Central Weather Bureau (CWB) of Taiwan for heavy rainfall events and is also used in Chen et al. (2007). In a broad sense, this index describes the "frequency" of heavy rainfall events. The third index, R5d, is the seasonal maximum consecutive 5-d rainfall amounts, with unit in mm. This index describes the "magnitude" of intense rainfall events. For the fourth index, the seasonal maximum number of consecutive dry days (CDD) is used. Long-term daily rainfall records from the CWB are used and there are 21 stations with data available for at least 41 years up to 2010, and seven out of these 21 stations have at least 60 years of records.

Figure 6.2a displays the trend pattern for SDII. All 21 stations across Taiwan and its outer isles show an increasing trend in rainfall intensity and trends at 7 out of these 21 stations and are statistically significant at the 5% level. Therefore, increasing rainfall intensity during the typhoon season is not a local phenomenon but occurs at an island-wide scale. Save for one station, the other six stations with significant upward trends are not located at high elevations, although, climatologically, the highest rainfall intensity is found in the mountainous region of the northern Taiwan and the CMR (Yu and Cheng, 2014). This is primarily because of the interaction between the typhoon's cyclonic circulation and the moisture supply associated with the prevailing low-level southwesterly monsoonal flow, which tends to strengthen the confluence flow and convection and then lead to heavy rainfall along the windward slope of the mountains (e.g., Wu et al., 2002; Chen and Chen, 2003; Ge et al., 2010; Yu and Cheng, 2013). Based on Doppler radar observations during Typhoon Morakot over the mountainous central and southern Taiwan in August 2009, Yu and Cheng (2013) found that upslope lifting accounts for the observed rainfall maxima along the windward slopes of the higher and wider barrier of the CMR. Morakot was a very slow-moving typhoon, lingering in the vicinity of Taiwan for 52 h, which is rather unusual. Wu et al. (2011) emphasized the monsoonal influence on maintaining the major rainfall area in southern Taiwan through the reduction of the translation speed and northward shifting of Morakot. For the southern barrier of the CMR, which is lower and narrower, the strongest rainfall tends to occur slightly downstream of the mountain crest. The degree of the observed orographic enhancement depends on typhoon

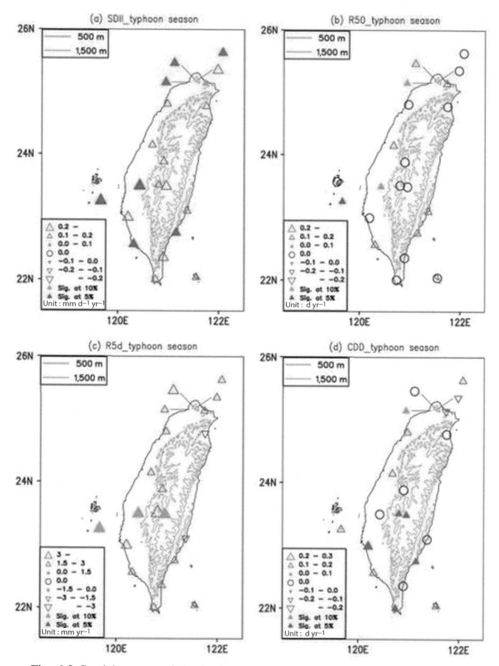

Fig. 6.2 Spatial pattern of the Sen's slopes for (a) SDII, (b) R50, (c) R5d, and (d) CDD during July–October, for 1950–2010. Triangles and circles denote the location of stations. Upward (downward) hollow triangles indicate positive (negative) trends and their size correspond to the magnitude of trends in the legend. Solid (light) filled-in triangles indicate trends significant at the 5% (10%) level. Source: Chu et al., 2013; © RightsLink/John Wiley & Sons. Used with permission.

background precipitation and the oncoming wind speed, which in turn can be interpreted as the seeder–feeder effect (Yu and Cheng, 2013).

For R50 (Fig. 6.2b), coastal and lower elevation stations experienced an increasing trend but only one offshore island (Penghu, Fig. 6.6a) showed a significant trend. The pattern of the trend in R5d is marked by positive trends in the entire region (Fig. 6.2c), except for two stations in eastern Taiwan (Ilan and Cheng-Kung). Changes in the plain stations are more uniform among the three rainfall-related indices but this is not the case for mountain stations (Fig. 6.2a–c). For CDD (Fig. 6.2d), the most pronounced feature is the prevailing upward trend across Taiwan, which is most significant in southern Taiwan and at high elevations in the CMR (Alishan and Yushan, Fig. 6.6a). That is, since 1950 consecutive dry days during the typhoon season in southern Taiwan and the CMR have become longer.

Rainfall in Taiwan during the typhoon season is contributed not only by typhoons, but also by mesoscale convective disturbances or local thunderstorms associated with the diurnal cycle of summer heating patterns embedded in the prevailing southwesterly monsoon (Chen and Chen, 2003; Chen et al., 2007). Chen et al. (2010) partitioned summer rainfall in Taiwan into two components: tropical cyclone (TC) rainfall and seasonal monsoonal rainfall. They define TC rainfall days as when a TC center is located near Taiwan within 2.5° in latitude and longitude (117.5°E–124.5°E, 19.5°N–27.5°N). The remainder of the rainfall systems that are not associated with TCs is termed as seasonal monsoon rainfall. Adopting their definition, we classify a wet day as either caused by a TC or monsoon systems. Here, we consider the maximum surface wind over 34 kts as a TC case. The TC thus includes both tropical storms and typhoons. Because most typhoons that impact Taiwan within the defined domain are short-lived, we will focus only on the SDII and R50.

For rainfall intensity induced by TCs, the majority of stations in Taiwan are marked by positive trends in SDII (Fig. 6.3a), suggestive of increasing rainfall intensity directly caused by typhoons since 1950. Although high rainfall is expected at high elevations of the CMR because of TC–topography interaction, the upward trend in rainfall intensity induced by TCs is generally not statistically significant in the CMR. One cautionary note is that the rainfall data associated with typhoons from conventional stations (only three in this case) may not be able to accurately depict the spatial rainfall distribution in the mountainous areas (Wu et al., 2016). In Fig. 6.3b, there is also an upward trend in rainfall intensity induced by monsoon systems over the last 60 years. It is likely that the positive contributions made by the two components in Fig. 6.3a and b enhance the overall signal, rendering a strong and prevailing upward trend in SDII during the typhoon season as seen in Fig. 6.2a.

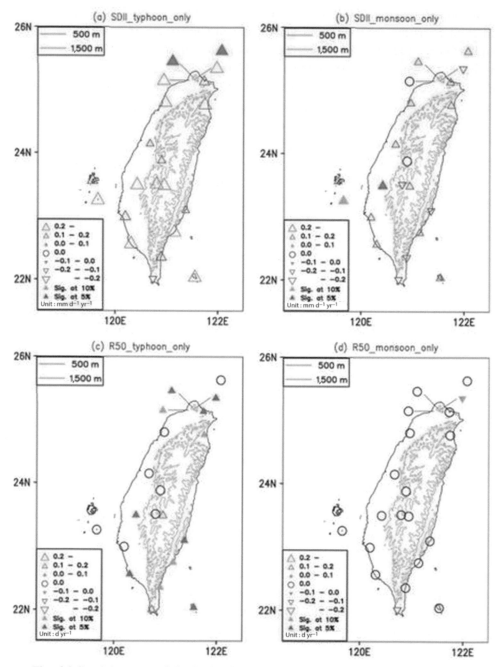

Fig. 6.3 Spatial pattern of the Sen's slopes for (a) SDII and (c) R50 from typhoon-induced precipitation, and (b) SDII and (d) R50 from monsoon-induced precipitation during July–October, for 1950–2010. Triangles and circles denote the location of stations. Upward (downward) hollow triangles indicate positive (negative)

In a different analysis, Tu and Chou (2013) used hourly rainfall data at 21 stations from 1970 to 2009 in Taiwan to examine changes in rainfall frequency and intensity associated with typhoon and non-typhoon events. They broke the entire record into two 20-year periods (P1, 1970–1989; P2, 1990–2009). For change in rainfall intensity, Tu and Chou (2013) chose the 99th percentile of rainfall distribution as the basis of comparison. At the top one percent, both typhoon and non-typhoon rainfall intensity is increased from P1 to P2 (Fig. 6.4b). Time series of rainfall intensity averaged over the entire 99th percentile shows a remarkable increase for typhoon rainfall events from 1970 to 2009, with a rate increase of 2.5 mm h^{-1} per decade, in contrast to a very small increase (0.6 mm h^{-1} per decade) for non-typhoon events (Fig. 6.4c). Although the indices used, methodology of analysis, and data period used in Tu and Chou (2013) and Chu et al. (2013) are different, the general results from these two independent studies are consistent and indicate that rainfall intensity induced by typhoons near Taiwan has increased over time.

Returning to Fig. 6.3c, for heavy rainfall frequency (R50), many stations show an increasing trend induced by TCs since 1950. This feature is particularly evident in northern and southeastern Taiwan. In Fig. 6.3d, there is essentially no change in the frequency of heavy precipitation days caused by monsoon systems over the last 60 years. According to Tu and Chou (2013), the frequency of typhoon rainfall increases in all intensity bins (e.g., 5 mm h^{-1},100 mm h^{-1}) from P1 to P2 (Fig. 6.5b); the increase is largest at the extremely high bins. In the meantime, the frequency tends to reduce at most bins for non-typhoon events, especially for lighter rain. These results suggest that increase in the frequency of heavier rainfall from P1 to P2 is attributed to typhoon events and the decrease in frequency of lighter rainfall is associated with non-typhoon events.

Chang et al. (2013) analyzed 84 typhoons that made landfall in Taiwan during 1960–2011 and found that nine out of 12 typhoons had the heaviest rainfall since 2000. They classified typhoons into three prevailing tracks across the island: northern (N) type with 26 cases; central (C) type with 23 cases; and southern (S) type with 14 cases. The N-type track is characterized by a strong confluence of low-level westerly and southwesterly flows, which impinge upon the western mountain range, producing the highest rainfall intensity (Fig. 6.6). This is followed by the C-type and the S-type, which has the least TC-topography interaction and smallest rainfall.

Caption for fig. 6.3 (*cont.*). trends and their size correspond to the magnitude of trends in the legend. Solid (light) filled-in triangles indicate trends significant at the 5% (10%) level.

Source: Chu et al., 2013; © RightsLink/John Wiley & Sons. Used with permission.

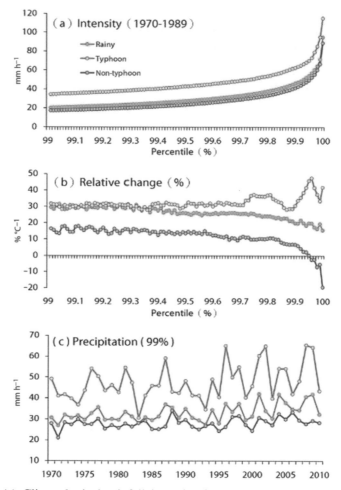

Fig. 6.4 (a) Climatological rainfall intensity for extremes (the 99th percentile) during 1970–1989 in Taiwan with a bin width of 0.01%. The unit for rainfall intensity is mm h^{-1}. (b) The relative changes in the intensity of rainfall extremes between 1990–2009 and 1970–1989, normalized by the global-mean surface temperature difference. (c) Time series of rainfall intensity averaged over the entire 99th percentile in 1970–2010. Green, red, and blue curves denote total, typhoon, and non-typhoon rainfall, respectively. A black and white version of this figure will appear in some formats. For the color version, refer to the plate section. Source: Tu and Chou, 2013; © CC By license.

The large increase in typhoon rainfall in the recent decade is caused by the change in storm duration and track types. For example, when the entire record is broken into two periods, 1960–2003 and 2004–2011, Chang et al. (2013) noted that the N-type typhoons have decreased in frequency (44.6% to 31.3%) but duration has increased substantially (16–24 h) from the first to the second period,

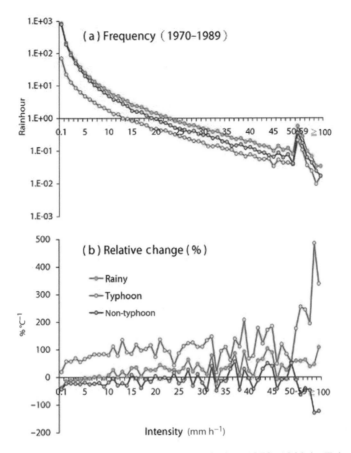

Fig. 6.5 (a) Climatological rainfall frequency during 1970–1989 in Taiwan. The bin width of rainfall intensity along the horizontal axis is 1 and 10 mm h^{-1} for intensity smaller and greater than 50 mm h^{-1}, respectively. The unit for frequency along the vertical axis is per station per year. (b) The relative changes in rainfall frequency between 1990–2009 and 1970–1989, normalized by the global-mean surface temperature difference between these two epochs. For both (a) and (b), green, red, and blue curves denote total, typhoon, and non-typhoon rainfall, respectively. A black and white version of this figure will appear in some formats. For the color version, refer to the plate section.
Source: Tu and Chou, 2013; © CC BY license.

implying slower TC motion and longer lingering time since 2004 (Table 6.1). For the C-type, its frequency has increased dramatically (31.9% to 50%) and its duration has remained essentially unchanged (23.9 to 23.3 h) from the first period to the second period. The S-type track exhibits a relatively small decrease in both frequency (23.4% to 18.8%) and duration (21.7 to 20.0 h). In short, a dramatic increase in the frequency of C-type track, substantial increase in duration of N-type typhoons, together with a decrease in both duration and frequency of S-type track

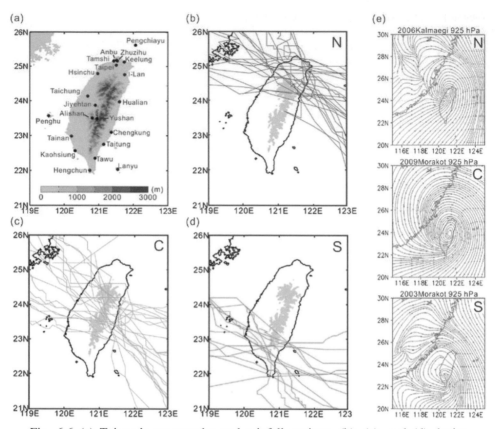

Fig. 6.6 (a) Taiwan's topography and rainfall stations. (b), (c), and (d) depict northern (N), central (C), and southern (S) track types, respectively. The typical 925-hPa streamlines for N, C, and S track types are shown in (e). Typhoon Kalmaegi in 2006 was chosen for the N-type, Typhoon Morakot in 2009 for the C-type, and Typhoon Morakot in 2003 for the S-type in (e).
Source: Chang et al., 2013; ©American Meteorological Society. Used with permission.

account for the large increase in typhoon rainfall intensity in Taiwan observed since 2004.

The recent increase in storm's duration in the vicinity of Taiwan is also observed in Tu and Chou (2013) and Liang et al. (2017), and is consistent with a weakening of the easterly steering flow over the western portion of the WNP and the northern portion of the South China Sea from 1958 to 2009 (Chu et al., 2012). A slower steering flow would weaken the speed of all three track types across Taiwan but it affects the N-type the most because of the topographic locking effect (Chang et al., 2013; Hsu et al., 2013). That is, besides the common influences of obstacles on the flow, such as lifting, sinking, blocking, and channeling effects, the topographically

Table 6.1. The frequency (number of typhoons) and duration (h) of the prevailing three track types across Taiwan before and after 2004. The numbers in the parentheses denote the percentage of each type to the total number of typhoons. Sixty-three typhoons that made landfall in Taiwan during 1960–2011 are used.

	1960–2003	2004–2011
Frequency		
N	21 (44.6%)	5 (31.3%)
C	15 (31.9%)	8 (50.0%)
S	11 (23.4%)	3 (18.8%)
Duration		
N	15.9	24.0
C	23.9	23.3
S	21.7	20.0

Source: Chang et al., 2013; © American Meteorological Society. Used with permission.

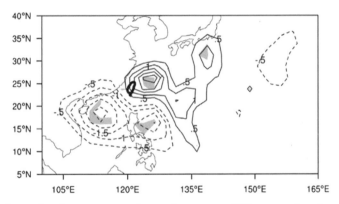

Fig. 6.7 June–October mean typhoon frequency differences for the period of 2000–2006 minus the period of 1970–1999. The contour interval is 0.5 and shading indicates the difference between the mean of the two epochs, which is statistically significant at the 5% level.
Source: Tu et al., 2009; ©American Meteorological Society. Used with permission.

phase-locked convection acts to slow down the N-type typhoons but speed up the S-type typhoons (Hsu et al., 2013). For slow-moving N-type storms, a positive feedback is suggested in the sense that the asymmetric convective heating pattern forced by topography reduces TC motion and slow-moving TCs then give rise to more overland durations, resulting in more precipitation and convective heating (Hsu et al., 2013). As a result, TC speed is further reduced.

Apart from an increase in storm duration or equivalently a slowdown in the translation speed, TC tracks in the western portion of the WNP have shifted in recent decades as TCs tend to take a more northward path (Tu et al., 2009; Liang et al., 2017; Shan and Yu, 2021). Specifically, an increase in typhoon frequency near Taiwan and the East China Sea is found since 2000 based on the Bayesian change-point analysis, concurrently with a decreasing trend in the northern South China Sea due to the weakening and eastward retreat of the WNP subtropical high (Fig. 6.7). As a result of the increased number of northward-moving typhoons, which induce cyclonic circulation around Taiwan and interact with the CMR to favor heavy rainfall along the western side of the island, long-term trends of typhoon-induced rainfall are increasing and statistically significant for northern and western stations in Taiwan (Liang et al., 2017). This is qualitatively consistent with Chang et al. (2013), who showed that the frequency of S-type tracks has decreased while the C-type storms have increased substantially since 2004.

6.3 Long-Term Changes in Return Levels of 24-h Rainfall Extremes during the Typhoon Season

Besides the four rainfall-related indices presented in Figs. 6.2 and 6.3, another popular index, annual maximum daily rainfall data, is also frequently used as a measure of extreme rainfall (e.g., Zwiers and Kharin, 1998; Min et al., 2011). This index, which is also advocated by WMO/CLIVAR/ETCCDI, is commonly used in estimating the amount of rainfall that can be expected to occur in a 24-h interval for the average of a period of many years (e.g., 20-, 100-, and 200-yr). The average "return periods" are derived from quantiles of a particular probability distribution on the basis of extreme value theory. The return period statistics are commonly used for hydro-meteorological and engineering designs, environmental regulations, risk analysis, and disaster prevention purposes. For the statistics of the maxima of long sequences of random variables, the generalized extreme value (GEV) distribution is often found to be a good approximation (Coles, 2001; Katz et al., 2002; Kharin and Zwiers, 2005; Wilks, 2011). This distribution is characterized by three parameters – location, scale, and shape. The data set used for the GEV distribution is often the annual or seasonal maximum daily value, R, which is the largest single 24-h value in each of n years, known as the block maximum series. This selection ensures that the underlying data are independent from year to year.

The aforementioned GEV model can be viewed as a stationary process because the model parameters do not change with time so the estimated return level is a constant. The return level is the extreme rainfall value (e.g., 489 mm within 24 h) that corresponds to any desired return period. Because the climate is changing, it is also reasonable to expect that the return level will change with time. By allowing the time-dependent change in the GEV parameters, the stationary model can be extended to a non-stationary process (Coles, 2001; Katz et al., 2002). Therefore, the statistical theory of extremes can be applied in the context of climate change. Moreover, besides examining changes in precipitation extremes with time, the non-stationary GEV (NGEV) model can also be used to investigate how extreme events will co-vary with external climate drivers such as the El Niño–Southern Oscillation (ENSO) phenomenon and others (Coles, 2001; Katz et al., 2002; Westra et al., 2013; Chen and Chu, 2014; Villafuerte et al., 2015; Chu et al., 2018; Lu et al., 2018). Because the variability of the shape parameter is small and allowing this parameter to vary may cause numerical problems, both the location and scale parameters co-vary with time and/or with ENSO. The NGEV parameters are fitted by maximum likelihood estimation, which maximizes the likelihood function. Specifically, they are estimated by the Extreme Toolkit using the R statistical programming language developed by University Corporation for Atmospheric Research (Gilleland and Katz, 2011).

The extreme-value data used here are the maximum 24-h rainfall (or block maximum) values during the typhoon season. It is worth noting that the true maximum 24-h rainfall event does not always fall in a fixed 24-h window. Extreme rainfall may start at any time of the day and sometimes spans two consecutive days. In this regard, a moving window covering the true maximum 24-h event is chosen, not rainfall accumulated for a fixed 24-h interval in each year. To ensure the rainfall extremes selected are only associated with typhoons, the official typhoon warnings from the CWB are considered. The CWB issued their first typhoon warning in 1958. There are two types of warnings issued by the CWB. The sea warning is issued when a TC with sustainable winds of 34 kts or greater is within the 100 km sea area of Taiwan in 24 h, and is updated every 3 h after the initial warning. The land warning is issued when a TC with sustained winds of 34 kts or more is projected to hit Taiwan and its offshore islets in 18 h. The warnings are updated every three hours while the center of the typhoon is located every hour. Only hourly rainfall data associated with these warnings are used.

Figure 6.8 shows the time series of the typhoon-induced seasonal maximum daily rainfall for Taipei and Taichung during 1958–2013 (see Fig. 6.6a for their locations). Also shown are three estimated return levels (2-, 20-, and 100-yr). For both stations, the return levels at both 20-yr and 100-yr increase with time and such results are common for other stations. Altogether, 20 stations are analyzed. Take Taichung as an example. The 20-yr return level was 381 mm in 1958 and increased to 585 mm by 2013. This amounts to a sizable 54% increase in 20-yr return level over a span of 56 years. Another way to interpret this result is that an event with a 20-yr return-interval threshold value in 1958 occurred on average once every

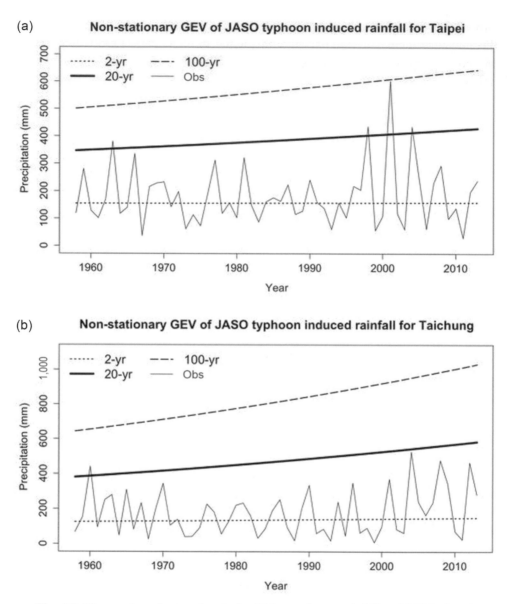

Fig. 6.8 Time series of return levels for 24-h maximum rainfall at (a) Taipei and (b) Taichung according to the non-stationary, time-varying GEV model (NGEV). The dotted, solid, and dashed curves denote the 2-yr, 20-yr, and 100-yr return levels, respectively. Solid thin line at the bottom represents observation data.
Source: Chu et al., 2018; © RightsLink/John Wiley & Sons. Used with permission.

13 years by 2013. Therefore, the waiting time for an extreme event that usually occurred on an average once in 20 years ($p = 0.05$) has shortened considerably to only once in 13 years in recent years. Also note that the slope of return levels becomes steeper as return periods increase from 2-yr to 100-yr because the

probability of annual occurrence (p) decreases from 0.5 (2-yr) to 0.01 (100-yr). Therefore, the increasing return level with time becomes more distinct as the probability of annual occurrence decreases from the 2-yr return period to the 100-yr return period.

Figure 6.9 shows the spatial patterns of the trends in the three return levels of the 24-h maximum rainfall during the typhoon season for the NGEV with time as a covariate for all 20 stations. A large majority of the stations, denoted by triangles pointing upward, have a positive slope in the 2-yr return level except for three stations (Fig. 6.9a). For the 20-yr and 100-yr return levels (Fig. 6.9b and c), all stations are positive except for two stations, which are characterized by a downward trend. Trends at all stations are significant at the 5% level. As expected, the slope of the 100-yr return level is larger than the corresponding 20-yr and 2-yr value. Combining results from Figs. 6.8 and 6.9, we can infer that an increasing trend in return levels associated with heavy precipitation events induced by typhoons is prevalent in Taiwan since 1958. A recent study also shows that Taiwan's extreme rainfall in typhoon season, defined using a 99th percentile threshold, has increased consistently over much of the island since 1960 (Henny et al., 2021). Alternatively, the return-interval threshold values shortened considerably throughout the last 56 years. That is, the frequency of typhoon-induced extreme rainfall has occurred more often in recent years.

Figure 6.8 focused on the changes in return levels with time. It is also informative to apply the NGEV analysis by considering both time and the Ocean Niño Index (ONI) as co-variates. By doing this, we can see how high-frequency interannual variability in rainfall extremes is modulated by time or a major climate mode in return levels. The ONI is a three-month running mean of the Extended Reconstructed Sea Surface Temperature version 4 (ERSSTv4) data set in the Niño 3.4 region (5°N–5°S, 120°W–170°W) and is available from the National Weather Service's Climate Prediction Center website (www.cpc.noaa.gov/products/analy sis_monitoring/). Figure 6.10 displays the changes in three return levels in Taipei and Taichung from 1958 to 2013. Take the 20-yr return level from the bi-covariate model as an example. Instead of a rather smooth and rising line as seen in Fig. 6.8, the return level fluctuates at an interannual time scale confounded by a slowly varying long-term rising trend in Fig. 6.10.

The high-frequency interannual variations in return levels are more distin-guished for Taichung, which are exemplified by the high return levels in the second year of a La Niña event (e.g., 2000, 2008) and low return levels in the second year of an El Niño event (e.g., 1998 and 2010). The results presented in Fig. 6.10 suggest that the short-term interannual variations influenced by ENSO are more dominant than the long-term time trend in return levels for Taichung and other stations located in western and central Taiwan (Chu et al., 2018).

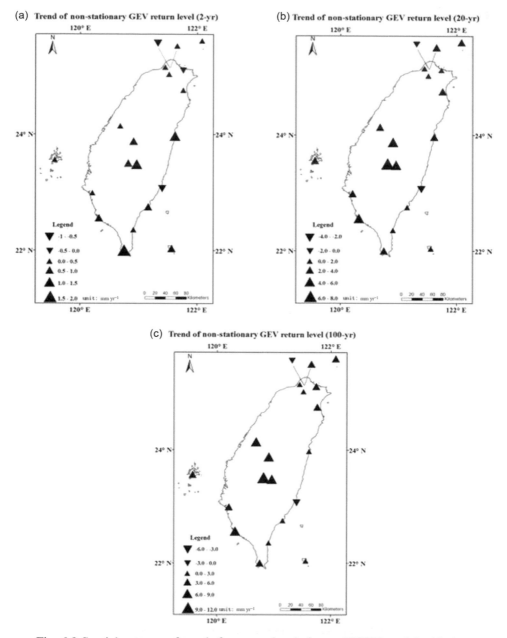

Fig. 6.9 Spatial patterns of trends for return levels from a NGEV model with time as covariate for maximum 24-h precipitation, (a) 2-yr, (b) 20-yr, and (c) 100-yr. Triangles denote the location of the individual stations. Upward (downward) triangles indicate positive (negative) direction of change; size corresponds to the magnitude of the trend in the legend. Note that all stations are significant at the 5% level.

Source: Chu et al., 2018; ©RightsLink/John Wiley & Sons. Used with permission.

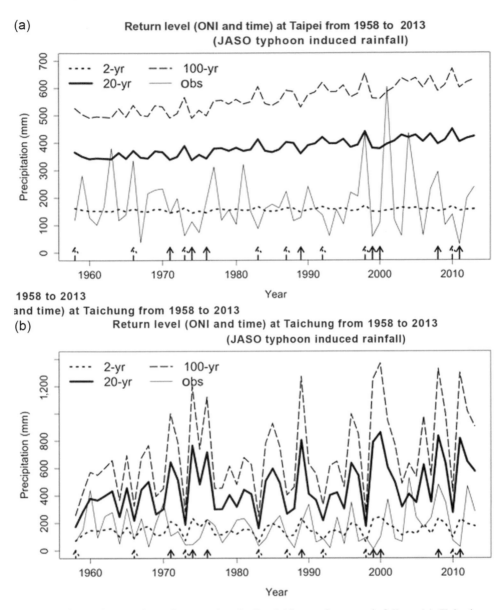

Fig. 6.10 Time series of return levels for 24-h maximum rainfall at (a) Taipei and (b) Taichung according to the non-stationary, bi-covariate GEV model. The short dashed, solid, and dashed curves denote the 2-yr, 20-yr, and 100-yr return levels, respectively. Thin solid line at the bottom denotes observations. Broken (solid) arrows along the abscissa indicate the second year of an El Niño (La Niña) episode.

Source: Chu et al., 2018; ©RightsLink/John Wiley & Sons. Used with permission.

6.4 Summary

This chapter describes the observed long-term changes in extreme TC rainfall. Direct measurements of TC rainfall are scanty, particularly on a climate time scale. Fortunately, Taiwan is one of the exceptions because of the availability of reliable, high temporal resolution (hourly) rainfall records, its unique location in relation to the prevailing typhoon path, and abundance of published literature on this topic. Rainfall extremes during the typhoon season (JASO) in Taiwan have indeed undergone a dramatic change over the last 60 years (1950–2010), which is clearly supported in various studies. For example, four climatic change indices recommended by WMO/CLIVAR/ETCCDI are used to investigate variations related to rainfall extremes. The first index describes daily rainfall intensity (mm d^{-1}), the second for the frequency of heavy rainfall events (days), the third for the magnitude of the maximum consecutive 5-d rainfall amounts (mm), and the fourth the maximum number of consecutive days within a season.

Upward trends are noted for rainfall related indices (e.g., rainfall intensity, magnitude of 5-d rainfall totals) and also for the drought duration index since 1950. This is indicative of more distinct dry-wet conditions during the typhoon season. An increase in rainfall intensity induced by TCs and monsoon is also noted. Heavy rainfall days caused by TCs have also increased but changes caused by southwesterly monsoons are rather flat. Independently, storm duration in the vicinity of Taiwan also increased. As a result of an increase in storm duration or a slowdown in TC translation speed together with the increased frequency of northward-moving typhoons, cyclonic circulation around Taiwan and interaction with the mountainous terrain favor heavy rainfall along the western side of the island. Long-term trends of typhoon-induced rainfall are increasing and statistically significant for northern and western portions of Taiwan.

Another popular index, annual maximum daily rainfall, is also used as a measure of extreme rainfall in climate research. This index is used to estimate the amount of rainfall expected to recur in a 24-h interval for the average of a period of many years (return periods). The return level is the extreme rainfall amount that corresponds to any desired return period (e.g., 576 mm within 24 h). Here, three return levels are chosen (2-yr, 20-yr, and 100-yr). The statistics of extreme value can be represented by the generalized extreme value distribution (GEV). In this chapter, a non-stationary GEV (NGEV) is applied to investigate how extreme events will co-vary with time or external climate drivers such as ENSO. To ensure the rainfall extremes selected are only associated with typhoons, the official typhoon warnings from the CWB are considered and hourly rainfall data associated with these warnings are used. Results indicate that an increasing trend in return levels associated with heavy rainfall events induced by typhoons has been prevalent in Taiwan since 1958. In other words, the return-interval threshold values shortened considerably or the frequency of typhoon-induced extreme rainfall has occurred more often in recent decades. The short-term interannual variations influenced by ENSO are more dominant than long-term rising trends for stations located in western and central Taiwan.

Exercises

6.1 The Mann–Kendall test is commonly used in trend analysis. What is this test?

6.2 (a) What is the generalized extreme value (GEV) distribution? (b) What kind of precipitation data are commonly used when applying this distribution? (c) What is the return period and return level?

6.3 (a) What is the non-stationary GEV distribution and its major difference from the stationary GEV distribution? (b) How can the non-stationary GEV distribution be used to study extreme events?

References

Alexander, L., and Coauthors, 2006: Global observed changes in daily climate extremes of temperature and precipitation. *J. Geophys. Res.*, **111**, D05109.

Allen, M. R., and W. J. Ingram, 2002: Constraints on future changes in the hydrological cycle. *Nature*, **419**, 224–228.

Chang, C.-P., Y.-T. Yang, and H.-C. Kuo, 2013: Large increasing trend of tropical cyclone rainfall in Taiwan and the roles of terrain. *J. Climate*, **26**, 4138–4147.

Chen, C.-S., and Y.-L. Chen, 2003: The rainfall characteristics of Taiwan. *Mon. Wea. Rev.*, **131**, 1323–1341.

Chen, C.-S., Y.-L. Chen, W. C. Chen, and P.-L. Lin, 2007: Statistics of rainfall occurrences in Taiwan. *Wea. Forecasting*, **22**, 981–1002.

Chen, J.-M., T. Li, and C.-F. Shih, 2010: Tropical cyclone- and monsoon-induced rainfall variability in Taiwan. *J. Climate*, **23**, 4107–4120.

Chen, Y. R., and P.-S. Chu, 2014: Trends in precipitation extremes and return levels in the Hawaiian Islands under a changing climate. *Int. J. Climatol.*, **34**, 3913–3925.

Chou, C., and C. A. Chen, 2010: Depth of convection and the weakening of tropical circulation in global warming. *J. Climate*, **23**, 3019–3030.

Chu, P.-S., Y. R. Chen, and T. A. Schroeder, 2010: Changes in precipitation extremes in the Hawaiian Islands in a warming climate. *J. Climate*, **23**, 4881–4900.

Chu, P.-S., J.-H. Kim, and Y. R. Chen, 2012: Have steering flows in the western North Pacific and the South China Sea changed over the last 50 years? *Geophys. Res. Lett.*, **39**, L10704.

Chu, P.-S., D.-J. Chen, and P.-L. Lin, 2013: Trends in precipitation extremes during the typhoon season in Taiwan over the last 60 years. *Atmos. Sci. Lett.*, **15**, 37–43.

Chu, P.-S., H. Zhang, H.-L. Chang, T.-L. Chen, and K. Tofte, 2018: Trends in return levels of 24-h precipitation extremes during the typhoon season in Taiwan. *Int. J. Climatol.*, **38**, 5107–5124.

Coles, S., 2001: *An Introduction to Statistical Modeling of Extreme Values*. Springer.

Frich, P., and Coauthors, 2002: Observed coherent changes in climatic extremes during the second half of the twentieth century. *Clim. Res.*, **19**, 193–212.

Ge, X., T. Li, S. Zhang, and M. Peng, 2010: What causes the extremely heavy rainfall in Taiwan during Typhoon Morakot (2009)? *Atmos. Sci. Lett.*, **11**, 46–50.

Gilleland, E., and R. W. Katz, 2011: New software to analyze how extremes change over time. *Eos, Tran. AGU*, **92**(2), 13–14.

Harr, P. A., and R. L. Elsberry, 1991: Tropical cyclone track characteristics as a function of large-scale circulation anomalies. *Mon. Wea. Rev.*, **119**, 1448–1468.

Henny, L., C. Thorncroft, H.-H. Hsu, and L. F. Bosart, 2021: Extreme rainfall in Taiwan: Seasonal statistics and trends. *J. Climate*, **34**, 4711–4731.

Hsu, L.-H., H.-C. Kuo, and R. G. Fovell, 2013: On the geographic asymmetry of typhoon translation speed across the mountainous island of Taiwan. *J. Atmos. Sci.*, **70**, 1006–1022.

Katz, R. W., M. B. Parlange, and Naveau, P., 2002: Statistics of extremes in hydrology. *Adv. Water Resour.*, **25**(8), 1287–1304.

Kharin, V. V., and F. W. Zwiers, 2005: Estimating extremes in transient climate change simulations. *J. Climate*, **18**, 1156–1173.

Knutson, T., and Coauthors, 2010: Tropical cyclones and climate change. *Nat. Geosci.*, **3**, 157–163.

Knutson, T., and Coauthors, 2019: Tropical cyclones and climate change assessment. Part I: Detection and attribution. *Bull. Amer. Meteorol. Soc.*, **100**, 1987–2007.

Knutson, T., and Coauthors, 2020: Tropical cyclones and climate change assessment. Part II: Projected response to anthropogenic warming. *Bull. Amer. Meteorol. Soc.*, **101**, 303–322.

Lau, W. K. M., and Y. P. Zhou, 2012: Observed recent trends in tropical cyclone rainfall over the North Atlantic and North Pacific. *J. Geophys. Res.*, **117**, D03104.

Liang, A., L. Oey, S. Huang, and S. Chou, 2017: Long-term trends of typhoon-induced rainfall over Taiwan: In situ evidence of poleward shift of typhoons in western North Pacific in recent decades. *J. Geophys. Res. Atmos.*, **122**, 2750–2765.

Liu, M., G. A. Vecchi, J. A. Smith, and H. Murakami, 2018: Projection of landfalling-tropical cyclone rainfall in the eastern United States under anthropogenic warming. *J. Climate*, **31**, 7269–7286.

Lu, M., and Coauthors, 2018: Changes in extreme precipitation in the Yangtze River basin and its association with global mean temperature and ENSO. *Int. J. Climatol.*, **38**, 1989–2005.

Min, S. K., F. Zwiers, and G. C. Hegerl, 2011: Human contribution to more-intense precipitation extremes. *Nature*, **470**, 378–381.

Shan, K., and X. Yu, 2021: Variability of tropical cyclone landfalls in China. *J. Climate*, **34**, 9235–9247.

Trenberth, K. E., A. Dai, R. M. Rasmussen, and D. B. Parsons, 2003: The changing character of precipitation. *Bull. Amer. Meteorol. Soc.*, **84**, 1205–1217.

Tu, J.-Y., and C. Chou, 2013: Changes in precipitation frequency and intensity in the vicinity of Taiwan: Typhoon versus non-typhoon events. *Environ. Res. Lett.*, **8**, 014023.

Tu, J.-Y., C. Chou, and P.-S. Chu, 2009: The abrupt shift of typhoon activity in the vicinity of Taiwan and its association with western North Pacific-East Asian climate change. *J. Climate*, **22**, 3617–3628.

Villafuerte, M. Q. II, J. Matsumoto, and H. Kubota, 2015: Changes in extreme rainfall in the Philippines (1911–2010) linked to global mean temperature and ENSO. *Int. J. Climatol.*, **35**, 2033–2044.

Wang, C.-C., B.-X. Lin, C.-T. Chen, and S.-H. Lo, 2015: Quantifying the effects of long-term climate change on tropical cyclone rainfall using a cloud-resolving model: Examples of two landfall typhoons in Taiwan. *J. Climate*, **28**, 66–85.

Westra, S., L. V. Alexander, and F. W. Zwiers, 2013: Global increasing trends in annual maximum daily precipitation. *J. Climate*, **26**, 3904–3918.

Wilks, D. S., 2011: *Statistical Methods in the Atmospheric Sciences*. Academic Press.

Wu, C.-C., T.-H. Yen, Y.-H. Kuo, and W. Wang, 2002: Rainfall simulation associated with Typhoon Herb (1996) near Taiwan. Part I: The topographic effect. *Wea. Forecasting*, **17**, 1001–1015.

Wu, C.-C., and Coauthors, 2016: Statistical characteristic of heavy rainfall associated with typhoons near Taiwan based on high-density automatic rain gauge data. *Bull. Amer. Meteorol. Soc.*, **97**, 1363–1375.

Wu, L., J. Liang, and C.-C. Wu, 2011: Monsoonal influence on Typhoon Morakot (2009). Part I: Observational analysis. *J. Atmos. Sci.*, **68**, 2208–2221.

Yu, C.-K., and L.-W. Cheng, 2008: Radar observations of intense orographic precipitation associated with Typhoon Xangsane (2000). *Mon. Wea. Rev.*, **136**, 497–521.

Yu, C.-K., and L.-W. Cheng, 2013: Distribution and mechanisms of orographic precipitation associated with Typhoon Morakot (2009). *J. Atmos. Sci.*, **70**, 2894–2915.

Yu, C.-K., and L.-W. Cheng, 2014: Dual-Doppler-derived profiles of the southwesterly flow associated with southwest and ordinary typhoons off the southwestern coast of Taiwan. *J. Atmos. Sci.*, **71**, 3202–3222.

Zhang, X., and Coauthors, 2011: Indices for monitoring changes in extremes based on daily temperature and precipitation data. *Wiley Interdiscip. Rev. Clim. Change*, **2**, 851–870.

Zwiers, F. W., and V. V. Kharin, 1998: Changes in the extremes of the climate simulated by CCC GCM2 under CO_2 doubling. *J. Climate*, **11**, 2200–2222.

7

Future Tropical Cyclone Projections and Uncertainty Estimates

7.1 Introduction

Tropical cyclones exert great impact on society. Therefore, for a long time an important scientific question has been whether changing the climate system affects tropical cyclone activity. Recent studies reported that global mean temperature has been rising since the mid-twentieth century, and the temperature rise is attributable to increases in emissions of greenhouse gasses (Bindoff et al., 2013; IPCC, 2013). An intuitive hypothesis under the warming trend at a global scale is that the mean global number of tropical cyclones would increase and mean storm intensity would be stronger because tropical cyclone activity could be favorable in a warmer environment. However, the science community has not yet reached a robust consensus on whether this hypothesis is true or not, especially for the effect of global warming on global tropical cyclone numbers (IPCC, 2013). The main sources of the uncertainties are as follows:

- Limited availability of reliable long-term observed records to infer the effect of global warming on tropical cyclone activity.
- Significant influence of intrinsic natural variability on tropical cyclone activity.
- Lack in theory of physical mechanisms controlling the number of global tropical cyclones.
- Diverse results of climate model simulations.

In this chapter, the possible effects of climate change on tropical cyclone activity will be introduced, based on observations and modeling studies associated with the above uncertainties.

7.2 Limited Availability of Long-Term Observations for Tropical Cyclones

The black line in Fig. 7.1a denotes the annual number of observed global tropical cyclones since 1958. Hereafter, a tropical cyclone is defined as a storm for which

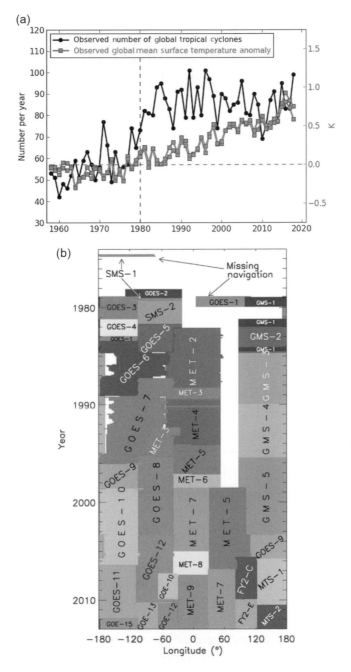

Fig. 7.1 (a) Observed annual number of global tropical cyclone (black, left axis) based on the IBtRACK best track data (Knapp et al., 2010) superimposed on the observed global mean surface temperature anomaly relative to the mean of 1961–1990 based on HadCRUT4 (Morice et al., 2012) and GISS Surface Temperature Analysis (GISTEMP v4; Lenssen et al., 2019) (gray, right axis, units: K). (b) Global geostationary satellite coverage over the past 40 years, indicating that the satellite-based tropical-cyclone observations are only available since 1980 at a global scale.

Source: Adapted from Kossin et al., 2013; ©American Meteorological Society. Used with permission.

its life-time maximum wind speeds exceed 17 m s^{-1}. It shows a significant positive trend since 1958 with a marked gap around 1980. Given the fact that the observed global mean surface temperatures have been rising since the mid-twentieth century due to the increases in anthropogenic forcing (gray lines in Fig. 7.1a), the observed trend in global mean tropical cyclone numbers appears to be linked to underlying climate change. However, this positive trend in the global number of tropical cyclones is artificial due to changes in data quality. As seen in Fig. 7.1b, satellite data was not available prior to 1980. Measurement of global tropical cyclones had been achieved by satellite visible and infrared images through analyzing cloud characteristics (the so-called Dvorak technique (Dvorak, 1973, 1984); see Velden et al., 2006 for a review). Before the satellite era, tropical cyclones were mainly observed by ships, land-based observations, and reconnaissance aircraft. Therefore, prior to the meteorological satellite era, tropical cyclones that did not approach land or encounter a ship or aircraft had a greater chance of not being detected, leading to unphysical upward trends in the record of storm frequency because more storms have been detected during the satellite era (Vecchi and Knutson, 2008). Therefore, the global tropical cyclone dataset prior to 1980 is inhomogeneous, and any trend of tropical cyclones including the pre-satellite era is unphysical.

Even after excluding the pre-1980 data from the analysis, the annual number of global tropical cyclones has been steady around 86 over the period 1980–2018 despite the significant increase in the global mean surface temperature (Fig. 7.1a). There is also significant interannual variation in the number of global tropical cyclones (one standard deviation is 8; Frank and Young, 2007). The short period for the reliable dataset, as well as the marked internal variation in the data, make it difficult to detect secular trends due to anthropogenic climate change from the observations for global tropical cyclone activity.

Despite the steady number of observed global tropical cyclones, there is no theory to support the evidence that the mean annual number of global tropical cyclones is steady around 86. In other words, why don't we observe a year with either 150 or 40 global tropical cyclones? A popular potential explanation for the steady global storm number is that there is a negative feedback that tends to limit the global number of tropical cyclones per year (e.g., Henderson-Sellers et al., 1998). In other words, an enhanced storm number in one ocean basin tends to suppress the storm activity in the other basins. This explanation implies the existence of significant relationships between tropical cyclones and climate. On the other hand, other studies have suggested that the significant negative correlations in the numbers of tropical cyclones among the ocean basins, such as a negative correlation between North Atlantic and Eastern North Pacific, are just a result of climate modes such as ENSO (e.g., Bell and Chelliah, 2006), and are not

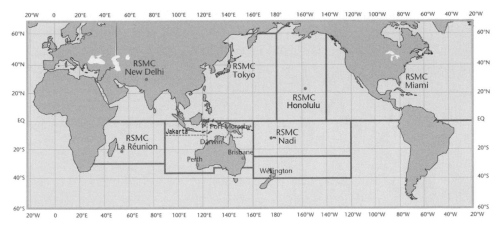

Fig. 7.2 The six tropical cyclone Regional Specialized Meteorological Centres (RSMCs) together with six Tropical Cyclone Warning Centres (TCWCs) having regional responsibility, provide advisories and bulletins with up-to-date first-level basic meteorological information on all tropical cyclones, hurricanes, and typhoons everywhere in the world.

Source: Adapted from www.wmo.int/pages/prog/www/tcp/Advisories-RSMCs.html.

relevant to the negative feedback induced by storms. Frank and Young (2007) also concluded that there is no observed tendency for above-normal cyclone activity in one basin to be compensated by below-normal cyclone activity over the other ocean basins, indicating no evidence to support the existence of negative feedback in the limited observed record.

As shown in Fig. 7.1b, the available satellites have been changing decade by decade. Also, because the analysis technique and protocols to extract tropical cyclones, the Dvorak Technique (Dvorak, 1973, 1984), had been changing, the data naturally contained temporal heterogeneities especially for the storm intensity record (Kossin et al., 2013). Tropical cyclones are also analyzed by different organizations from basin by basin by applying different methodology (Fig. 7.2).

Therefore, there is an inherent lack of temporal and spatial consistency in the global tropical cyclone datasets. This evolution of the available data and analysis techniques, even after the satellite era, is also one of the critical sources causing artificial trends in tropical cyclone activity. Specifically, the observed cyclone datasets contain uncertainty in the storm intensity so that they also contain uncertainty in the genesis number, specifically for the intense storms. Webster et al. (2005) claimed that both the number and the percentage of category 4 and 5 hurricanes for all global basins increases using the best track data issued by the US Department of Defense's Joint Typhoon Warning Center (JTWC, Chu et al., 2002) (Fig. 7.3). The largest increases have occurred in the western North Pacific, the Indian Ocean, and the South Pacific, with smaller increases in the eastern North

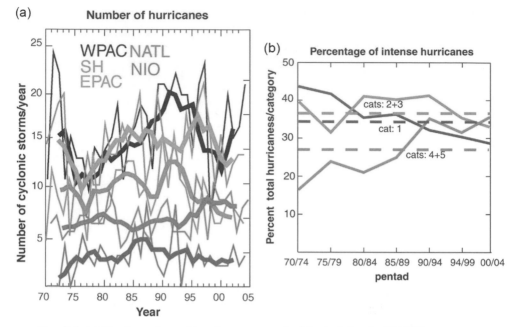

Fig. 7.3 (a) Total number of hurricanes over the North Atlantic (NATL), western North Pacific (WPAC), eastern North Pacific (EPAC), North Indian Ocean (NIO), and Southern Hemisphere (South Indian Ocean plus South Pacific) for the period 1970–2004. (b) The percentage of the total number of hurricanes in each category class at the global scale in 5-yr periods. The dashed lines show the 1970–2004 average numbers in each category. A black and white version of this figure will appear in some formats. For the color version, refer to the plate section.
Source: Adapted from Webster et al., 2005; ©Science. Used with permission.

Pacific and the North Atlantic. However, several studies since Webster et al. (2005) have questioned these trends. Wu et al. (2006) reported that there was no increase in the category 4 and 5 typhoons over the western North Pacific if the different best track data from the Regional Specialized Meteorological Center (RSMC) Tokyo (RSMC, 2020) and that of the Hong Kong Observatory were analyzed. A few studies have questioned the accuracy of the analyzed data in the earlier portion of the record (Landsea et al., 2006; Kuleshov et al., 2010; Kossin et al., 2013). Klotzbach and Landsea (2015) investigated whether the trends reported by Webster et al. (2005) have continued, by including 10 additional years of data, showing insignificant trends in the data. They concluded that the observed trends in the category 4 and 5 hurricanes between 1970 and 2004 are attributed to observational improvements at the various global tropical cyclone warning centers.

There are also uncertainties in the attributes of observed trends to global warming. As shown in Fig. 7.3a, the observed number of hurricanes shows significant multi-decadal oscillations in the western North Pacific and North

Atlantic, with significant changes around 1998. Coincidentally, the Pacific Decadal Oscillation (PDO; Mantua et al., 1997; Lupo et al., 2008; Wang et al., 2010) and Atlantic Multidecadal Oscillation (AMO; Delworth and Mann, 2000), which are the known internal variability at a decadal time scale, changed their sign around 1998. A few studies have argued that the observed trends shown in Webster et al. (2005) were mainly due to the intrinsic natural variability (Pielke et al., 2005; Bell and Chelliah, 2006), although other studies insisted that these trends are due to global warming (Emanuel, 2005; Anthes et al., 2006; Hoyos et al., 2006; Mann and Emanuel, 2006; Trenberth and Shea, 2006; Holland and Webster, 2007; Mann et al., 2007a, 2007b). Overall, up to now, no consensus has yet been reached in the science community regarding the cause of the observed trends. (Note that changes in the observed TC activity are described separately in Section 4.7.)

7.3 Reconstructing Observed Tropical Cyclone Data

As indicated in Section 7.2, evolution of the available data and analysis techniques is one of the causes for the inhomogeneities, leading to artificial trends in tropical cyclone activity especially for tropical cyclone intensity. In order to alleviate the inhomogeneities, Kossin et al. (2013) developed the Advanced Dvorak Technique-Hurricane Satellite-B1 (ADT-HURSAT) in order to maintain the same protocol to determine tropical cyclone intensities throughout all ocean basins over the period 1982–2009. In short, geostationary satellite imagery is analyzed around the tropical cyclones archived in the best track data (IBTrACS, Knapp et al., 2010), and subsampled to be both spatially and temporally homogeneous. Finally, a simplified version of the advanced Dvorak technique (Olander and Velden, 2007) is applied to determine a maximum storm wind speed. Therefore, ADT-HURSAT is a spatially and temporally homogeneous dataset and suitable to analyze the observed trend in tropical cyclone activity. The ADT-HURSAT data was utilized for analyzing the observed trends in mean location of lifetime maximum intensity of tropical cyclones (Section 4.7.2) and mean translation speed of tropical cyclones (Section 4.7.3). Also note that changes in the observed TC activity in the context of frequency of occurrence, motion, intensity, and meridional migration of lifetime maximum intensity are described separately in Section 4.7.

7.4 Projected Future Changes in Tropical Cyclone Activity

When the Intergovernmental Panel on Climate Change (IPCC) was established in 1988, the scientific community held several discussions regarding the effect of global warming on atmospheric phenomena. Because a tropical cyclone gains energy when water vapor condenses in its convection and releases energy in the air (i.e., diabatic heating), scientists argued whether tropical cyclones would be more intense and frequent due to increasing water vapor in the air by global warming. In this section, the history of research on potential future changes in tropical cyclones will be introduced.

7.4.1 Early Studies Using Low- and Medium-Resolution Global Climate Models

Broccoli and Manabe (1990) was the first study to utilize a climate dynamical model to elucidate the impact of global warming on tropical cyclone numbers. The authors utilized the global climate model with two versions with different horizontal resolutions (600 km and 300 km). They conducted two experiments with CO_2 at the present-day level and doubling CO_2 relative to the present-day level. They also conducted two experiments by prescribing climatological clouds and varying clouds. Regardless of the model resolution, they found that the projected number of tropical cyclones decreases in the case of varying clouds and increases in the case of fixed clouds. The scientific community (e.g., Evans, 1992) was skeptical about these results because the simulated mean number of tropical cyclones is too low in the Eastern Pacific in the present-day simulation. Evans (1992) also requested a modeling study to evaluate model fidelity in simulating the annual cycle of the number of tropical cyclones compared with observations before conducting an experiment with increasing CO_2. Similar to Broccoli and Manabe (1990), Haarsma et al. (1993) also conducted a doubling CO_2 experiment using a 300-km mesh global climate model, showing a 50% increase in global tropical cyclone numbers in a warming climate. Although their model showed a realistic annual cycle in the number and spatial distribution of tropical cyclones, their model had no storm in the Southern Hemisphere, which was unrealistic. In 1993, The World Meteorological Organization Third International Workshop on Tropical Cyclone (IWTC-III) was held in Mexico to discuss the effect of global warming on tropical cyclones (Lighthill et al., 1994). The workshop concluded that it is unfeasible to estimate the impact of global warming on tropical cyclone activity by climate models because the 300-km resolution in the climate models is too coarse to resolve tropical cyclone inner structure. However, it also concluded that a climate model would be a useful tool when sophisticated physical processes are incorporated in a model as well as using a higher resolution.

Since the late 1990s, several studies have used climate models with horizontal resolution higher than a 300-km mesh. Bengtsson et al. (1995, 1996) reported the effect of increases in CO_2 on tropical cyclones using 120-km mesh atmospheric global circulation model (AGCM). In the studies, the so-called time-slice method was first applied. Ideally, it is preferable to use a high-resolution atmosphere–ocean coupled global circulation model (CGCM) because the physical processes of both atmosphere and ocean are important for simulating tropical cyclones. However, it was computationally very expensive to conduct multi-year climate simulations with a high-resolution CGCM. In the first step in the time-slice method, multi-year long-term climate simulations are preliminarily conducted with a low-resolution CGCM. Thereafter, the simulated sea surface temperatures (SSTs) for a specific period in the long-term climate simulations are used as boundary conditions in a simulation with a high-resolution AGCM.

In Bengtsson et al. (1995, 1996), a long-term simulation for 100 years from 1985 to 2085, assuming approximately 1% annual increase of CO_2 (IPCC Scenario A, IPCC, 1990), was conducted using a low-resolution 600-km mesh CGCM. Using the simulated mean SST in the last five years when CO_2 doubled as lower boundary conditions, a five-year simulation with a high-resolution 100-km AGCM was conducted. Although the low-resolution CGCM cannot resolve tropical cyclones due to the coarse horizontal resolution, a high-resolution AGCM can simulate tropical cyclones in a specific period (time-slice). The results of Bengtsson et al. (1995, 1996) indicated for the first time that the global number of tropical cyclones would decrease by 37% due to increases in CO_2. Although the time-slice method could save computational resources because it only requires simulations using a high-resolution AGCM for the specific period of interest, rather than simulations for the full period, the caveat is that the AGCM simulation ignores an important physical process of air–sea interactions. In general, SST decreases along tropical cyclone tracks due to cold-wakes induced by wind-induced ocean mixing (Lin et al., 2009; Lloyd and Vecchi, 2011), which serves to weaken tropical cyclone intensity and suppress subsequent tropical cyclone genesis (Shade and Emanuel, 1999; Bender and Ginis, 2000; Knutson et al., 2001). Because SST is fixed in an AGCM simulation, the cold-wake effect is ignored, leading to overestimation of the mean intensity of tropical cyclones. In fact, Landsea (1997) discusses problems in the experimental settings in Bengtsson et al. (1995, 1996).

Since Bengtsson et al. (1995, 1996), several studies have conducted time-slice experiments to estimate the impact of CO_2 increases on tropical cyclone numbers. Sugi et al. (2002) conducted experiments similar to Bengtsson et al. but with a different 120-km mesh AGCM. They reported projected decreases in tropical cyclone numbers by 28%, 39%, and 34%, respectively, for the Northern Hemisphere, the Southern Hemisphere, and globally. McDonald et al. (2005) also conducted a time-slice experiment using a 100-km mesh AGCM (Hadley Center HadAM3), showing a 6% decrease in global tropical cyclone numbers in the case of doubling CO_2. Similar projected decreases in global tropical cyclone numbers were also reported by Hasegawa and Emori (2005) and Yoshimura et al. (2006). In those experiments, however, the horizontal resolution of an AGCM is still around a 100-km mesh, which is not high enough to simulate intense tropical cyclones such as category 3–5 (major) hurricanes.

7.4.2 Dynamical Downscaling

As reviewed in Section 7.4.1, the horizontal resolution of an AGCM was around a 100-km mesh, which is not high enough to estimate intensity changes in tropical cyclones by increases in CO_2. To overcome the resolution problem in an AGCM, a new technique, the so-called dynamical downscaling, was introduced. In this method, tropical cyclones simulated by a low- or medium-resolution AGCM or CGCM are further downscaled using a finer regional model or hurricane model. For example, Knutson et al. (1998) simulated 51 tropical cyclones using a present-day simulation and a high-CO_2 climate simulation with a 300-km CGCM. In order to estimate tropical cyclone intensity more accurately, the selected storm cases were then rerun as five-day "forecast" experiments using the high-resolution (18-km mesh) Hurricane Prediction System, which is used in the US National Centers for Environmental Prediction (NCEP) as an operational hurricane prediction model. They showed 6% increases in maximum wind speed around the storm center for the high-CO_2 climate compared with the present-day climate. Knutson and Tuleya (2004) further repeated similar experiments but using nine CGCMs as input along with a 9-km mesh hurricane model as an inner model. They found increases in storm intensity in terms of a 14% decrease in central pressure and a 6% increase in maximum wind speed. They also added 18% projected increases in precipitation around a storm center.

Similar dynamical downscaling simulations to Knutson et al. (1998) were conducted by Walsh and Ryan (2000) and Walsh et al. (2004), targeting tropical cyclones near Australia. They generally showed an increase in tropical cyclone intensity in terms of a 15–25% increase in maximum wind speeds.

7.4.3 High-Resolution Global and Regional Climate Models

Based on the research introduced in Sections 7.4.1 and 7.4.2, the World Meteorological Organization Sixth International Workshop on Tropical Cyclone (IWTC-VI; WMO, 2006) was held in Costa Rica in 2006. The workshop concluded that it is uncertain whether tropical cyclone numbers would decrease under a warming environment as simulated by AGCMs. The main reason for the uncertainty is that the 100-km mesh horizontal resolution is still not high enough to reproduce the observed storm genesis process. Consequently, the Intergovernmental Panel on Climate Change (IPCC) Fourth Assessment Report (IPCC, 2007) concluded in its summary chapter for policymakers that there is "*less confidence*" in projections of a global decrease in the frequency of tropical cyclones. Researchers felt that increasing horizontal resolution was the priority in order to improve model confidence.

Oouchi et al. (2006) used a 20-km mesh AGCM, which was the highest resolution for an AGCM for climate simulations at that time. They showed that the total number of global tropical cyclone numbers decreases in a warming world as in the previous studies. However, they also found that intense cyclones (maximum surface winds >45 m s^{-1}) would increase, while weaker cyclones would decrease. Bengtsson et al. (2007) utilized T63 (200-km mesh), T213 (60-km mesh), and T319 (40-km mesh) AGCMs to conduct time-slice experiments for future projections under the IPCC A1B scenario, showing a projected decrease in global tropical cyclones by 10–20%. The T213 and T319 AGCMs showed a projected increase in intense cyclones, but the T63 AGCM did not, which underlines the importance of using a high-resolution AGCM for reliable future projections. Gualdi et al. (2008) used a high-resolution CGCM (120 km in atmosphere and $2° \times 2°$ ocean components) for a $4 \times CO_2$ warming experiment compared with a present-day experiment. They showed a significant decrease in the global number of tropical cyclones, especially in the western North Pacific and North Atlantic. Zhao et al. (2009) also used a 50-km mesh AGCM developed at the Geophysical Fluid Dynamics Laboratory (GFDL) in the United States, showing a decrease in global tropical cyclone numbers. Since 2007, there have been a number of experiments conducted using high-resolution AGCMs (finer than 50-km mesh) (e.g., Murakami and Sugi, 2010; Yamada et al., 2010; Murakami et al., 2012a, 2012b). Most of the studies using high-resolution AGCMs and CGCMs generally supported the results by Oouchi et al. (2006). The mean number of global tropical cyclones was projected to decrease and the projected decrease in total storm number is mainly caused by significant decreases in weaker storms; however, intense storms are projected to increase.

Refining horizontal resolution in dynamical downscaling also helped to improve the fidelity of tropical cyclone simulations (Chauvin et al., 2006; Leslie et al., 2007; Stowasser et al., 2007; Knutson et al., 2008; Semmler et al., 2008; Lavender and Walsh, 2011). For example, Bender et al. (2010) conducted an experiment similar to Knutson et al. (2008) but used a fine-resolution 8-km mesh hurricane model. Bender et al. (2010) reported that the model was able to reproduce category 4 and 5 (maximum wind speeds >50 m s^{-1}) hurricanes. In their warming experiment, they projected a 100% increase in category 4 and 5 hurricanes in the North Atlantic at the end of the twenty-first century. In general, most of the dynamical downscaling models support the conclusion that the mean intensity of tropical cyclones in terms of wind speeds and precipitation would increase as the climate warms.

Based on the studies since IWTC-VI held in 2006, Knutson et al. (2010a, 2010b) summarized and reviewed the potential influence of human activity on tropical cyclones as follows:

- It is uncertain whether human activity affected tropical cyclones in the past due to limited record of observations and natural variability.
- Theory and high-resolution models project an increase in storm intensity by 2–20% at the end of the twenty-first century.
- Dynamical models project a decrease in the number of global tropical cyclones by 6–34% by the end of the twenty-first century, under the IPCC A1B future scenario.
- High-resolution models project increases in intense tropical cyclones and 20% increases in mean precipitation within 100 km from a storm center by the end of the twenty-first century.
- Regional changes in tropical cyclone activity remain uncertain in the future.

7.4.4 Statistical-Dynamical Downscaling Approach

One of the problems for a climate modeling study is its computational cost. As the typical horizontal scale of a tropical cyclone ranges from 100 km to 1,000 km, a horizontal resolution finer than a 25-km mesh is required to simulate realistic tropical cyclones. However, running multi-ensemble and multi-decadal simulations using such a high horizontal resolution is very expensive. In order to save on computational cost, a new downscaling method, statistical-dynamical downscaling, has been implemented to quantify projected future changes in tropical cyclone activity.

Emanuel (2006) was the first to develop a new statistical-downscaling approach. Put simply, this approach includes three processes: *Genesis*; *Tracks*; and *Intensity*. In general, the downscaling technique applies a storm intensity model (*Intensity*) to tropical cyclone tracks initiated by random seeding in space and time (*Genesis*), and propagates it forward using a beta-and-advection model driven by winds derived from the output by dynamical climate models (*Tracks*). More specifically, in the approach of Emanuel (2006), about 1,000 weak vortices (12 m s^{-1} in terms of maximum wind speed) were randomly placed over the global tropics (i.e., *Genesis*). Second, these vortices were propagated following large-scale flows using an advection model (Marks, 1992; i.e., *Tracks*). The large-scale flows were derived from existing climate simulations, which were not required to be high-resolution models. Third, a hurricane intensity model computed the development (or decay) of each vortex along the vortex propagation (i.e., *Intensity*), forced with thermodynamic and dynamic large-scale parameters derived from the same existing climate simulations. Most vortices decay due to strong vertical wind shear or dry conditions in the mid-troposphere. Only a vortex that attains 21 m s^{-1} in the hurricane intensity model is considered a tropical cyclone. Emanuel (2006) utilized a hurricane intensity model called the Coupled Hurricane Intensity Prediction System (CHIPS). CHIPS is based on a simple axisymmetric hurricane model that can compute the attainable tropical cyclone intensity given large-scale environmental conditions, such as SST, and the atmospheric vertical structure of moisture and temperature. CHIPS is also coupled to a very simple ocean model consisting of a series of independent one-dimensional ocean columns strung out along a storm track to represent the impact of air–sea interactions on storm intensity. CHIPS is very simple and computationally cheap, and even a personal computer can run the model. The source code is available online at ftp://texmex.mit.edu/pub/emanuel/HURRICANE/.

Emanuel et al. (2008) applied this downscaling approach to quantify projected changes in tropical cyclone activity. They utilized large-scale parameters derived from the seven climate models archived in the World Climate Research Program (WCRP) third Climate Model Intercomparison Project (CMIP3) as the inputs for the downscaling model. Two sets of climate models were compared. The first set is based on simulations of the climate of the twentieth century and the second set is based on IPCC scenario A1B, for which atmospheric CO_2 concentration increases to 720 ppm by the year 2100. The projected future changes in storm genesis density are shown in Fig. 7.4, revealing overall decreases in tropical cyclone genesis on a global scale except for the western North Pacific. In addition, the downscaling model also projects increases in mean storm intensity. These results are in line with the statement in the previous studies reviewed by Knutson et al. (2010a, 2010b; see Section 7.4.3) noting that the mean number of tropical cyclones is projected to decrease, whereas mean storm intensity is projected to increase in the future.

Similar downscaling approaches have been also developed by Satoh et al. (2011) and Lee et al. (2018). Lee et al. (2018) incorporated a statistical assumption that the seeding rate varies with thermodynamic and dynamic large-scale conditions. In this way, the downscaling approach is also called "statistical-dynamical downscaling" because it incorporates both the dynamical (e.g., intensity model like the CHIPS model) and statistical (e.g., seeding based on statistics) approaches. As will be discussed in Section 7.6, new studies applying this downscaling approach to the large-scale parameters by the new climate models showed considerable uncertainty in the projected future changes in the number of global storms.

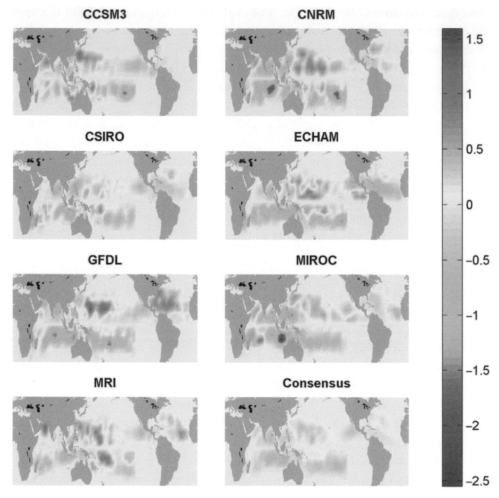

Fig. 7.4 Projected future change in storm genesis density by a downscaling model by Emanuel et al. (2008) using seven different large-scale parameters from different climate models. The ensemble mean is shown in the bottom right panel. Units: TC number per 5° latitude-longitude square per year. A black and white version of this figure will appear in some formats. For the color version, refer to the plate section.

Source: Adapted from Emanuel et al., 2008; ©American Meteorological Society. Used with permission.

7.5 Physical Mechanisms behind the Projected Decrease in Tropical Cyclones in the Future

Previous modeling studies commonly reported projected decreases in the number of global tropical cyclones in a warmer climate. However, it is still unclear why the number of tropical cyclones is projected to decrease in a warmer climate. In this section, a few hypotheses for the projected decrease in the number of global tropical cyclones are discussed.

7.5.1 Weakening of Tropical Circulation

Bengtsson et al. (1996) suggested that one of the probable causes for the projected decrease in global tropical cyclones was due to large-scale changes in the tropical circulation. Sugi et al. (2002) also proposed a hypothesis that the projected weakening of the tropical overturning circulation (e.g., weakening of the Hadley or Walker circulations) is the main reason for the decrease in the global storm number. In fact, several climate models reported a projected weakening of tropical overturning circulation by global warming (Knutson and Manabe, 1995; Sugi et al., 2002; Held and Soden, 2006; Vecchi and Soden, 2007).

Figure 7.5 shows a schematic diagram for a tropical circulation. A tropical circulation consists of a strong upward branch associated with deep convection and a weak downward branch as well as upper and lower flows bridging the branches. In general, adiabatic heating associated with vertical air motion balances the total diabatic heating (e.g., condensation heating or radiative cooling) in a tropical circulation (Holton, 1990; Knutson and Manabe, 1995):

$$M \left| \frac{\partial \theta}{\partial p} \right| \approx \frac{\theta}{T} \frac{Q}{C_p}, \tag{7.1}$$

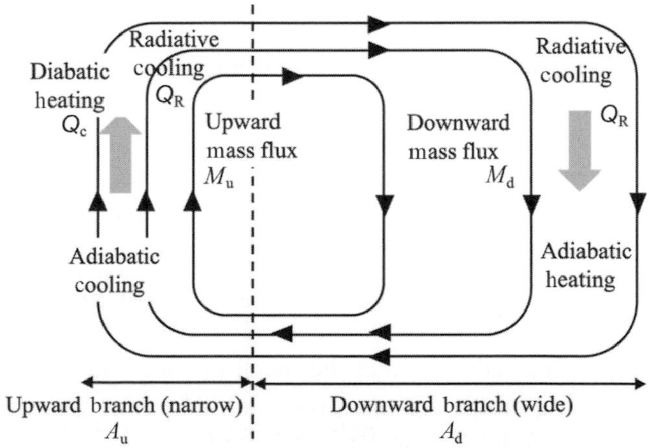

Fig. 7.5 A schematic diagram showing the energy balance of tropical circulation. M_u (M_d) and A_u (A_d) are the mass flux, and fractional area of the upward (downward) motion branch. Q_c and Q_R are the condensation heating rate and radiative cooling rate, respectively.
Source: Adapted from Sugi et al., 2002.

where M denotes the vertical mass flux; θ is the time-mean potential temperature at a specific pressure level; T is the time-mean temperature at a specific pressure level; p denotes the pressure level; Q denotes the total diabatic heating; C_p is the heat capacity at constant pressure; and $\partial\theta/\partial p$ is atmospheric stability. The left side of Eq. 7.1, a product of vertical mass flux and atmospheric stability, represents adiabatic heating associated with vertical motion, whereas the right side of Eq. 7.1 represents the diabatic heating associated with condensation heating or/and radiative cooling.

For the upward branch, the condensation heating associated with convective activity along with radiative cooling balances the adiabatic cooling associated with upward motion:

$$M_u \left|\frac{\partial\theta}{\partial p}\right| \approx \frac{\theta}{T}\frac{1}{C_p}(Q_c - Q_R), \tag{7.2}$$

where M_u denotes the upward mass flux, Q_c and Q_R denote the condensation heating associated with convection and radiative cooling, respectively. Q_c is actually proportional to the total precipitation. In contrast, at the downward branch, radiative cooling balances the adiabatic heating associated with downward motion. Moreover, total condensation heating at the upward branch balances the total radiative cooling at the downward branch:

$$M_d \left|\frac{\partial\theta}{\partial p}\right| \approx \frac{\theta}{T}\frac{1}{C_p}Q_R, \tag{7.3}$$

where M_d denotes the downward mass flux.

Sugi et al. (2012) found in their global warming experiment that the fractional changes in total precipitation in the tropics was about +1% relative to the present-day climate experiment, indicating a +1% increase in Q_c. They also found that the projected increase in Q_R was +1% relative to the present-day experiment. On the other hand, the atmospheric stability ($\partial\theta/\partial p$; defined as the potential temperature difference at 200 hPa and 100 hPa) increased by +6% in the warming experiment. To keep the balance in Eqs. 7.1, 7.2, and 7.3, the mean mass flux should decrease by about 6%. Thus, the increase in static stability and decrease in mass flux nearly cancel each other and balance with the small increase in the heating terms. A decrease in mass flux indicates that the mean upward motion decreases in the tropics. The weakening of upward mass flux and increasing atmospheric stability are both unfavorable conditions for tropical cyclone genesis, which may lead to a decrease in global tropical cyclone numbers in a warming climate. Projected weakening of the tropical circulation due to global warming is consistent among several studies (e.g., Knutson and Manabe, 1995; Bengtsson et al., 1996; Sugi et al., 2002; Held and Soden, 2006; Vecchi and Soden, 2007).

7.5.2 Increasing of Entropy Deficit

Emanuel et al. (2008) hypothesized that the projected increase in a thermo-dynamical parameter, "entropy deficit," was the main driver for decreases in global tropical cyclone numbers in a warming climate. The entropy deficit is defined as

$$\chi_m \equiv \frac{s_m - s_m^*}{s_o^* - s_b}, \tag{7.4}$$

where s_m is the environmental moist entropy at 600 hPa, s_m^* is the saturation entropy at 600 hPa in the inner core of a tropical cyclone; s_o^* is the moist entropy of air saturated at SST and pressure; and s_b is the moist entropy of the boundary layer. To calculate the moist entropy, the pseudo-adiabatic entropy from Bryan (2008) can be applied:

$$s = c_p \log (T) - R_d \log (p_d) + \frac{L_{vo} r_v}{T} - R_v r_v \log (H), \tag{7.5}$$

where c_p is the specific heat at constant pressure for dry air; T is the temperature; R_d is the gas constant for dry air; p_d is the partial pressure of dry air; L_{vo} is the latent heat of vaporization (set at 2.555×10^6 J kg^{-1}); r_v is the water vapor mixing ratio; R_v is the gas constant for water vapor; and H is the relative humidity.

The entropy deficit is non-positive and its magnitude measures the degree of thermodynamic inhibition of tropical cyclone formation (Emanuel et al., 2008). It also represents the amount by which the water vapor in the air must be increased to achieve saturation without changing temperature and pressure. The denominator of Eq. 7.4 represents the air–sea thermodynamic disequilibrium, which is proportional to the surface turbulent energy flux into the atmosphere at constant surface wind speed. In a warming climate, the denominator would not change much or increase more slowly relative to a present-day climate because it is tied to the surface enthalpy fluxes at the potential intensity, which is projected to only increase slightly. On the other hand, the numerator of Eq. 7.4 is proportional to increases in saturation specific humidity in free atmosphere.

As dictated by Clausius–Clapeyron, the saturation specific humidity exponentially increases along with increases in air temperature under the assumption of constant relative humidity (Soden and Held, 2006). Although previous studies reported that relative humidity in free atmosphere would not change much under a warming climate (e.g., Wetherald and Manabe, 1980; Mitchell and Ingram, 1992; Soden and Held, 2006), the numerator of Eq. 7.4 would become much larger in a warmer climate due to a large increase in s_m^*. Therefore, the entropy deficit, Eq. 7.4, would increase markedly due to a larger increase in the numerator than the denominator in a warmer climate relative to a present-day climate. This indicates that a larger volume of water vapor would be required for saturation in a warmer climate than in a present-day climate. Because an energy source for a tropical cyclone is latent heat release in the air, the increase in entropy deficit would lead to a decrease in the probability of developing deep convection, which in turn inhibits the development of tropical cyclones. Therefore, an increase in entropy deficit could be an important thermodynamic factor that potentially leads to decreases in global tropical cyclone numbers in a warmer climate.

7.5.3 Increase in Ventilation Index

As discussed in Section 7.5.2, entropy deficit influences the probability of tropical cyclone genesis in terms of thermodynamical conditions. On the other hand, there are a number of other environmental factors that potentially control tropical cyclone genesis. One factor is environmental vertical wind shear, which is defined as the wind speed difference between low-level troposphere (e.g., 850 hPa) and upper-level troposphere (e.g., 200 hPa). Strong vertical wind shear is detrimental to tropical cyclone genesis and intensity (McBride and Zehr, 1981; Zehr, 1992) because it discourages storm development by ventilating the incipient disturbance with low-entropy air (Simpson and Riehl, 1958). Gray (1968) proposed that advection by the environmental flow removes the condensation heat from a vortex, preventing it from developing. The environmental flow also advects dry air into the disturbance, inhibiting the formation of a deep convection. Therefore, vertical wind shear acts to prohibit development of tropical cyclones by ventilating a storm with low-entropy air at the mid-level troposphere.

Figure 7.6 illustrates the ventilation process in a sheared tropical cyclone (Tang and Emanuel, 2012). The vertically sheared flow tilts the vortex inner core and causes convective asymmetries, which excite mesoscale eddies at the mid-level troposphere. These eddies transport surrounding low-entropy air into the inner core of a storm. The low-entropy air then mixes into the inner core and undercuts updrafts, where evaporation of rain triggers downdrafts that flush the inflow layer with low-entropy air. These processes act to inhibit the generation of available potential energy by surface fluxes and weaken a tropical cyclone because ventilation keeps energy away from the tropical cyclone.

Tang and Emanuel (2010) developed a new index for tropical cyclone genesis, the so-called ventilation index, which is a combination of entropy deficit and vertical wind shear. The ventilation index is defined as follows:

$$\Lambda \equiv \frac{u_s \chi_m}{u_{pi}}, \tag{7.6}$$

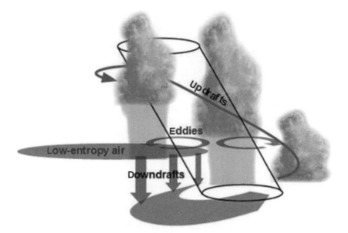

Fig. 7.6 An illustration of a tropical cyclone undergoing ventilation.
Source: Adapted from Tang and Emanuel, 2012; ©American Meteorological Society. Used
with permission.

where Λ is the non-dimensional ventilation index; u_s is vertical wind shear; χ_m is the entropy deficit defined as Eq. 7.4; and u_{pi} is the maximum potential intensity (MPI) based on Bister and Emanuel (1998). The MPI is defined as follows:

$$u_{pi} \equiv \sqrt{\frac{C_k}{C_D} \frac{T_s - T_o}{T_o}} \left(h_o^* - h^* \right), \tag{7.7}$$

where C_k is the surface enthalpy exchange coefficient; C_D is the surface momentum exchange coefficient; T_s is the SST; T_o is the outflow temperature; h_o^* is the saturation moist static energy of the sea surface; and h^* is the saturation moist static energy of the free troposphere. The MPI consists of two key factors: the thermodynamic efficiency, $(T_s - T_o)/T_o$, and an air–sea disequilibrium term, $h_o^* - h^*$.

The MPI represents a metric that a tropical cyclone can achieve at steady state in a given thermodynamic environment (Emanuel, 1986; Bister and Emanuel, 1998). Emanuel (2013) showed that most increases in thermodynamic efficiency are due to decreases in outflow temperature (T_o) rather than increases in SST. The SST mainly influences u_{pi} through the air–sea disequilibrium term. Therefore, the difference between the SST and free troposphere along with the outflow temperature is key to the variation of u_{pi}. A larger ventilation index (Eq. 7.6) implies an unfavorable large-scale environment for tropical cyclogenesis and tropical cyclone intensification. As indicated by Eq. 7.6, large values of vertical wind shear and entropy deficit lead to an increase in the ventilation index. The idea of the ventilation index is that vertical wind shear and entropy deficit play a role of "efficiency" to attain the theoretical maximum storm intensity.

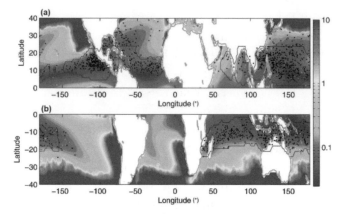

Fig. 7.7 (a) July–October mean ventilation index for the Northern Hemisphere and (b) December–March ventilation index for the Southern Hemisphere averaged over 1990–2009 using interim European Center for Medium-Range Weather Forecasts (ECMWF) Re-Analysis (ERA-Interim) data (Dee et al., 2011). Black dots are tropical cyclogenesis points over the same period. The black outline demarcates the main genesis regions, which are considered to be equatorward of 25°. A black and white version of this figure will appear in some formats. For the color version, refer to the plate section.
Source: Adapted from Tang and Emanuel, 2012; ©American Meteorological Society. Used with permission.

Figure 7.7 displays the seasonally averaged (1990–2009) ventilation index calculated from interim European Center for Medium-Range Weather Forecasts (ECMWF) Re-Analysis (ERA-Interim) data (Dee et al., 2011) (shading), which reveals good correspondence with the observed tropical cyclogenesis (dots). The seasonally averaged ventilation index is less than 0.1 over the tropics where tropical cyclogenesis is abundant. In the subtropics on the eastern side of the ocean basins, the ventilation index is relatively high due to lower potential intensity, high vertical wind shear, and high entropy deficit. The high entropy deficit in these regions is caused by dry air at the mid-level troposphere. The ventilation index increases quickly poleward of 30° due to increasing upper-level westerlies and a sharp decrease in potential intensity.

Tang and Camargo (2014) applied the ventilation index to the model outputs from the eight Coupled Model Intercomparison Project 5 (CMIP5) models. The projected ventilation index for the present-day climate (1981–2000) was compared with that for the future climate (2081–2100) under the RCP8.5 scenario. The RCP8.5 experiment assumes a high-end emissions scenario in which anthropogenic emissions lead to a radiative forcing of 8.5 W m^{-2}. Tang and Camargo (2014) showed projected future changes in the ventilation index by the eight CMIP5 models. Most of the tropical oceans, except for the North Indian Ocean, show positive changes in the ventilation index in the future. The projected increases in the ventilation index were mostly due to a projected increase in entropy deficit, although the projected increase in potential intensity offsets the increase in entropy deficit. The vertical wind shear was not as robust, although the models project slight decreases in the vertical wind shear, which also acts to compensate a small amount for the increase in the entropy deficit.

Overall, the projected increase in ventilation index is consistent with the projected decrease in global tropical cyclone numbers. However, when the results of individual models rather than the ensemble mean of the models are examined more closely, it is not always true that the ventilation index represents projected changes in global tropical cyclone numbers. Indeed, unlike the conventional results of decreases in global tropical cyclone numbers, some of the CMIP5 models project an increase in global tropical cyclone numbers under the RCP8.5 scenario (e.g., Camargo, 2013; Tory et al., 2013; Murakami et al., 2014).

7.6 Studies Showing Projected Increases in Tropical Cyclones in the Future

Although a number of previous studies reported projected decreases in the number of tropical cyclones in a warmer climate, a few studies since 2013 have reported an increasing number of global tropical cyclones under a specific future scenario and experimental design, leaving substantial inconsistencies and uncertainties among the climate research.

Emanuel (2013) applied the statistical-dynamical downscaling approach (see Section 7.4.4) developed by Emanuel (2006) to the outputs of the CMIP5 climate models, showing a decreasing number of global tropical cyclones in the future under the RCP4.5 scenario (medium emission scenario), but an increasing number of global tropical cyclones under the RCP8.5 scenario (high emission scenario). Using the same downscaling approach but applying it to the output of the CMIP6 climate models, Emanuel (2021) also showed an increasing number of global tropical cyclones in a warmer climate. Figure 7.8a shows the simulated number of global tropical cyclones by Emanuel (2021). The blue line shows the simulated number during the historical period (1850–2014), whereas the red line shows the scenarios in which the CO_2 level is increased 1% per year from the present-day condition.

The figure shows significant increases in the number of global tropical cyclones under the increased CO_2. Although it is unclear why the downscaling approach by Emanuel (2021) projected increases in global tropical cyclone numbers, there is excellent correspondence between the simulated number of global tropical cyclones and the genesis potential index (GPI) calculated from the large-scale

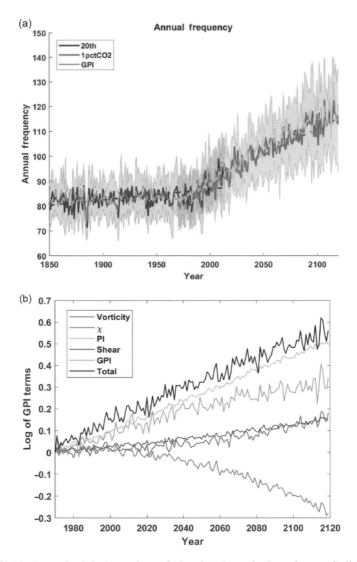

Fig. 7.8 (a) Annual global number of simulated tropical cyclones. Solid curves represent multi-model means and shading indicates one standard deviation up and down. Dashed lines show linear regression trends. Blue indicates the historical period 1850–2014 while red shows the 1% yr^{-1} CO_2 increase experiment beginning in 1970. Green curves show the multi-model mean, globally summed genesis potential index (GPI). (b) Each term contribution to the total GPI in terms of the right side of Eq. 7.9. The contributions are from vorticity, χ, potential intensity (PI), and vertical wind shear. The black curve shows their sum. A black and white version of this figure will appear in some formats. For the color version, refer to the plate section.

Source: Adapted from Emanuel, 2021; ©American Meteorological Society. Used with permission.

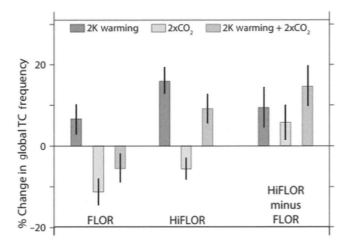

Fig. 7.9 Response of global tropical cyclone numbers in the idealized experiments: leftmost bars are for the FLOR model; the second set of bars for the HiFLOR model; and rightmost bars show the difference between HiFLOR and FLOR. The blue bars show the response to a combined uniform 2K warming and a CO_2 doubling, the gray bars show the response to CO_2 doubling with fixed SST, and the red bars show the response to uniform 2K warming. Black lines show the 95% confidence interval on the change. A black and white version of this figure will appear in some formats. For the color version, refer to the plate section.
Source: Adapted from Vecchi et al., 2019; ©creativecommons.org/licenses/by/4.0/.

outputs by the CMIP6 models (the green line in Fig. 7.8a). The definition of GPI used here is from Emanuel (2010):

$$GPI = |\eta|^3 \chi^{-4/3} \, \text{MAX} \left(\left(V_{pot} - 35 \right), 0 \right)^2 \left(25 + V_{shear} \right)^{-4}, \tag{7.8}$$

where η is the absolute vorticity of the 850 hPa, capped by $5 \times 10^{-5} \, s^{-1}$; V_{pot} is the potential intensity defined as Eq. 7.7; V_{shear} is the magnitude of the 850 hPa–250 hPa wind shear; and χ is the entropy deficit defined in Eq. 7.4. The GPI is summed over the globe and averaged among the models to produce the green line in Fig. 7.9 (top), which shows good agreement with the increasing number of global tropical cyclones. In order to estimate each term contribution to the GPI increase, logarithm is taken in both sides in Eq. 7.8:

$$\log{(GPI)} = 3 \log{(|\eta|)} - \frac{4}{3} \log{(\chi)} + 2 \log{\left(\text{MAX} \left(\left(V_{pot} - 35 \right), 0 \right) \right)} \tag{7.9}$$
$$- 4 \log{\left(25 + V_{shear} \right)}.$$

Figure 7.8b shows each term contribution of Eq. 7.9 to the GPI time evolution. As in the previous studies discussed in Section 7.4.2, an increase in the entropy deficit (red line) could serve as a negative factor for tropical cyclone genesis.

However, the negative effect from the entropy deficit is offset by the positive effect by potential intensity (yellow line), in addition to vertical wind shear (magenta line) and vorticity (blue line). The potential intensity (see Eq. 7.7) becomes larger when the outflow temperature (i.e., temperature at the upper troposphere) becomes lower and/or SST is higher. Given the projected increases in static stability in the future by the increased temperature at the upper troposphere, the projected increased SST dominates the increased potential intensity. In Emanuel (2021)'s downscaling approach, tropical cyclone activity appears to be more sensitive to the thermodynamic parameter, especially to SST, although a number of dynamical climate models, as introduced in Section 7.4.3, showed a decreased number of global tropical cyclones despite the increased SST under global warming.

There is also a dynamical climate model that does not show a decrease in the number of global tropical cyclones in a warmer climate. Vecchi et al. (2019) utilized a coupled global climate model with increasingly refined atmosphere/land horizontal grids (FLOR [~50 km; Vecchi et al., 2014] and HiFLOR [~25 km; Murakami et al., 2015]). In a warmer climate, owing to the doubling of CO_2 along with the increased SST, the simulated number of global number of tropical cyclones markedly decreases in the FLOR model relative to the present-day climate, while the HiFLOR model shows an increasing change (blue bars in Fig. 7.9). Vecchi et al. (2019) also conducted additional idealized experiments in which either CO_2 (gray bars in Fig. 7.9) or SST (red bars in Fig. 7.10) is increased separately. Both models show an increasing number of tropical cyclones in the increased SST only and decreasing number of tropical cyclones in the increased CO_2 only. However, the HiFLOR model shows larger (smaller) increases (decreases) by the increased SST (CO_2) than FLOR does (Fig. 7.9). In other words, the changes in the number of global tropical cyclones depends on the model's sensitivity to increasing CO_2 and SST.

Vecchi et al. (2019) also investigated why the HiFLOR model projected an increased number of global tropical cyclones in a warmer climate associated with the large-scale parameter changes. They first computed a few large-scale parameters: GPI developed by Emanuel (2013, Eq. 7.8); the ventilation index developed by Tang and Emanuel (2012; see Eq. 7.6 in Section 7.5.3); and the vertical *p*-velocity with respect to the mass flux as discussed in Section 7.5.1. First, the changes in GPI (Fig. 7.10a) show positive correlations with the model simulated changes in global tropical cyclone numbers within the individual models (red and gray lines). However, it cannot fully explain the differences between models. For example, even with the same degree of the changes in GPI, HiFLOR tends to project to increased storm numbers, while FLOR tends to project decreased storm numbers. Second, the changes in ventilation index (Fig. 7.10b) do

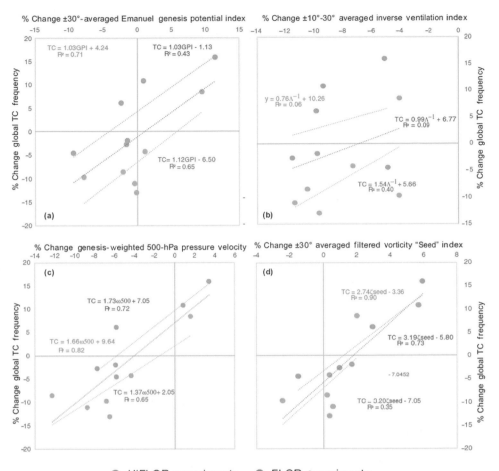

Fig. 7.10 Fractional response in global number of tropical cyclones versus fractional response in spatially averaged large-scale parameters. Orange symbols show the response of HiFLOR and gray symbols denote those for FLOR. The dotted lines indicate the linear least-squares regression fit with the covariance indicated by R^2. Orange lines show regression for HiFLOR points, gray for FLOR points, and blue for all data combined. Each symbol is the response of one idealized experiment relative to the present-day control experiment (e.g., doubling CO_2 experiments, +2K uniform SST experiments, and the combined experiments). Fractional response of simulated global tropical cyclone numbers is compared with (a) tropical-mean, Emanuel (2013) GPI (Eq. 7.8) response , (b) ± 10–30° averaged inverse Tang and Emanuel ventilation index (Eq. 7.6 with minus sign), (c) 500 hPa pressure velocity, and (d) tropical cyclones "seed" index (Li et al., 2010). A black and white version of this figure will appear in some formats. For the color version, refer to the plate section.

Source: Adapted from Vecchi et al., 2019; ©creativecommons.org/licenses/by/4.0/.

not account for both the intra- and inter-model differences in the changes in storm number. Third, it appears that the changes in vertical p-velocity (Fig. 7.10c) is highly correlated with the changes in storm number relative to the GPI and ventilation index. However, the regression lines show different offsets between FLOR and HiFLOR, revealing different sensitivities of p-velocity to storm number in the models. Therefore, none of the large-scale parameters perfectly reflect both intra- and inter-model differences in the changes in storm number.

Vecchi et al. (2019) further hypothesized that the projected changes in tropical cyclone genesis number is influenced by the rate of pre-storm synoptic-scale disturbances, so-called tropical cyclones seeds. Because tropical cyclones originate from tropical disturbances such as synoptic wave trains and easterly waves (e.g., Lau and Lau, 1990; Li et al., 2003; Tam and Li, 2006; Fu et al., 2007), the number of tropical cyclones could depend on the likelihood that these seeds develop into tropical cyclones. Moreover, the projected changes in the number of tropical cyclone seeds could be independent from these in the large-scale parameters. Following Li et al. (2010), Vecchi et al. (2019) analyzed a "seed index" in which the variance of 3–10-day relative vorticity fields at 850 hPa is computed after removing the tropical cyclones from the original vorticity fields. Figure 7.10d reveals that the seed index represents both the intra- and inter-model differences among the model experiments. Moreover, the regression lines are similar for the two models, indicating a potential to be a universal index that quantifies the projected changes in global tropical cyclone numbers in any climate models.

Vecchi et al. (2019) noted that tropical cyclone genesis is a binomial process, in which the expected number of tropical cyclones depends on the product of the number of seeds (i.e., trials) and the probability of success of each trial developing into tropical cyclone genesis. The large-scale parameters could influence the latter. This hypothesis indicates that even in a situation where the probability of a seed developing into tropical cyclone genesis by negative large-scale conditions is small, the number of tropical cyclone genesis could increase in the case of a larger number of seeds. Indeed, the HiFLOR model projected an increase in seeds in a warmer climate, which was the main reason for the projected increases in tropical cyclone numbers, despite the unfavorable large-scale conditions such as increases in the mean ventilation index and decreases in the mean upward p-velocity. Although the HiFLOR shows an increased number of seeds in a warmer climate, it is not clear what controls the number of seeds. In contrast to Vecchi et al. (2019), Sugi et al. (2020) reported that two different dynamical models showed projected decreases in the number of seeds in a warmer climate. Further studies are needed in the future to clarify the projected changes in seeds and the physical mechanisms of future changes in tropical cyclones.

7.7 Recent Scientific Views on the Potential Future Changes in Tropical Cyclones

As reviewed in Sections 7.4–7.6, a number of studies attempted to address possible future changes in tropical cyclone activity. These results were discussed at the World Meteorological Organization Seventh International Workshop on Tropical Cyclone (IWTC-VII; WMO, 2010; Knutson et al., 2010b) and reviewed by scientists and reflected in the IPCC Fifth Assessment Report (AR 5; IPCC, 2013). Knutson et al. (2020) also updated the studies after IPCC AR 5 and reported the scientific viewpoints of the potential future changes in tropical cyclones. In this section, up-to-date knowledge on the potential future changes in the tropical cyclone activity is introduced.

Because model projections often vary in the details of the models, it is difficult to objectively assess their combined results to form a consensus. In IPCC (2013), the modeling results were normalized using a combination of objective and subjective expert judgements. The results are summarized in Fig. 7.11. Potential changes in tropical cyclone activity were evaluated for each ocean basin as well as the hemispheric and global scale in terms of: annual frequency of tropical cyclones (Metric I); annual frequency of category 4–5 tropical cyclones (Metric II); mean life-time maximum storm intensity (Metric III); and mean precipitation rate induced by tropical cyclones (Metric IV). Figure 7.11 reveals that Metric I is generally projected to decrease or remain unchanged in the next century, globally as well as in most regions, although confidence in the projections is lower in specified regions than global projections. Specifically, the models consistently project decreased frequency in tropical cyclones in the ocean basins in the South Hemisphere (e.g., South Indian Ocean and South Pacific), while the projected future changes in the ocean basins in the North Hemisphere vary markedly. As for Metric II, most of the modeling studies showed increases in the intense storms at the global scale; however, there are large uncertainties among the ocean basins. In contrast to Metrics I and II, Metrics III and IV, which represent the mean tropical cyclone intensity and tropical cyclone precipitation rate, respectively, are relatively consistent among the modeling studies. Metrics III and IV generally show projected increases in most ocean basins as well as at the global scale. Based on the results, IPCC (2013) concluded:

It is *likely* that the global frequency of occurrence of tropical cyclones will either decrease or remain essentially unchanged, concurrent with a *likely* increase in both global mean tropical cyclone maximum wind speed and precipitation rates. But the specific characteristics of the changes are not yet well quantified and there is *low confidence* in region-specific projections of frequency of tropical cyclones.

Knutson et al. (2020) updated the new modeling studies since IPCC (2013) and reported the expert viewpoints on the potential future changes in tropical cyclone

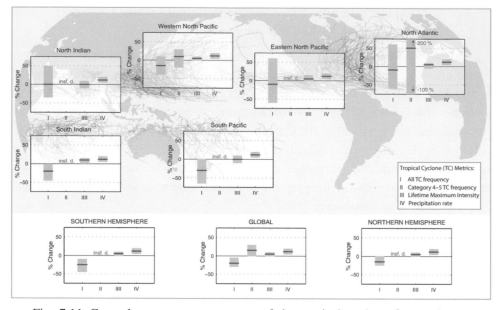

Fig. 7.11 General consensus assessment of the tropical cyclone future change made by the IPCC AR 5 (IPCC, 2013). This consensus was made by reviewing the numerical experiments available up to 2013. All values represent the expected percentage change in the average over period 2081–2100 relative to 2000–2019, under an IPCC A1B-like scenario, based on expert judgement after subjective normalization of the model projections. Four metrics were considered: percentage change in (I) the total annual frequency of tropical storms; (II) annual frequency of category 4 and 5 storms; (III) mean Lifetime Maximum Intensity (LMI; the maximum intensity achieved during a storm's lifetime); and (IV) precipitation rate within 200 km of storm center at the time of LMI. For each metric plotted, the solid line is the best guess of the expected percentage change, and the gray bar provides the 67% (likely) confidence interval for this value (note that this interval ranges across –100% to +200% for the annual frequency of category 4 and 5 storms in the North Atlantic). Where a metric is not plotted, there are insufficient data (denoted "insf. d.") available to complete an assessment. A randomly drawn (and colored) selection of historical storm tracks is underlain to identify regions of tropical cyclone activity.
Source: Adapted from IPCC, 2013.

activity. The results are summarized in Fig. 7.12. Overall consensus has not been changed from IPCC (2013). However, a major difference from IPCC (2013, Fig. 7.11) can be recognized at the global scale, showing a larger uncertainty range in the number of global tropical cyclones in Knutson et al. (2020) than in IPCC (2013). This is because the experts considered new studies showing projected increases in the number of global tropical cyclones (e.g., Emanuel, 2013). Knutson et al. (2020) had lower confidence about the decrease in global tropical cyclone numbers. Instead, they showed *medium-to-high confidence* in an increase

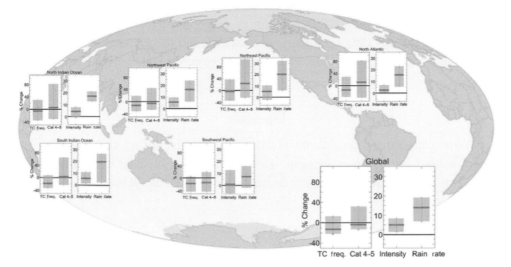

Fig. 7.12 Summary of tropical cyclone projections for a 2°C global anthropogenic warming reported in Knutson et al. (2020). Similar to Fig 7.11, shown for each basin and globally are the median and percentile ranges for projected percentage changes in tropical cyclone frequency, category 4–5 tropical cyclone frequency, mean tropical cyclone intensity, and tropical cyclone rain rates. For storm frequency, the 5th–95th-percentile range across published estimates is shown. For category 4–5 storm frequency, mean tropical cyclone intensity, and tropical cyclone rain rates, the 10th–90th-percentile range is shown.
Source: Adapted from Knutson et al., 2020; ©American Meteorological Society. Used with permission.

in the proportion of tropical cyclones that reach very intense levels as an increase in the proportion metric is projected by almost all modeling studies. As in IPCC (2013), Knutson et al. (2020) showed higher confidence in increases in mean storm intensity and related storm hazard in the future. Specifically, the most confident storm-related projection is that sea-level rise over the coming century will lead to higher storm surge levels for the tropical cyclones that do occur.

7.8 Summary

Future projections of tropical cyclone activity remain a challenging scientific topic despite the great societal impact and interest. There are large uncertainties in the future projections of tropical cyclone numbers as well as regional changes in tropical cyclone activity. Reducing uncertainties in climate model projections in tropical cyclones is important. Understanding the physical mechanisms controlling the observed number of tropical cyclones is also an important topic.

It is expected that accumulating long-term homogeneous observations will help the climatic change in tropical cyclones. At the time of writing this chapter, we are in the beginning era for tropical cyclone future projections. New works using high-resolution climate models, long-term observations, and theories are desired to shed further light on the uncertainties in the future projections of tropical cyclones.

Exercises

7.1 What causes uncertainty in observed trends in tropical cyclone activity?
7.2 What is a time-slice experiment? Describe the strengths and weaknesses.
7.3 Why is the vertical wind shear detrimental for tropical cyclone genesis?

References

Anthes, R. A., R. W. Corell, G. Holland, J. W. Hurrell, M. C. MacCracken, and K. E. Trenberth, 2006: Hurricanes and global warming–potential linkages and consequences. *Bull. Amer. Meteorol. Soc.*, **87**, 623–628.

Bell, G. D., and M. Chelliah, 2006: Leading tropical modes associated with interannual and multidecadal fluctuations in North Atlantic hurricane activity. *J. Climate*, **19**, 590–612.

Bender, M. A., and I. Ginis, 2000: Real-case simulations of hurricane-ocean interaction using a high-resolution coupled model: Effects on hurricane intensity. *Mon. Wea. Rev.*, **128**, 917–946.

Bender, M. A., T. R. Knutson, R. E. Tuleya, J. J. Sirutis, G. A. Vecchi, S. T. Garner, and I. M. Held, 2010: Modeled impact of anthropogenic warming on the frequency of intense Atlantic Hurricanes. *Science*, **327**, 454–458.

Bengtsson, L., M. Botzet, and M. Esch, 1995: Hurricane-type vortices in a general circulation model. *Tellus*, **47A**, 175–196.

Bengtsson, L., M. Botzet, and M. Esch, 1996: Will greenhouse gas-induced warming over the next 50 years lead to higher frequency and greater intensity of hurricanes? *Tellus*, **48A**, 57–73.

Bengtsson, L., K. I. Hodges, M. Esch, N. Keenlyside, L. Kornblueh, J.-J. Luo, and T. Yamagata, 2007: How may tropical cyclones change in a warmer climate? *Tellus*, **59A**, 539–561.

Bindoff, N. L., and Coauthors, 2013: Detection and attribution of climate change: From global to regional. In *Climate Change 2013: The Physical Science Basis. Contribution of Working Group I to the Fifth Assessment Report of the Intergovernmental Panel on Climate Change*, T. F. Stocker, D. Qin, G.-K. Plattner, et al., Eds., Cambridge University Press, 867–952.

Bister, M., and K. A. Emanuel, 1998: Dissipative heating and hurricane intensity. *Meteorol. Atmos. Phys.*, **52**, 233–240.

Broccoli, A. J., and S. Manabe, 1990: Can existing climate models be used to study anthropogenic changes in tropical cyclone climate? *Geophys. Res. Lett.*, **17**, 1917–1920.

Bryan, G., 2008: On the computation of pseudoadiabatic entropy and equivalent potential temperature. *Mon. Wea. Rev.*, **136**, 5239–5245.

Camargo, S. J., 2013: Global and regional aspects of tropical cyclone activity in the CMIP5 models, *J. Climate*, **26**, 9880–9902, doi:10.1175/JCLI-D-12–00549.1.

Chauvin, F., J.-F. Royer, and M. Deque, 2006: Response of hurricane-type vortices to global warming as simulated by ARPEGE-Climate at high resolution. *Clim. Dyn.*, **24**, 377–399.

Chu, J.-H., C. R. Sampson, A. S. Levin, and E. Fukada, 2002: The Joint Typhoon Warning Center tropical cyclone best tracks 1945–2000. Joint Typhoon Warning Center Rep., Pearl Harbor, HI. Available at: www.usno.navy.mil/NOOC/nmfc-ph/RSS/jtwc/best_tracks/TC_bt_report.html.

Dee, D., and Coauthors, 2011: The ERA-Interim reanalysis: Configuration and performance of the data assimilation system. *Quart. J. Roy. Meteorol. Soc.*, **137**, 553–597.

Delworth, T. L., and M. E. Mann, 2000: Observed and simulated multidecadal variability in the Northern Hemisphere. *Clim. Dyn.*, **16**, 661–676.

Dvorak, V. F., 1973: A technique for the analysis and forecasting of tropical cyclone intensities from satellite pictures. *NOAA Technical Memorandum NESS*, **45**.

Dvorak, V. F., 1984: Tropical cyclone intensity analysis using satellite data. *NOAA Technical Report NESDIS*, **11**.

Emanuel, K., 2006: Climate and tropical cyclone activity: A new model downscaling approach. *J. Climate*, **19**, 4797–4802.

Emanuel, K., R. Sundararajan, and J. Williams, 2008: Hurricanes and global warming: Results from downscaling IPCC AR4 simulations. *Bull. Amer. Meteorol. Soc.*, **89**, 347–367.

Emanuel, K. A., 1986: An air–sea interaction theory for tropical cyclones. Part I: Steady state maintenance. *J. Atmos. Sci.*, **43**, 585–604.

Emanuel, K. A., 2005: Increasing destructiveness of tropical cyclones over the past 30 years. *Nature*, **436**, 686–688.

Emanuel, K. A., 2010: Tropical cyclone activity downscaled from NOAA-CIRES reanalysis, 1908–1958. *J. Adv. Model. Earth Sys.*, **2**, 1–12.

Emanuel, K. A., 2013: Downscaling CMIP5 climate models shows increased tropical cyclone activity over the 21st century. *PNAS*, **110**, 12219–12224.

Emanuel, K., 2021: Response of global tropical cyclone activity to increasing CO2: Results from downscaling CMIP6 models. *J. Climate*, **34**, 57–70.

Evans, J. L., 1992: Comment on "Can existing climate models be used to study anthropogenic changes in tropical cyclone climate". *Geophys. Res. Lett.*, **19**, 1523–1524.

Frank, W. M., and G. S. Young, 2007: The interannual variability of tropical cyclones. *Mon. Wea. Rev.*, **135**, 3587–3598.

Fu, B., T. Li, M. Peng, and F. Weng, 2007: Analysis of tropical cyclone genesis in the western North Pacific for 2000 and 2001. *Wea. Forecasting*, **22**, 763–780.

Gray, W., 1968: Global view of the origin of tropical disturbances and storms. *Mon. Wea. Rev.*, **96**, 669–700.

Gualdi, S., E. Scoccimarro, and A. Navarra, 2008: Changes in tropical cyclone activity due to global warming: Results from a high-resolution coupled general circulation model. *J. Climate*, **20**, 5204–5228.

Haarsma, R. J., J. F. B. Mitchell, and C. A. Senior, 1993: Tropical disturbances in a GCM. *Clim. Dyn.*, **8**, 247–257.

Hasegawa, A., and S. Emori, 2005: Tropical cyclones and associated precipitation over the western North Pacific: T106 atmospheric GCM simulation for present day and doubled CO_2 climate. *SOLA*, **1**, 145–148.

Held, I. M., and B. J. Soden, 2006: Robust responses of the hydrological cycle to global warming. *J. Climate*, **19**, 5686–5699.

Henderson-Sellers, A., and Coauthors, 1998: Tropical cyclones and global climate change: A post-IPCC assessment. *Bull. Amer. Meteorol. Soc.*, **79**, 19–38.

Holland, G. J., and P. J. Webster, 2007: Heightened tropical cyclone activity in the North Atlantic: Natural variability or climate trend? *Philos. Trans. R. Soc. A*, **365**, 2695–2716.

Holton, J. R., 1990: On the global exchange of mass between the stratosphere and the troposphere. *J. Atmos. Sci.*, **47**, 392–395.

Hoyos, C. D., P. A. Agudelo, P. J. Webster, and J. Acurry, 2006: Deconvolution of the factors contributing to the increase in global hurricane intensity. *Science*, **312**, 94–97.

IPCC, 1990: *Climate Change: The IPCC Scientific Assessment*. J. T. Houghton, G. J. Jenkins and J. J. Ephraums, Eds. Cambridge University Press.

IPCC, 2007: *Climate Change 2007: The Physical Science Basis*. Contribution of Working Group I to the Fourth Assessment Report of the Intergovernmental Panel on Climate Change. Cambridge University Press.

IPCC, 2013: *Climate Change 2013: The Physical Science Basis*. Cambridge University Press, doi:10.1017/CBO9781107415324.

Klotzbach, P. J., and C. W. Landsea, 2015: Extremely intense hurricanes: Revisiting Webster et al. (2005) after 10 years. *J. Climate*, **28**, 7621–7629, doi:10.1175/JCLI-D-15-0188.1.

Knapp, K. R., M. C. Kruk, D. H. Levinson, H. J. Diamond, and C. J. Neuman, 2010: The International Best Track Archive for Climate Stewardship (IBTrACS): Unifying tropical cyclone best track data. *Bull. Amer. Meteorol. Soc.*, **91**, 363–376.

Knutson, T. R., and S. Manabe, 1995: Time-mean response over the tropical Pacific to increased CO_2 in a coupled ocean-atmosphere model. *J. Climate*, **8**, 2181–2199.

Knutson, T. R., and R. E. Tuleya, 2004: Impact of CO_2-induced warming on simulated hurricane intensity and precipitation: Sensitivity to the choice of climate model and convective parameterization. *J. Climate*, **17**, 3477–3495.

Knutson, T. R., R. E. Tuleya, and Y. Kurihara, 1998: Simulated increase of hurricane intensities in a CO_2-warmed climate. *Science*, **279**, 1018–1020.

Knutson, T. R., R. E. Tuleya, W. Shen, and I. Ginis, 2001: Impact of CO_2-induced warming on hurricane intensities as simulated in a hurricane model with ocean coupling. *J. Climate*, **14**, 2458–2468.

Knutson, T. R., J. J. Sirutis, S. T. Garner, G. A. Vecchi, and I. Held, 2008: Simulated reduction of Atlantic hurricane frequency under twenty-first-century warming condition. *Nat. Geosci.*, **1**, 359–364.

Knutson, T. R., and Coauthors, 2010a: Tropical cyclones and climate change. *Nat. Geosci.*, **3**, 157–163.

Knutson, T. R., C. Landsea, and K. Emanuel, 2010b: Tropical cyclones and climate change: A review. In *Global Perspectives on Tropical Cyclones*, J. C. L. Chan and J. D. Kepert, Eds. World Scientific, 243–284.

Knutson, T. R., and Coauthors, 2020: Tropical cyclones and climate change assessment: Part II: Projected response to anthropogenic warming. *Bull. Amer. Meteorol. Soc.*, **101**, E303–E322.

Kossin, J. P., T. L. Olander, and K. R. Knapp, 2013: Trend analysis with a new global record of tropical cyclone intensity. *J. Climate*, **26**, 9960–9976, doi:10.1175/JCLI-D-13-00262.1.

Kuleshov, Y., and Coauthors, 2010: Trends in tropical cyclones in the South Indian Ocean and the South Pacific Ocean. *J. Geophys. Res.*, **115**, D01101, doi:10.1029/2009JD012372.

Landsea, C. W., 1997: Comments on "Will greenhouse gas-induced warming over the next 50 years lead to higher frequency and greater intensity of hurricanes?" *Tellus*, **49A**, 622–623.

Landsea, C. W., B. A. Harper, K. Hoarau, and J. A. Knaff, 2006: Can we detect trends in extreme tropical cyclones? *Science*, **313**, 452–453, doi:10.1126/science.1128448.

Lau, K.-H., and N.-C. Lau, 1990: Observed structure and propagation characteristics of tropical summertime synoptic-scale disturbances. *Mon. Wea. Rev.*, **118**, 1888–1913, doi:10.1175/1520-0493(1990)118<1888:OSAPCO>2.0.CO;2.

Lavender, S. L., and K. J. E. Walsh, 2011: Dynamically downscaled simulations of Australian region tropical cyclones in current and future climates. *Geophys. Res. Lett.*, **38**, L10705., doi:10.1029/2011GL047499.

Lee, C.-Y., M. K. Tippett, A. H. Sobel, and S. J. Camargo, 2018: An environmentally forced tropical cyclone hazard model. *J. Adv. Model. Earth Syst.*, **10**, 233–241.

Lenssen, N., G. Schmidt, J. Hansen, M. Menne, A. Persin, R. Ruedy, and D. Zyss, 2019: Improvements in the GISTEMP uncertainty model. *J. Geophys. Res. Atmos.*, **124**(12), 6307–6326, doi:10.1029/2018JD029522.

Leslie, L. M., D. J. Karoly, M. Leplastrier, and B. W. Buckley, 2007: Variability of tropical cyclones over the southwest Pacific Ocean using a high-resolution climate model. *Meteorol. Atmos. Phys.*, **97**, 171–180.

Li, T., B. Fu, X. Ge, B. Wang, and M. Peng, 2003: Satellite data analysis and numerical simulation of tropical cyclone formation. *Geophys. Res. Lett.*, **30**, 2122–2126.

Li, T., M. Kwon, M. Zhao, J.-J. Kug, J.-J. Luo, and W. Yu, 2010: Global warming shifts Pacific tropical cyclone location. *Geophys. Res. Lett.*, **37**, L21804, doi:10.1029/2010GL045124.

Lighthill, J., G. Holland, W. Gray, C. Landsea, G. Craig, J. Evans, Y. Kurihara, and C. Guard, 1994: Global climate change and tropical cyclones. *Bull. Amer. Meteorol. Soc.*, **75**, 2147–2157.

Lin, I.-I., I.-F. Pun, and C.-C. Wu, 2009: Upper-ocean thermal structure and the western North Pacific category-5 typhoons. Part II: Dependence on translation speed. *Mon. Wea. Rev.*, **137**, 3744–3757.

Lloyd, I. D., and G. A. Vecchi, 2011: Observational evidence for oceanic controls on hurricane intensity. *J. Climate*, **24**, 1138–1153.

Lupo, A. R., T. K. Latham, T. Magill, J. V. Clark, C. J. Melick, and P. S. Market, 2008: The interannual variability of hurricane activity in the Atlantic and East Pacific regions. *Nat. Wea. Dig.*, **32**(2), 119–135.

Mann, M. E., and K. A. Emanuel, 2006: Atlantic hurricane trends linked to climate change. *Eos Trans. AGU*, **87**(24), 233–241.

Mann, M. E., K. A. Emanuel, G. J. Holland, and P. J. Webster, 2007a: Atlantic tropical cyclones revisited. *Eos Trans. AGU*, **88**(36), 349–350.

Mann, M. E., T. A. Sabbatelli, and U. Neu, 2007b: Evidence for a modest undercount bias in early historical Atlantic tropical cyclone counts. *Geophys. Res. Lett.*, **34**, L22707.

Mantua, N. J., S. R. Hare, Y. Zhang, J. M. Wallace, and R. C. Francis, 1997: A Pacific interdecadal climate oscillation with impacts on salmon production. *Bull. Amer. Meteorol. Soc.*, **78**, 1069–1079.

Marks, D. G., 1992: The beta and advection model for hurricane track forecasting. NOAA Technical Memorandum NWS NMC 70, National Meteorological Center, Camp Spring, MD.

McBride, J., and R. Zehr, 1981: Observational analysis of tropical cyclone formation. Part II: Comparison of non-developing versus developing systems. *J. Atmos. Sci.*, **38**, 1132–1151.

McDonald, R. E., D. G. Bleaken, D. R. Cresswell, V. D. Pope, and C. A. Senior, 2005: Tropical storms: Representation and diagnosis in climate models and the impacts of climate change. *Clim. Dyn.*, **25**, 19–36.

Mitchell, J. F. B., and W. J. Ingram, 1992: Carbon dioxide and climate: Mechanisms of changes in cloud. *J. Climate*, **5**, 5–21.

Morice, C. P., J. J. Kennedy, N. A. Rayner, and P. D. Jones, 2012: Quantifying uncertainties in global and regional temperature change using an ensemble of observational estimates: The HadCRUT4 dataset. *J. Geophys. Res. Atmos.*, **117**, D08101, doi:10.1029/2011JD017187.

Murakami, H., and M. Sugi, 2010: Effect of model resolution on tropical cyclone climate projections. *SOLA*, **6**, 73–76.

Murakami, H., R. Mizuta, and E. Shindo, 2012a: Future changes in tropical cyclone activity projected by multi-physics and multi-SST ensemble experiments using the 60-km-mesh MRI-AGCM. *Clim. Dyn.*, **39(9–10)**, 2569–2584, doi:10.1007/s00382-011-1223-x.

Murakami, H., et al., 2012b: Future changes in tropical cyclone activity projected by the new high-resolution MRI AGCM. *J. Climate*, **25**, 3237–3260.

Murakami, H., P.-C. Hsu, O. Arakawa, and T. Li, 2014: Influence of model biases on projected future changes in tropical cyclone frequency of occurrence. *J. Climate*, **27**, 2159–2181.

Murakami, H., et al., 2015: Simulation and prediction of category 4 and 5 hurricanes in the high-resolution GFDL HiFLOR coupled climate model. *J. Climate*, **28**, 9058–9079.

Olander, T. L., and C. S. Velden, 2007: The advanced Dvorak technique: Continued development of an objective scheme to estimate tropical cyclone intensity using geostationary infrared satellite imagery. *Wea. Forecasting*, **22**, 287–298.

Oouchi, K., J. Yoshimura, H. Yoshimura, R. Mizuta, S. Kusunoki, and A. Noda, 2006: Tropical cyclone climatology in a global-warming climate as simulated in a 20km-mesh global atmospheric model: Frequency and wind intensity analysis. *J. Meteorol. Soc. Japan*, **84**, 259–276.

Pielke, R. A., C. Landsea, M. Mayfield, J. Laver, and R. Pasch, 2005: Hurricanes and global warming. *Bull. Amer. Meteorol. Soc.*, **11**, 1571–1575.

RSMC, 2020: Regional Specialized Meteorological Centers – Tokyo Typhoon Center tropical cyclone data. Available at: www.jma.go.jp/jma/jma-eng/jma-center/rsmc-hp-pub-eg/trackarchives.html.

Satoh, T., A. Juri, K. Masuyama, E. Imakita, and M. Kimoto, 2011: Verification of downscaling framework for interannual variation of tropical cyclone in Western North Pacific. *SOLA*, **7**, 169–177.

Semmler, T., S. Varghese, R. McGrath, P. Nolan, S. Wang, P. Lynch, and C. O'Dowd, 2008: Regional climate model simulations of North Atlantic cyclones: Frequency and intensity changes. *Clim. Res.*, **36**, 1–16.

Shade, L., and K. Emanuel, 1999: The ocean's effect on the intensity of tropical cyclones: Results from a simple coupled atmosphere-ocean model. *J. Atmos. Sci.*, **56**, 642–651.

Simpson, R., and R. Riehl, 1958: Mid-tropospheric ventilation as a constraint on hurricane development and maintenance. Preprints, Technical Conference on Hurricanes, Miami Beach, FL, *Amer. Meteorol. Soc.*, D4-1–D4-10.

Soden, B. J., and I. M. Held, 2006: An assessment of climate feedbacks in coupled ocean-atmosphere models. *J. Climate*, **19**, 3354–3360.

Stowasser, M., Y. Wang, and K. Hamilton, 2007: Tropical cyclone changes in the western North Pacific in a global warming scenario. *J. Climate*, **20**, 2378–2396.

Sugi, M., A. Noda, and N. Sato, 2002: Influence of the global warming on tropical cyclone climatology. *J. Meteorol. Soc. Japan*, **80**, 249–272.

Sugi, M., H. Murakami, and J. Yoshimura, 2012: On the mechanism of tropical cyclone frequency changes due to global warming. *J. Meteorol. Soc. Japan*, **90A**, 397–408.

Sugi, M., Y. Yamada, K. Yoshida, R. Mizuta, M. Nakano, C. Kodama, and M. Satoh, 2020: Future changes in the global frequency of tropical cyclone seeds. *SOLA*, **60**, 70–74.

Tam, C.-Y., and T. Li, 2006: The origin and dispersion characteristics of the observed summertime synoptic-scale waves over the western Pacific. *Mon. Wea. Rev.*, **134**, 1630–1646.

Tang, B., and S. J. Camargo, 2014: Environmental control of tropical cyclones in CMIP5: A ventilation perspective. *J. Adv. Model. Earth Syst.*, **6**, 115–128, doi:10.1002/2013MS000294.

Tang, B., and K. Emanuel, 2010: Midlevel ventilation's constraint on tropical cyclone intensity. *J. Atmos. Sci.*, **67**, 1817–1830.

Tang, B., and K. Emanuel, 2012: A ventilation index for tropical cyclones. *Bull. Amer. Meteorol. Soc.*, **93**, 1901–1912.

Tory, K. J., S. S. Chand, J. L. McBride, H. Ye, and R. A. Dare, 2013: Projected changes in late-twenty-first-century tropical cyclone frequency in 13 coupled climate models from phase 5 of the Coupled Model Intercomparison Project. *J. Climate*, **26**, 9946–9959, doi:10.1175/JCLI-D-13-00010.1.

Trenberth, K. E., and D. J. Shea, 2006: Atlantic hurricanes and natural variability in 2005. *Geophys. Res. Lett.*, **33**, L12704.

Vecchi, G. A., and T. R. Knutson, 2008: On estimates of historical North Atlantic tropical cyclone activity. *J. Climate*, **21**, 9960–9976.

Vecchi, G. A., and B. J. Soden, 2007: Global warming and the weakening of the tropical circulation. *J. Climate*, **20**, 4316–4340.

Vecchi, G. A., et al., 2014: On the seasonal forecasting of regional tropical cyclone activity. *J. Climate*, **27**, 7994–8016.

Vecchi, G. A., and Coauthors, 2019: Tropical cyclone sensitivities to CO_2 doubling: Roles of atmospheric resolution, synoptic variability and background climate changes. *Clim. Dyn.*, **53(9–10)**, 5999–6033.

Velden, C., and Coauthors, 2006: The Dvorak tropical cyclone intensity estimation technique: A satellite-based method that has endured for over 30 years. *Bull. Amer. Meteorol. Soc.*, **87**, 1195–1210.

Walsh, K. J. E., and B. F. Ryan, 2000: Tropical cyclone intensity increase near Australia as a result of climate change. *J. Climate*, **13**, 3029–3036.

Walsh, K. J. E., K.-C. Nguyen, and J. L. McGregor, 2004: Fine-resolution regional climate model simulations of the impact of climate change on tropical cyclones near Australia. *Clim. Dyn.*, **22**, 47–56.

Wang, B., Y. Yang, Q.-H. Ding, H. Murakami, and F. Huang, 2010: Climate control of the global tropical storm days (1965–2008). *Geophys. Res. Lett.*, **37**, L07704.

Webster, P. J., G. J. Holland, J. A. Curry, and H.-R. Chang, 2005: Changes in tropical cyclone number, duration and intensity in a warming environment. *Science*, **309**, 1844–1846.

Wetherald, R. T., and S. Manabe, 1980: Cloud cover and climate sensitivity. *J. Atmos. Sci.*, **37**, 1485–1510.

WMO, 2006: Statement on tropical cyclones and climate change. In *The 6th International Workshop on Tropical Cyclones of the World Meteorological Organization*

(WMO IWTC-VI). Available at: www.wmo.int/pages/prog/arep/tmrp/documents/iwtc_statement.pdf.

WMO, 2010: *7th International Workshop on Tropical Cyclones (ITWC-VII).* Available at: www.library.wmo.int/index.php?lvl=notice_display&id=10562#.X77cW2j0lQU.

Wu, M.-C., K.-H. Yeung, and W.-L. Chang, 2006: Trends in western North Pacific tropical cyclone intensity. *Eos, Trans. Amer. Geophys. Union,* **87**, 537–538, doi:10.1029/2006EO480001.

Yamada, Y., K. Oouchi, M. Satoh, H. Tomita, and W. Yanase, 2010: Projection of changes in tropical cyclone activity and cloud height due to greenhouse warming: Global cloud-system-resolving approach. *Geophys. Res. Lett.,* **37**, L07709, doi:10.1029/2010GL042518.

Yoshimura, J., M. Sugi, and A. Noda, 2006: Influence of greenhouse warming on tropical cyclone frequency. *J. Meteorol. Soc. Japan,* **84**, 405–428.

Zehr, R., 1992: Tropical cyclogenesis in the western north Pacific. *NOAA Tech. Rep. NOAA Technical Report NESDIS,* **61**, 181pp.

Zhao, M., I. Held, S.-J. Lin, and G. A. Vecchi, 2009: Simulations of global hurricane climatology, interannual variability, and response to global warming using a 50 km resolution GCM. *J. Climate,* **22**, 6653–6678.

Index